EQUILIBRIUM CONFIGURATIONS OF DEGENERATE GASEOUS MASSES

G. S. Saakyan

EQUILIBRIUM CONFIGURATIONS OF DEGENERATE GASEOUS MASSES

Edited by V. A. Ambartsumyan

Translated from Russian by C. F. Hall, Ph. D

A HALSTED PRESS BOOK

JOHN WILEY & SONS
New York · Toronto

ISRAEL PROGRAM FOR SCIENTIFIC TRANSLATIONS
Jerusalem · London

Sole distributors for the Western Hemisphere
HALSTED PRESS, a division of
JOHN WILEY & SONS, INC., NEW YORK

Library of Congress Cataloging in Publication Data

Saakiân, Gurgen Serbovich.
Equilibrium configurations of degenerate gaseous masses.

Translation of *Ravnovesnye konfiguratsii vyrozhdennykh gazovykh mass.*
"A Halsted Press book."
Bibliography: p.
1. Rotating masses of fluid. 2. Neutron stars. 3. White dwarfs. 4. Gravitation. I. Title.
QB410.S213 523.8 74-13583
ISBN 0-470-74805-2

Distributors for the U. K., Europe, Africa and the Middle East
JOHN WILEY & SONS, LTD., CHICHESTER

Distributors for Japan, Southeast Asia and India
TOPPAN COMPANY, LTD., TOKYO AND SINGAPORE

Distributed in the rest of the world by
KETER PUBLISHING HOUSE JERUSALEM LTD.
ISBN 0 7065 1376 2
IPST cat. no. 22095

This book is a translation from Russian of
RAVNOVESNYE KONFIGURATSII VYROZHDENNYKH
GAZOVYKH MASS
Izdatel'stvo "Nauka"
Moscow, 1972

Printed and bound by Keterpress Enterprises, Jerusalem
Printed in Israel

CONTENTS

EDITOR'S PREFACE

The latest discoveries in astrophysics have attracted universal interest to new and previously unknown phenomena taking place in the stellar world. In particular, the discovery of pulsars has demonstrated directly that the Galaxy contains superdense bodies of near-nuclear average density. The properties of superdense matter and the possible equilibrium configurations of stellar masses consisting almost exclusively of baryons have been discussed in a large number of papers in the specialized literature. These papers are not easily unified. Moreover, the research on which they are based is carried out at different levels, often repeating, in a coarser or less rigorous form, material discussed in earlier papers. In view of this situation, the publication of an authoritative monograph reviewing a significant part of the field is most welcome.

The present book, written by Professor G. S. Saakyan, Corresponding Member of the Academy of Sciences of the Armenian SSR, considers the theory of the equation of state of extremely degenerate matter at nuclear density and the resulting theory of equilibrium configurations of degenerate stellar masses.

Saakyan is an eminent authority in the field. Many of the results presented in the book are to his credit. As in his other work, he maintains consistent rigor in his treatment of the subject. As a result, his monograph should considerably contribute toward somehow ordering the informational chaos obtaining in the field, which includes many harmful, disturbing factors.

The last two chapters of the book are extrapolatory in nature and study the possible consequences of varying the equations or constants. These chapters are interesting and useful, but the reader should be forewarned and adopt a critical approach to their contents.

V. A. Ambartsumyan

AUTHOR'S PREFACE

This monograph presents a systematic exposition of the theory of stellar configurations consisting of degenerate gaseous masses (white dwarfs and baryon stars). The foundations of the theory of superdense celestial bodies appeared in the thirties. The latest advances in this field are related to the discovery of hyperons and a host of other particles. At the present time, the existence of white dwarfs is established beyond doubt, since many stars of this type have actually been observed. As to baryon (neutron) stars, until quite recently they seemed to be highly theoretical constructions. Although the theory of these celestial bodies involves no hypothetical points, it was nevertheless necessary to obtain direct or indirect observational data pointing to their actual existence. Such data appeared quite unexpectedly in radio-astronomical observations made in 1968 which revealed pulsating sources of radio emission, known as pulsars. Because of various special features, it became obvious almost immediately that the pulsating radiation is generated by rotating baryon stars. This remarkable discovery stimulated a new and considerable interest in superdense celestial bodies.

At present, the theory of superdense celestial bodies is one of the most interesting and rapidly developing fields of modern astrophysics. Much material has been accumulated, which needs collection, critical analysis and a systematic exposition of what has been achieved. In the rather sparse literature available, not all aspects of the theory of superdense configurations have been covered. We hope that this book will in some degree fill this gap.

The last two chapters are devoted to the generalized theory of gravitation; underlying this theory is the conjecture that the gravitational constant is variable, and certain assumptions have been made concerning the choice of the action for the gravitational field. These assumptions are as yet unsubstantiated by experimental evidence, and so the results presented in these two chapters are in a sense debatable.

In the bibliography we have tried as far as possible to cover all publications with a bearing on the problem of superdense stars. However, we make no claim to completeness.

I am indebted to all my co-workers at the Faculty of Theoretical Physics of Erevan State University for their help in the preparation of the monograph.

G. S. Saakyan

Chapter I

FUNDAMENTALS OF THE RELATIVISTIC THEORY OF GRAVITATION

1. FOUR-DIMENSIONAL TENSORS

An event is defined in space and time by a quadruple of numbers $ct \equiv x^0$, x^1, x^2, x^3, the first of which represents time and the last three represent space coordinates, i. e. the place of occurrence. * This also implies a concrete frame of reference with a set of synchronized clocks. In the language of geometry, the continuum of these numbers forms a four-dimensional space. In the case of inertial frames of reference this is called the Minkowski space.

From special theory of relativity it is known that the space and time coordinates are mixed in a transition between two frames of reference. Moreover, in strong gravitational fields, i. e. in the presence of large accumulations of mass, the time and spatial properties of the space will differ at different points. Thus the nature of the space—time continuum is determined by matter, and therefore the Newtonian concept of absolute space and an independent absolute time is only an approximation. In the general case, space and time are continuously interconnected, comprising a single entity. The concept of a four-dimensional space is therefore more accurate than that of independent absolute space and time. In this four-dimensional space an event corresponds to a point, and the history of an event is represented by a line.

Let x^i and x'^i $(i=0, 1, 2, 3)$ be the coordinates of a single event in two arbitrary frames of reference. These coordinates are then related through the transformations

$$x^i = f^i(x'^0, \quad x'^1, \quad x'^2, \quad x'^3). \tag{1.1}$$

We shall now examine the behavior of physical quantities with respect to this transformation.

* In the general theory of relativity the coordinates are denoted by upper indices.

A scalar is defined as a function φ which remains unchanged through any coordinate transformation, i. e.

$$\varphi(x^0, x^1, x^2, x^3) = \varphi(x'^0, x'^1, x'^2, x'^3). \tag{1.2}$$

In the transformations (1.1), the coordinate differentials transform as

$$dx^i = \frac{\partial x^i}{\partial x'^k} dx'^k. \tag{1.3}$$

Here, as well as below, repeated indices always signify summation. A set of four quantities A^i which transform according to (1.3), i. e. as the coordinate differentials, is called a contravariant vector:

$$A^i = \frac{\partial x^i}{\partial x'^k} A'^k. \tag{1.4}$$

A different type of quantity is obtained when one examines the gradient of a scalar

$$\frac{\partial \varphi}{\partial x^i} = \frac{\partial \varphi}{\partial x'^k} \cdot \frac{\partial x'^k}{\partial x^i}. \tag{1.5}$$

A set of four quantities which transform as the gradient of a scalar is called a covariant vector:

$$A_i = \frac{\partial x'^k}{\partial x^i} A'_k. \tag{1.6}$$

A set of 16 quantities A^{ik} which transform as the product of the components of two contravariant vectors A^i and B^k, i. e. according to

$$A^{ik} = \frac{\partial x^i}{\partial x'^l} \cdot \frac{\partial x^k}{\partial x'^m} A'^{lm}, \tag{1.7}$$

is called a contravariant tensor of the second rank. Similarly, one also defines a covariant tensor of second rank:

$$A_{ik} = \frac{\partial x'^l}{\partial x^i} \cdot \frac{\partial x'^m}{\partial x^k} A'_{lm}, \tag{1.8}$$

a mixed tensor of second rank A_k^i,

$$A_k^i = \frac{\partial x^i}{\partial x'^l} \cdot \frac{\partial x'^m}{\partial x^k} A'^l_m, \tag{1.9}$$

as well as tensors of higher ranks. It would be logical to regard a scalar as a tensor of rank zero, and a vector as a tensor of rank one.

From the above definitions it follows that tensors of any higher rank can be obtained through multiplication of the components of covariant and contravariant vectors and tensors. One can also obtain tensors of lower rank from tensors of higher rank. Thus, for example, it is easy to prove that the product $A^{ikl}B_{kl}$ transforms as a contravariant vector. In fact

$$A^{ikl}B_{kl} = \frac{\partial x^i}{\partial x'^m} \cdot \frac{\partial x^k}{\partial x'^n} \cdot \frac{\partial x^l}{\partial x'^p} \cdot \frac{\partial x'^q}{\partial x^k} \frac{\partial x'^r}{\partial x^l} A'^{mnp}B'_{qr} =$$

$$= \frac{\partial x^i}{\partial x'^m} \cdot \frac{\partial x'^q}{\partial x'^n} \cdot \frac{\partial x'^r}{\partial x'^p} A'^{mnp}B'_{qr} = \frac{\partial x^i}{\partial x'^m} \delta^q_n \delta^r_p A'^{mnp}B'_{qr} =$$

$$= \frac{\partial x^i}{\partial x'^m} A'^{mnp} B'_{np}.$$

Here δ_i^k is the Kronecker symbol which is a unit mixed tensor of the second rank, as follows directly from the obvious identity $A^i\delta_i^k = A^k$.

It can be easily proven that the expressions A_iB_i and A^iB^i are not scalars, whereas A_iB^i is an invariant quantity which represents the generalization of a scalar product of two vectors in the case of a curvilinear coordinate system. In fact,

$$A_iB^i = \frac{\partial x'^l}{\partial x^i} \cdot \frac{\partial x^i}{\partial x'^m} A'_l B'^m = \frac{\partial x'^l}{\partial x'^m} A'_l B'^m = A'_m B'^m.$$

Invariant expressions can also be formed from tensors of higher rank; thus, for example,

$$A^l_{ik} B^{ik}_l = \text{invariant}.$$

Similarly, one can also prove that A^{ikl}_{kl} is a covariant vector, and that A^i_i and A^{ik}_{ik} are scalars.

Two tensors A_{ik} and B^{ik} which satisfy the condition

$$A_{ik}B^{kl} = \delta^l_i,$$

are said to be mutually inverse. It can be shown directly that

$$B^{ik} = \frac{a^{ik}}{|A_{ik}|}, \tag{1.10}$$

where $|A_{ik}|$ is the determinant formed by the elements of the tensor A_{ik}, and a^{ik} is the cofactor of the element A_{ik}.

2. NONINERTIAL FRAMES OF REFERENCE

Let us examine the Minkowski space. The "distance" between two points (i. e. the interval between two events) in it is determined by the expression

$$ds^2 = dx_0^2 - (dx^2 + dy^2 + dz^2), \tag{2.1}$$

where $x_0 = ct$. Here the three-dimensional space is purely Euclidean, which cannot be said with respect to the four-dimensional space. The latter would have been Euclidean, if the sum of the squares of the differentials had appeared instead of (2.1). The above distance can, however, be given a Euclidean form by introducing the imaginary coordinate $x_4 = ict$. Thus the space—time coordinates x, y, z, t form not a true Euclidean space, but one similar to it, called pseudo-Euclidean.

Special theory of relativity is concerned only with inertial frames of reference, i. e. such that the bodies in them are free from the action of forces and therefore move uniformly and in straight lines. The coordinates of a single event in two inertial systems moving relative to each other along the x axis are related through the Lorentz transformations

$$t = \frac{t' + \frac{v}{c^2}x'}{\sqrt{1 - \frac{v^2}{c^2}}}, \quad x = \frac{x' + vt'}{\sqrt{1 - \frac{v^2}{c^2}}}, \quad y = y', \quad z = z'. \tag{2.2}$$

The quadratic form (2.1) is invariant with respect to these transformations.

Introducing the notations $x_0 = ct$, $x_1 = x$, $x_2 = y$, $x_3 = z$, (2.2) can be written in the form

$$x_i = \alpha_{ik} x_k', \quad x_k' = \alpha_{ik} x_i, \tag{2.3}$$

where the α_{ik} form the matrix

$$(\alpha_{ik}) = \left\{ \begin{array}{cccc} \dfrac{1}{\sqrt{1-\dfrac{v^2}{c^2}}} & \dfrac{v/c}{\sqrt{1-\dfrac{v^2}{c^2}}} & 0 & 0 \\ \dfrac{v/c}{\sqrt{1-\dfrac{v^2}{c^2}}} & \dfrac{1}{\sqrt{1-\dfrac{v^2}{c^2}}} & 0 & 0 \\ 0 & 0 & 1 & 0 \\ 0 & 0 & 0 & 1 \end{array} \right\}. \tag{2.4}$$

The Lorentz transformations are linear and homogeneous, and the coefficients α_{ik} satisfy the condition

$$\alpha_{il}\alpha_{im} = \delta_{lm}. \tag{2.5}$$

The quadratic form (2.1) can be rewritten in the following way:

$$ds^2 = \overset{0}{g}_{ik} dX^i dX^k, \tag{2.6}$$

where the notations $x_0 = X^0$, $x = X^1$, $y = X^2$, $z = X^3$ are used for the coordinates in an inertial system, and the $\overset{0}{g}_{ik}$ form the matrix

$$(\overset{0}{g}_{ik}) = \begin{pmatrix} 1 & 0 & 0 & 0 \\ 0 & -1 & 0 & 0 \\ 0 & 0 & -1 & 0 \\ 0 & 0 & 0 & -1 \end{pmatrix}. \tag{2.7}$$

Let us now examine a noninertial frame of reference, which performs an arbitrary motion with respect to some inertial system. These are assumed to be related through the coordinate transformations

$$X^i = F^i(x^0, x^1, x^2, x^3), \tag{2.8}$$

where the small letters denote the coordinates of an event in the noninertial system. Since ds is a scalar, its value will not change under the transformations (2.8). Substituting (2.8) in (2.6), we find the expression for ds^2 in the new coordinates:

$$ds^2 = g_{ik} dx^i dx^k, \tag{2.9}$$

where

$$g_{ik} = \frac{\partial X^l}{\partial x^i} \frac{\partial X^m}{\partial x^k} \overset{0}{g}_{lm}. \tag{2.10}$$

In general, g_{ik} are complicated functions of the coordinates x^i. The sum of the products of the functions g_{ik} and the coordinate differentials, which constitute a contravariant vector, produces a scalar ds^2. Consequently, the g_{ik} must be the components of a covariant tensor of second rank, and the same can certainly also be said of the $\overset{0}{g}_{ik}$. Comparison between (2.10) and (1.8) then shows that g_{ik} and $\overset{0}{g}_{ik}$ are the components of the same tensor in the noninertial and inertial systems, respectively, and that (2.10) is none other than the formula of their transformation. The above tensor, called the metric tensor, is of fundamental importance, and determines the geometric properties of the space.

The tensor g_{ik} is obviously symmetric with respect to its indices, namely $g_{ik}=g_{ki}$. Consequently, in the general case there are 10 different g_{ik} values in all. According to (2.7), in an inertial system with Cartesian coordinates

$$g_{00} = 1, \quad g_{11} = g_{22} = g_{33} = -1, \quad g_{ik} = 0 \qquad \text{if} \qquad i \neq k.$$

This special form of g_{ik} corresponds to a flat Minkowski world with Euclidean geometry. In a noninertial frame of reference, the four-dimensional coordinate system necessarily becomes curvilinear. This follows from the fact that the quadratic form ds^2 (i.e. the square of the distance between two infinitely close points of the four-dimensional world) is no longer expressed through the simple formula (2.1), which is characteristic of an orthogonal coordinate system with rectilinear axes. The expression for ds^2 now includes the coefficients $g_{ik}(x^l)$, as well as terms with products of the differentials of different coordinates. Figuratively speaking, one can say that space and time have become curved, and in such a system the geometry is obviously not Euclidean.

As an example of a noninertial system, let us consider a disk (system K') which rotates uniformly relative to a stationary plane (the inertial system K). A point on the disk will describe a circle, whose projection on the stationary plane is again a circle. An observer on K will measure the circle's radius as r and its circumference as l, with the relationship $l/r=2\pi$. Let us now analyze the measurements carried out in K' with measuring scales which are at rest in it. Viewed from K, one observes that a scale applied along the circumference undergoes a Lorentz contraction, as a result of which it fits into this circumference more times than in the measurement from the inertial system K. In measurements along the radius the length of the scale remains unchanged. Thus, one obtains $r'=r$, $l'=l\left(1-\frac{\omega^2r^2}{c^2}\right)^{-1/2}$, where ω is the angular velocity. Consequently,

$$\frac{l'}{r'}=\frac{2\pi}{\sqrt{1-\frac{\omega^2r^2}{c^2}}}>2\pi. \tag{2.11}$$

We thus see that the geometry of the rotating disk is not only non-Euclidean, but it also changes with distance from the center. Moreover, with a variable rotation velocity the geometry will also change with time. In going over to noninertial systems we will therefore have to deal with non-Euclidean geometries, in which an important role is played by the components of the metric tensor g_{ik}.

The tensor g^{ik} which is the inverse of g_{ik} is called the contravariant metric tensor:

$$g^{ik}g_{kl}=\delta^i_l, \qquad g^{ik}=\frac{b^{ik}}{g}, \tag{2.12}$$

where $g=|g_{ik}|$ is the determinant of g_{ik}, and b^{ik} is the cofactor of the element g_{ik}. It should be noted that, for a real four-dimensional space, always $g<0$.

With the help of the metric tensors g_{ik} and g^{ik} one can obtain covariant quantities from contravariant ones and vice versa, for example

$$g_{ik}A^k=A_i, \qquad g_{il}A^{lk}=A^k_i, \qquad g_{il}g_{km}A^{lm}=A_{ik},$$

$$g^{ik}A_k=A^i, \qquad g^{il}A_{kl}=A^i_k, \qquad g^{il}g^{km}A_{lm}=A^{ik}.$$

Here A^i and A_i are obviously the contravariant and covariant components of the same physical quantity. This, of course, also applies to the tensors A^{ik}, A_{ik}, A^k_i of the second rank, as well as to those of higher rank. All tensorial quantities have contravariant, covariant, and mixed components. The mixed components of the metric tensor coincide with the Kronecker symbol, $g^k_i=\delta^k_i$.

The metric tensor enables one not only to raise or lower the indices of tensorial quantities, but also to form tensors of lower rank, as well as scalars:

$$g^{lm}A_{iklm} = A_{ik}, \qquad g^{lm}\delta^n_k A_{inlm} = A_{ik},$$
$$A^2 = A_i A^i = g_{ik}A^i A^k = g^{ik}A_i A_k,$$
$$A_i B^i = A^i B_i = g_{ik}A^i B^k = g^{ik}A_i B_k,$$
$$A^{ik}B_{ik} = A_{ik}B^{ik} = A_i^k B_k^i = g_{ik}g_{lm}A^{ik}B^{lm} = g^{ik}g^{lm}A_{ik}B_{lm}.$$

The scalar $g_{ik}A^{ik} = g^{ik}A_{ik} = A_i^i$ is called the trace of the tensor.

The quadratic form (2.9) can also be written in the form

$$ds^2 = g^{ik}dx_i dx_k, \tag{2.13}$$

where $dx_i = g_{ik}dx^k$.

In special theory of relativity, where only the linear transformations (2.3) are permitted, there is no fundamental difference between the contravariant and covariant components of the same quantity:

$$A_0 = \overset{0}{g}_{0i} A^i = A^0, \quad A_\alpha = \overset{0}{g}_{\alpha i} A^i = -A^\alpha \quad (\alpha = 1,2,3). \tag{2.14}$$

The above was based on (2.7). In a three-dimensional Euclidean space, using rectangular Cartesian coordinates, the distance between two infinitely close points is

$$dl^2 = \overset{0}{\gamma}_{\alpha\beta} dx^\alpha dx^\beta \qquad (\alpha,\beta = 1,2,3). \tag{2.15}$$

Here γ^0 is the three-dimensional analogue of the metric tensor; $\overset{0}{\gamma}_{\alpha\beta} = 0$ for $\alpha \neq \beta$, $\overset{0}{\gamma}_{11} = \overset{0}{\gamma}_{22} = \overset{0}{\gamma}_{33} = 1$, and $A_\alpha = \overset{0}{\gamma}_{\alpha\beta}A^\beta = A^\alpha$. As a result, there is no difference between the covariant and contravariant components of tensors.

3. REAL WORLD METRIC

So far we considered only an empty space, but the real space is certainly not like this. In fact, it is filled with celestial objects, cosmic dust, gas, neutrinos, radiation, and the gravitational field created by them. We shall examine the actual properties of the space—time universe, but let us first recall one important property of the motion of particles in a gravitational field. In the nonrelativistic case

$$\ddot{\mathbf{r}} = -\frac{m_g}{m} \operatorname{grad} \varphi(x, y, z, t,), \tag{3.1}$$

where φ is the gravitational potential, and m, m_g are the inertial and gravitational mass respectively.* The first of these is a measure of the inertia of the body and equals the ratio between force and acceleration, and the second is the source of the gravitational field.

High-accuracy experiments show that in a gravitational field the same acceleration is imparted to all bodies, independent of their mass. This important fact was first established by Galileo for the motion of particles in the earth's field. From this Galilean law it follows that the ratio between gravitational and inertial mass is a universal constant, the value of which can depend only on the choice of measurement units. This constant is usually chosen to be unity, i.e. the gravitational and inertial masses are assumed to be identical:

$$m_g = m, \qquad \ddot{\mathbf{r}} = -\operatorname{grad} \varphi. \tag{3.2}$$

This fundamental property of the motion of bodies in a gravitational field enables one to establish a far-reaching analogy between noninertial systems and gravitational fields. Thus in a system moving in a straight line and with uniform acceleration relative to an inertial system, the free motion of objects is uniformly accelerated. An observer in this system who knows nothing about its motion can reach the logical conclusion that in this system there is a uniform gravitational field. On the other hand, an observer in a closed box (Einstein elevator), falling freely in the gravitational field, notes that the objects which experience no force are at rest or move uniformly and in a straight line, i.e. that there is no gravity field present. Systems moving in a straight line but with varying acceleration are equivalent to uniform gravitational fields which vary with time.

The real gravitational fields in nature are not uniform, but decrease with distance from mass and vanish at infinity. Moreover, these fields also change with time due to the movement of mass. It is obvious that such a field does not correspond to one equivalent noninertial reference system, and cannot be derived by means of

* In principle, one could also have introduced the concept of a passive gravitational mass m'_g expressing the ability of a body to receive the effect of an external gravitational field.

a single coordinate transformation. However, the field for infinitely small regions at any point in space can be considered uniform and derived by introducing a free-falling reference system (a local inertial system, also called Galilean), or by specifying an equivalent noninertial system (the principle of local equivalence). Thus, the field at each point in space will be represented by its particular frame of reference, and the whole field by an infinite number of different noninertial systems. Because of the difference in the motion of these noninertial systems at different points, these will obviously also differ in their geometries and clock behavior. We have already shown that in noninertial systems the space—time metric is non-Euclidean, and that the distance between two events is given by a quadratic form of the general type (2.9). From the physical equivalence between a field and a collection of noninertial systems it then follows that in a gravitational field the metric should also be non-Euclidean, and that

$$ds^2 = g_{ik}dx^i dx^k, \tag{3.3}$$

where $g_{ik}(x^0, x^1, x^2, x^3)$ is a function of the space—time coordinates.

We thus reach the conclusion that the gravitational field is determined by the quantities g_{ik} and serves as a measure of the curvature of the four-dimensional world. The real world in which we live is a Riemannian manifold with variable curvature, as opposed to the flat Minkowski world. All this points to the fact that the metric of the world is determined by matter and is not a fixed property of space—time. There is no world without matter. In principle, then, geometry becomes a division of physics, and this, in fact, is the main aspect of Einstein's theory of gravitation.

From the above it is clear that, as opposed to the flat Minkowski world, in an Einstein world with varying gravitational fields there is not just one single frame of reference with rigidly connected objects and a fixed time scale. Here the distance between objects, the length of measuring scales, the flow of time vary from point to point as well as with time. Consequently, in the case of an arbitrary gravitational field the frame of reference is not stable but related to a collection of an infinite number of bodies with their own arbitrary clocks, i. e. the frame of reference itself becomes arbitrary.

In connection with random frames of reference, the question arises: what are the mathematical equations that express the laws of nature? Einstein was convinced that these should have the same form in all systems (the principle of general relativity). This requirement of general covariance is automatically satisfied by tensorial equations, whose form is preserved under any coordinate transformations. The theory of space—time and gravitation which also incorporates the above principle was developed by Einstein and is called the general theory of relativity.

4. TRUE TIME AND DISTANCES

Let us assume that two events take place at some point of space—time, with an interval of coordinate time dx^0 between them. Since $dx^\alpha = 0$, the four-dimensional interval will be $ds^2 = g_{00}dx^{0^2}$. The true, or proper, time between these events is then the ratio $\frac{ds}{c}$, i. e.

$$d\tau = \frac{1}{c}\sqrt{g_{00}}\,dx^0, \quad \tau = \frac{1}{c}\int\limits_{x_1^0}^{x_2^0}\sqrt{g_{00}}\,dx^0, \tag{4.1}$$

where x_1^0 and x_2^0 are the times of occurrence of the two events, which take place at the same point.

We shall now define simultaneity and true (proper) distance. This can be done with the help of a light signal. Let us consider two neighboring points $A(x^\alpha)$ and $B(x^\alpha + dx^\alpha)$ in three-dimensional space, their clocks indicating x_A^0 and $x_B^0 + dx_B^0$. A light signal is sent out from B at the time $x_B^0 + \Delta x_B'^0$, reaches A at x_A^0, and after reflection returns to B at a time $x_B^0 + \Delta x_B''^0$. Clearly, an event simultaneous to the one taking place at A at a time x_A^0, should be one taking place at B at a time

$$x_B^0 = x_A^0 + \frac{1}{2}\left(\Delta x_B'^0 + \Delta x_B''^0\right). \tag{4.2}$$

The true distance between points A and B is half the product of the velocity of light and the interval of proper time, which is the time interval between sending and receiving of the signal at point B:

$$dl = \frac{1}{2}c\,d\tau = \frac{1}{2}\sqrt{g_{00}}\left(\Delta x_B''^0 - \Delta x_B'^0\right). \tag{4.3}$$

For a light signal $ds = 0$, therefore, $\Delta x_B'^0$ and $\Delta x_B''^0$ are determined from the equation

$$ds^2 = g_{00}dx^{0^2} + 2g_{0\alpha}dx^0dx^\alpha + g_{\alpha\beta}dx^\alpha dx^\beta = 0.$$

Solving the above for dx^0 we obtain

$$\left.\begin{aligned}\Delta x'^0_B &= \frac{-g_{0\alpha}dx^\alpha - \sqrt{(g_{0\alpha}g_{0\beta} - g_{00}g_{\alpha\beta})\,dx^\alpha dx^\beta}}{g_{00}},\\ \Delta x''^0_B &= \frac{-g_{0\alpha}dx^\alpha + \sqrt{(g_{0\alpha}g_{0\beta} - g_{00}g_{\alpha\beta})\,dx^\alpha dx^\beta}}{g_{00}},\end{aligned}\right\} \tag{4.4}$$

and from substitution of (4.4) into (4.2),

$$x^0_B = x^0_A - \frac{g_{0\alpha}dx^\alpha}{g_{00}}. \tag{4.5}$$

Thus, by moving from point to point, one can sychronize the clocks along any curve and for any distance between two points A and B:

$$x^0_B = x^0_A - \int\limits_L \frac{g_{0\alpha}}{g_{00}}\,dx^\alpha . \tag{4.6}$$

The above shows that the synchronization of clocks is not a unique procedure, since it depends on the path taken by the light signal. The clocks can be uniquely synchronized only when $g_{0\alpha} = 0$, i. e. in the case of static fields. The quantity x^0 is then usually called the world time.

From (4.3) and (4.4) we find for the true distance

$$dl^2 = \gamma_{\alpha\beta}dx^\alpha dx^\beta, \tag{4.7}$$

where

$$\gamma_{\alpha\beta} = -g_{\alpha\beta} + \frac{g_{0\alpha}g_{0\beta}}{g_{00}}. \tag{4.8}$$

It is easily shown that in transformations of the type $x^\alpha = f^\alpha(x'^1, x'^2, x'^3)$ the quantity g_{00} behaves as a scalar, and the $g_{0\alpha}$ behave as a three-dimensional vector. The product $g_{0\alpha}g_{0\beta}$ is then a tensor of the second rank, and as a result $\gamma_{\alpha\beta}$ acts as a metric tensor for the three-dimensional space. Consequently, (4.7) behaves as a scalar with respect to transformations of the spacelike coordinates. On the other hand, dl^2 is no longer a scalar when the transformations include time. It is easily shown that

$$\gamma^{\alpha\beta} = -g^{\alpha\beta}. \tag{4.9}$$

5. COVARIANT DIFFERENTIATION

The most general laws of nature are expressed in the form of differential equations, and in relativistic gravitation theory there is the additional requirement of general covariance. We shall see below that ordinary differentiation destroys the tensorial property of variables. In order to combine these two properties, it is therefore necessary to reconsider and extend the concepts of differential and derivative. To this end, let us consider the differential of a vector; according to (1.6),

$$dA_i = \frac{\partial x'^k}{\partial x^i} dA'_k + \frac{\partial^2 x'^k}{\partial x^i \partial x^l} A'_k dx^l,$$

where $dA_i = A_i(x^m + dx^m) - A_i(x^m)$. Because of the presence of the second term, dA_i does not transform as (1.6) and is therefore not a covariant vector. This is due to the fact that the transformation coefficients for vectors are themselves functions of the coordinates, thus transforming differently at different points, and that dA_i is the difference between the vector values at points $x^m + dx^m$ and x^m. In the case of linear transformations (e. g. the Lorentz transformations in the Minkowski world) the second term vanishes and dA_i is a vector. The tensorial character of A_i can be maintained in the differential if we first transfer $A_i(x^m)$ to point $x^m + dx^m$ and only then calculate the vector difference. This difference, which we shall call DA_i, is a vector, since the subtraction now involves only values pertaining to a single point. The problem, though, is how to transfer a vector from one point to another. In a flat world the vector is moved parallel to itself, i. e. the angle formed with the direction of movement (or the scalar product between the vector and a unit vector along this direction) remains constant. This direction being a straight line, we now pass a plane through it and then bend it into an arbitrary shape. The straight line thus becomes a geodesic of the new surface, and the scalar product $A_i dl^i$, where dl^i are the components of an arc element along the geodesic, should remain invariant. In other words, in a parallel transfer of a vector along a geodesic, the angle between the vector and the geodesic should remain fixed. This is the generalization of the concept of parallel displacement to the case of a non-Euclidean space.

In a Galilean system, parallel displacement of a vector does not affect its projections along the axes, whereas in a curvilinear coordinate system these do change as a result of the changes in the directions of the axes. With accuracy to terms of the order of dx^i, the change in $A_i(x^m)$ during parallel displacement is proportional to both A_i itself and the displacement dx^l:

$$\delta A_i = \Gamma^k_{il} A_k dx^l, \tag{5.1}$$

where Γ^k_{il} are functions of the coordinates, called the Christoffel symbols. Thus the covariant differential is

$$\begin{aligned} DA_i = A_i(x^m + dx^m) - \left[A_i(x^m) + \Gamma^k_{il} A_k dx^l\right] = \\ = dA_i - \Gamma^k_{il} A_k dx^l = \left(\frac{\partial A_i}{\partial x^l} - \Gamma^k_{il} A_k\right) dx^l, \end{aligned} \tag{5.2}$$

where DA_i is a vector but dA_i is not, since clearly Γ^k_{il} is not a tensor. Therefore the expression

$$\frac{DA_i}{Dx^l} \equiv A_{i;l} = \frac{\partial A_i}{\partial x^l} - \Gamma^k_{il} A_k \tag{5.3}$$

is a tensor and represents the natural extension of the concept of a derivative to the case of curvilinear coordinate systems.

We shall now find the covariant derivative of a contravariant vector A^i. To this end, we make use of the fact that a scalar does not change under parallel displacement, i. e. $\delta(A^k B_k) = 0$, where B_k is an arbitrary covariant vector:

$$\begin{aligned} \delta(A^k B_k) = B_k \delta A^k + A^k \delta B_k = B_i \delta A^i + A^k \Gamma^i_{kl} B_i dx^l = \\ = \left(\delta A^i + \Gamma^i_{kl} A^k dx^l\right) B_i = 0. \end{aligned}$$

From this we obtain

$$\delta A^i = -\Gamma^i_{kl} A^k dx^l \tag{5.4}$$

and consequently

$$DA^i = \left(\frac{\partial A^i}{\partial x^l} + \Gamma^i_{kl} A^k\right) dx^l, \quad A^i_{;l} = \frac{\partial A^i}{\partial x^l} + \Gamma^i_{kl} A^k. \tag{5.5}$$

It is easily shown that the rule for covariant differentiation of vector-component products, such as $A_i B^k C_l$ ..., is the same as that of ordinary differential calculus. Using this, one can easily calculate the covariant derivatives of tensors of any rank, covariant, contravariant, or mixed:

$$\left.\begin{aligned} A_{ik;l} &= \frac{\partial A_{ik}}{\partial x^l} - \Gamma^m_{il} A_{mk} - \Gamma^m_{kl} A_{im}, \\ A^{ik}_{;l} &= \frac{\partial A^{ik}}{\partial x^l} + \Gamma^i_{ml} A^{mk} + \Gamma^k_{ml} A^{im}, \\ A^k_{i;l} &= \frac{\partial A^k_i}{\partial x^l} + \Gamma^k_{ml} A^m_i - \Gamma^m_{il} A^k_m. \end{aligned}\right\} \tag{5.6}$$

By raising the indices of covariant differentiation, we obtain the contravariant derivatives of tensors:

$$g^{kl} A_{i;l} = A_i^{;k}, \qquad g^{kl} A^i_{;l} = A^{i;l}. \tag{5.7}$$

Sometimes Γ^i_{kl} is replaced by the following Christoffel symbol:

$$g_{im} \Gamma^m_{kl} = \Gamma_{i;kl}.$$

An explicit expression for Γ^i_{kl} will be derived in the next section.

In a Galilean reference system $\Gamma^i_{kl} = 0$, therefore covariant differentiation reduces to the usual operation.

6. SOME ESSENTIAL EQUATIONS

It was earlier mentioned that the equations describing the laws of nature are in tensor form and therefore do not change with the frame of reference.* This fact enables one to write these laws in curvilinear coordinates in a universal covariant form once they have been formulated in tensor form for a flat world.

To begin with, we shall use these to obtain explicit expressions for the Christoffel symbols. Let us consider the gradient $A_i = \frac{\partial \varphi}{\partial x^i}$ of an arbitrary scalar; since $\frac{\partial A_i}{\partial x^k} = \frac{\partial A_k}{\partial x^i}$, it follows that

$$A_{i;k} - A_{k,i} = (\Gamma^l_{ki} - \Gamma^l_{ik}) A_l \equiv B_{ik}.$$

In a local Galilean system all Christoffel symbols vanish and $B^0_{ik} = 0$. Consequently, in an arbitrary coordinate system also $B_{ik} = 0$,

* This statement is too rigid; generally speaking, invariant formulation of certain laws can be achieved without using tensors.

which is possible only when

$$\Gamma^l_{ik} = \Gamma^l_{ki}, \tag{6.1}$$

which means that the Christoffel symbols are symmetrical in their lower indices.

According to (2.7), in a locally Galilean system $g^0_{ik;l} = \frac{\partial g^0_{ik}}{\partial x^l} = 0$. Therefore in an arbitrary curvilinear coordinate system also

$$g_{ik;\,l} = \frac{\partial g_{ik}}{\partial x^l} - \Gamma^m_{il}\, g_{mk} - \Gamma^m_{kl}\, g_{im} = 0, \tag{6.2}$$

from which follows the identity

$$\frac{\partial g_{ik}}{\partial x^l} + \frac{\partial g_{lk}}{\partial x^i} - \frac{\partial g_{il}}{\partial x^k} = 2 g_{mk} \Gamma^m_{il}.$$

Finally, multiplication by g^{kn} leads to the desired expressions for the Christoffel symbols:

$$\Gamma^n_{il} = \frac{1}{2} g^{nk} \left(\frac{\partial g_{ki}}{\partial x^l} + \frac{\partial g_{kl}}{\partial x^i} - \frac{\partial g_{il}}{\partial x^k} \right). \tag{6.3}$$

Let us now consider the "free" motion of particles in a gravitational field. In a Galilean system, where there are no forces, a particle moves uniformly in a straight line; therefore $du^i = 0$, where $u^i = \frac{dx^i}{ds}$ is the four-dimensional velocity. According to our earlier discussion, in curvilinear coordinates one should use $Du^i = du^i + \Gamma^i_{kl}\, u^k\, dx^l$, i. e.

$$\frac{du^i}{ds} + \Gamma^i_{kl} u^k\, u^l = 0$$

or

$$\frac{d^2 x^i}{ds^2} + \Gamma^i_{kl} \frac{dx^k}{ds} \frac{dx^l}{ds} = 0. \tag{6.4}$$

It is known from Riemannian geometry that the latter is at the same time also the equation of a geodesic line. Thus in a space bent by gravitation, the "free" motion of a particle takes place along a geodesic line. In terms of the Newtonian theory of gravitation, the expression $m\Gamma^i_{kl} u^k u^l$ should be seen as a component of a four-dimensional force acting on the particle. As a matter of fact, in the Newtonian theory this becomes $m\frac{\partial\varphi}{\partial x^i}$.

In the presence of an electromagnetic field, a charged particle moves according to

$$\frac{du^i}{ds}+\Gamma^i_{kl}u^k u^l=\frac{e}{mc^2}F^{ik}u_k, \tag{6.5}$$

where $F_{ik}=A_{k;i}-A_{i;k}=\frac{\partial A_k}{\partial x^i}-\frac{\partial A_i}{\partial x^k}$, A_i is the four-dimensional potential of the electromagnetic field, and e is the charge of the particle.

Finally, let us examine the equations of hydrodynamics. In special theory of relativity, these equations are written as $\frac{\partial T_{ik}}{\partial x^k}=0$, where T_{ik} is the energy-momentum tensor of a continuous medium:

$$T_{ik}=(P+\rho)u_i u_k-Pg_{ik}. \tag{6.6}$$

Here P is pressure and ρ is the density of total energy (including the particles' rest energy) in the proper (accompanying) reference system, i. e. in the system attached to the moving mass. In a curved space, the equations of hydrodynamics assume the following form:

$$T^k_{i;k}=\frac{\partial T^k_i}{\partial x^k}+\Gamma^k_{lk}T^l_i-\Gamma^l_{ik}T^k_l=0.$$

With the help of a few transformations and the symmetry $T_{ik}=T_{ki}$, this equation can be written as

$$\frac{1}{\sqrt{-g}}\frac{\partial(\sqrt{-g}T^k_i)}{\partial x^k}-\frac{1}{2}T^{kl}\frac{\partial g_{kl}}{\partial x^i}=0. \tag{6.7}$$

7. THE CURVATURE TENSOR

In a space—time bent by gravitation, the higher covariant derivatives of vectors and tensors are dependent upon the order of differentiation. Using the rules of covariant differentiation presented in Section 5, it is easily shown that

$$\left.\begin{aligned} A_{i;\,k;\,l} - A_{i;\,l;\,k} &= R^m_{ikl} A_m, \\ A^i_{;k;\,l} - A^i_{;l;\,k} &= -R^i_{mkl} A^m, \end{aligned}\right\} \tag{7.1}$$

where R^m_{ikl} is called the curvature, or Riemann, tensor:

$$R^i_{klm} = \frac{\partial \Gamma^i_{km}}{\partial x^l} - \frac{\partial \Gamma^i_{kl}}{\partial x^m} + \Gamma^i_{nl}\Gamma^n_{km} - \Gamma^i_{nm}\Gamma^n_{kl}. \tag{7.2}$$

In a Galilean coordinate system of Minkowski space, an arbitrary vector obeys $A_{i;\,k;\,l} = A_{i;\,l;\,k} = \frac{\partial^2 A_i}{\partial x^k \partial x^l}$, which means that $R^m_{ikl} = 0$. However, if all components of a tensor are equal to zero in a particular coordinate system of a flat space, they have to remain so in any other reference system in that space. From this it follows that in a pseudo-Euclidean (flat) space—time all components of the curvature tensor have to vanish.

The tensor R^i_{klm} is involved even in parallel translations of vectors along infinitesimal closed contours. Thus, in a parallel translation of vectors A_k and A^k along a closed contour bounding a small two-dimensional surface Δf^{lm} one obtains /66/

$$\left.\begin{aligned} \Delta A_k &= \frac{1}{2} R^i_{klm} A_i \,\Delta f^{lm}, \\ \Delta A^k &= -\frac{1}{2} R^k_{ilm} A^i \,\Delta f^{lm}. \end{aligned}\right\} \tag{7.3}$$

Let all components of the tensor R^i_{klm} be zero everywhere in four-dimensional space. The components of vectors will then remain unchanged during parallel displacements. In particular, one can thus construct an inertial coordinate system at some point of space—time, and then fill the space with identical systems by means of parallel displacement. Therefore $R^i_{klm} = 0$ is not only a necessary condition, but also a sufficient condition, for the flatness of space—time. This means that, in this case, the metric tensor g_{ik} can be reduced to the form (2.7) through an appropriate coordinate transformation.

We should note the following symmetry properties of the curvature tensor, which follow from (7.2):

$$\left.\begin{aligned} &R^i_{klm} = -R^i_{kml},\\ &R^i_{klm} + R^i_{mkl} + R^i_{lmk} = 0,\\ &R_{iklm} = -R_{kilm} = -R_{ikml} = R_{lmik},\\ &R_{iklm} + R_{imkl} + R_{ilmk} = 0,\\ &R^n_{ikl;m} + R^n_{imk;l} + R^n_{ilm;k} = 0, \end{aligned}\right\} \tag{7.4}$$

where

$$R_{iklm} = g_{in} R^n_{klm}. \tag{7.5}$$

The fourth identity holds for any fixed index. The fifth is the Bianchi identity, from which one can obtain the relatively simple identity

$$R^k_{i;k} = \frac{1}{2}\frac{\partial R}{\partial x^i}. \tag{7.6}$$

through contraction of indices. Because of these symmetry properties, the number of independent components of R^i_{klm} is 20 instead of the expected 256.

In addition to the curvature tensor, an important quantity is also the tensor

$$R_{ik} = \delta^l_m R^m_{i.k} = R^l_{ilk}, \tag{7.7}$$

which is called the Ricci tensor. From (6.2) it follows that

$$R_{ik} = \frac{\partial \Gamma^l_{ik}}{\partial x^l} - \frac{\partial \Gamma^l_{il}}{\partial x^k} + \Gamma^l_{ik}\Gamma^m_{lm} - \Gamma^m_{il}\Gamma^l_{km}. \tag{7.8}$$

It is immediately evident that this tensor is symmetrical, i. e. $R_{ik} = R_{ki}$. Contraction of the Ricci tensor yields the invariant

$$R = g^{ik} R_{ik} = R^i_i, \tag{7.9}$$

which is called the scalar curvature, in analogy with the Gaussian curvature of a surface in ordinary three-dimensional space.

In contrast to the Riemann tensor, $R_{ik}=0$ and $R=0$ does not mean that space—time is flat — these conditions are necessary but not sufficient for flatness. In fact, it will be shown in the next section that $R=0$ and $R_{ik}=0$ even in curved space at points not occupied by matter.

8. THE EINSTEIN EQUATIONS

In mathematical field theory, a logically elegant system may be set up by first adopting a suitable expression for the action, and then using the extremum principle for the action to obtain the field equations and the equations of motion of the particles. In relativistic gravitation theory, the basic expression for the action is

$$S=\frac{1}{c}\int\left(\Lambda+\frac{c^2}{2\chi_0}R\right)\sqrt{-g}\,d\Omega, \tag{8.1}$$

where Λ is the Lagrangian density for the substance, R is the scalar curvature of space—time, $\chi_0=\frac{8\pi k_0}{c^2}$, and k_0 is Newton's gravitation constant.

In order to obtain the equations for the gravitation field, the trajectories of the particles are fixed and the varying quantities are those characterizing the field, namely g_{ik}. To obtain the particle-motion equations, the g_{ik} functions remain fixed and the coordinates x^i are subjected to virtual variation. From $\delta S=0$ and the fact that the variations δg_{ik} and δx^i are arbitrary one then obtains the equations

$$R_{ik}-\frac{1}{2}Rg_{ik}=\frac{\chi_0}{c^2}T_{ik}, \tag{8.2}$$

$$T^k_{i;k}=0, \tag{8.3}$$

where T_{ik} is the energy-momentum tensor for the substance; it is related to the scalar Λ through the formula

$$T_{ik}=\frac{2}{\sqrt{-g}}\left(\frac{\partial(\sqrt{-g}\Lambda)}{\partial g^{ik}}-\frac{\partial\partial(\sqrt{-g}\Lambda)}{\partial x^l\,\partial\left(\frac{\partial g^{ik}}{\partial x^l}\right)}\right). \tag{8.4}$$

In the case of a continuous medium this yields expression (6.6). In the presence of an electromagnetic field one has to replace Λ by

$\Lambda+\Lambda^{(n)}$ and T_{ik} by $T_{ik}+T_{ik}^{(n)}$, where

$$\left.\begin{aligned} \Lambda^{(n)} &= -\frac{1}{16\pi} g^{il} g^{km} F_{ik} F_{lm}, \\ T_{ik}^{(n)} &= \frac{1}{4\pi}\left(-F_{il}F_{k}^{l} + \frac{1}{4} g_{ik} F_{lm} F^{lm}\right). \end{aligned}\right\} \tag{8.5}$$

Equation (8.3), which is the hydrodynamics equation, was already given in Section 6.

Equations (8.2) are the Einstein equations. They can also be written in the equivalent forms

$$\left.\begin{aligned} R_{i}^{k} - \frac{1}{2} R\delta_{i}^{k} &= \frac{\chi_0}{c^2} T_{i}^{k}, \\ T_{ik} - \frac{1}{2} T g_{ik} &= \frac{c^2}{\chi_0} R_{ik}, \\ R &= -\frac{\chi_0}{c^2} T, \end{aligned}\right\} \tag{8.6}$$

where $T=T_{k}^{k}$ is the trace (sum of the diagonal elements) of the energy-momentum tensor. Outside the mass distribution (with no magnetic field),

$$R_{ik}=0. \tag{8.7}$$

The Einstein equations have an important property, namely that they also contain the equations of motion for matter. In fact, covariant differentiation of the first (8.6) equation with respect to k, together with the identity (7.6), immediately yields the relationship $T_{i;k}^{k}=0$.

It is not difficult to show that the Einstein equations also contain the Maxwell equations (as well as the equations for other fields). In this respect, relativistic gravitation theory occupies a special place among the theories for other physical fields. In the case of electromagnetic, meson, and weak forces, the field equations do not incorporate the equations for the motion of matter.

In equations (8.2) for the gravitational field, the unknown functions are the elements of the metric tensor, velocity, pressure, energy density, and temperature. Not all these functions, however, are independent. Of the ten g_{ik}, four are determined through proper choice of the coordinates x^i; the components of velocity obey the relationships $u^k u_k=1$; and the pressure P, the energy density ρ, and the temperature T are related through the equation of state. Thus only 11 unknown functions remain. In theoretical calculations for celestial bodies (polytropic models or the configurations of degenerate gaseous masses), temperature can often be excluded. In general, the Einstein equations are of the second order in the time—space coordinates. The choice of initial and boundary conditions will be considered later on during the discussion of specific problems.

Let us consider a weak gravitation field created by masses moving with relatively low velocities. Obviously, in this case one can choose the reference system so that it will be very close to an inertial system:

$$g_{ik} = g^0_{ik} + h_{ik}, \tag{8.8}$$

where h_{ik} are small corrections, and the g^0_{ik} are the elements of the matrix (2.7):

$$g^0_{00} = 1, \quad g^0_{11} = g^0_{22} = g^0_{33} = -1, \quad g^0_{0\alpha} = g^0_{\alpha\beta} = 0, \quad \alpha \neq \beta.$$

In all discussions of the general theory of relativity it is shown that when (8.8) is substituted in (8.2) and some additional restrictions are imposed on the choice of coordinates (see /66/ and /53/), then in the first approximation in h_{ik} one obtains

$$\Delta h_{ik} - \frac{1}{c^2}\frac{\partial^2 h_{ik}}{\partial t^2} = -g^0_{ik}\chi_0\rho, \tag{8.9}$$

where Δ is the Laplace operator. The solution of (8.9) is

$$h_{ik} = -g^0_{ik}\frac{\chi_0}{4\pi}\int\left(\frac{\rho dv}{r}\right)_{t-r/c} = -g^0_{ik}\frac{2\varphi}{c^2},$$

where φ is the Newtonian potential of the gravitational field. From this,

$$g_{00} = 1 + \frac{2\varphi}{c^2}, \quad g_{11} = g_{22} = g_{33} = -1 + \frac{2\varphi}{c^2}. \tag{8.10}$$

Using the above result, we can rewrite (8.9) for $i=k=0$ in the form

$$\Delta\varphi - \frac{1}{c^2}\frac{\partial^2\varphi}{\partial t^2} = 4\pi k_0\rho. \tag{8.11}$$

This is the well-known wave equation of the nonrelativistic theory of gravitation. Thus, in the limiting case of a weak gravitational field, the Einstein equations yield Newton's law of gravitation.

9. SPHERICALLY SYMMETRICAL GRAVITATIONAL FIELDS

An important type of gravitational field is one which is centrally symmetrical. It is created by a spherically symmetrical distribution of mass, which is at rest or performs an arbitrary radial motion. There is a certain idealization in assuming this model of the physical situation, which implies a celestial body not rotating and remote from other bodies. Often, however, these two conditions are fulfilled with sufficiently high accuracy. In the case of central symmetry the Einstein equations are considerably simplified, and in a number of interesting cases they provide both analytical and numerical solutions. As a result, this enables us to work out the main features of the relativistic theory of gravitation and to arrive at a number of new effects, which permit its experimental verification.

It is here convenient to use the spherical coordinates r, θ, φ, and we shall introduce the notation $r=x^1, \theta=x^2$, $\varphi=x^3$. Using the property of central symmetry, as well as the arbitrariness involved in choosing the coordinates x^0 and r, it is possible to show that the metric assumes the following form:

$$dS^2 = e^{\varphi}\, dx^{0^2} - e^{\lambda}\, dr^2 - r^2\,(d\theta^2 + \sin^2\theta\, d\varphi^2), \tag{9.1}$$

where ν and λ are unknown functions of r and x^0. The components of the metric tensor are

$$\left.\begin{aligned} g_{00} &= e^{\nu}, \; g_{11} = -e^{\lambda}, \; g_{22} = -r^2, \; g_{33} = -r^2\sin^2\theta, \\ g^{00} &= e^{-\nu}, g^{11} = -e^{-\lambda}, g^{22} = -r^{-2}, g^{33} = -r^{-2}\sin^{-2}\theta. \end{aligned}\right\} \tag{9.2}$$

The contravariant components were calculated from equation (2.12). The tensor R_{ik} is calculated from (7.8), and the energy-momentum tensor from (6.6). Finally, from (8.2) we obtain the following system of equations:

$$\left.\begin{aligned} &-e^{-\lambda}\left(\frac{\nu'}{r} + \frac{1}{r^2}\right) + \frac{1}{r^2} = \frac{\chi_0}{c^2} T^1_1, \\ &-\frac{1}{2}e^{-\lambda}\left(\nu'' + \frac{\nu'^2}{2} + \frac{\nu' - \lambda'}{r} - \frac{\nu'\lambda'}{2}\right) + \\ &+ \frac{1}{2}e^{-\nu}\left(\ddot{\lambda} + \frac{\dot{\lambda}^2}{2} - \frac{\dot{\lambda}\,\dot{\nu}}{2}\right) = \frac{\chi_0}{c^2} T^2_2 = \frac{\chi_0}{c^2} T^3_3, \\ &-e^{-\lambda}\left(\frac{1}{r^2} - \frac{\lambda'}{r}\right) + \frac{1}{r^2} = \frac{\chi_0}{c^2} T^0_0, \\ &-e^{-\lambda}\frac{\dot{\lambda}}{r} = \frac{\chi_0}{c^2} T^1_0, \end{aligned}\right\} \tag{9.3}$$

where the prime and dot indicate differentiation with respect to r and x^0, respectively.

In the general case, for equations of state which present a physical interest one can find an interior solution only through numerical integration. Analytical solutions exist for an incompressible fluid ($\rho=\text{const}$) and for a region where $P=\rho=0$. In the latter case, it is known as the Schwarzschild solution:

$$ds^2 = \left(1 - \frac{r_g}{r}\right) dx^{0^2} - \frac{dr^2}{1 - \frac{r_g}{r}} - r^2 (d\theta^2 + \sin^2 d\varphi^2), \qquad (9.4)$$

where $r_g = 2k_0 M/c^2$ is the so-called gravitational radius of the body. According to Newton's gravitation theory, at a radius $R \approx r_g$ the gravitational energy of a celestial body becomes of the order of its total energy: $Mc^2 \approx K_0 M^2/R$. The gravitational radii of the known celestial bodies are smaller by many orders of magnitude than their actual radii, e.g. $r_g = 2.956$ km for the sun and $r_g = 0.443$ cm for the earth.

Expression (9.4) constitutes the metric for the region of space outside a spherical celestial body. When $R < r_g$, there exist a singular region $0 \leqslant r \leqslant r_g$ and a singular surface $r = r_g$ (the singular Schwarzschild surface). When $r < r_g$, the character of the metric changes: x^0 becomes spacelike, r becomes timelike, $g_{00}(r_g) = 0$, and $g_{11}(r_g) = \infty$. In the case of real celestial bodies this situation does not occur, since always $R \gg r_g$. For a finite coordinate time dt, the proper time $d\tau = \left(1 - \frac{r_g}{r}\right)^{1/2} dt$ tends to 0 when $r \to r_g$; on the other hand, a finite $d\tau$ on the singular Schwarzschild surface corresponds to an infinite time for a distant observer:

$$dt = d\tau \left(1 - \frac{r_g}{r}\right)^{-1/2} \to \infty \quad \text{when } r \to r_g.$$

This peculiarity of the metric can, however, be eliminated through special choice of the frame of reference. This was first done by Lemaître with the following coordinate transformation:

$$\left.\begin{aligned} x_0 &= \pm c\tau - 2\sqrt{r_g r} + r_g \ln\left|\frac{\sqrt{r} + \sqrt{r_g}}{\sqrt{r} - \sqrt{r_g}}\right|, \\ r &= r_g \left[\frac{3}{2r_g}(R \mp c\tau)\right]^{2\,3}. \end{aligned}\right\} \qquad (9.5)$$

In the new system

$$ds^2 = c^2 d\tau^2 - \left[\frac{3}{2r_g}(R \mp c\tau)\right]^{-2/3} dR^2 - \\ - \left[\frac{3}{2r_g}(R \mp c\tau)\right]^{4/3} - r_g^2 (d\theta^2 + \sin^2\theta d\varphi^2). \qquad (9.6)$$

As can be seen, the new expression for ds^2 has no singularity at $r = r_g$, the singularity at $r = 0$ remains, but the metric has now become nonstationary. In the Lemaître reference system the clocks can be unequivocally synchronized, since now also $g_{0\alpha} = 0$. Moreover, the time coordinate now coincides everywhere with the proper time ($g_{00} = 1$). Such a reference system is called synchronous. The time lines (i.e. the world lines with $x^\alpha = \text{const}$, $\alpha = 1, 2, 3$, and consequently $u^\alpha = 0$, $u^0 = 1$, where $u^k = \frac{dx^k}{ds}$ is the four-dimensional velocity) in this system are geodesic. Indeed, they obey the equation

$$\frac{du^i}{ds} + \Gamma^i_{kl} u^k u^l = 0.$$

Of course, a geodesic line remains such in all frames of reference. From this it follows that an object which is at rest with respect to a Lemaître system (including, of course, the system itself) will move freely in a Schwarzschild system, i.e. it will fall freely (along geodesics) toward the center or escape to infinity along the same lines. Simple analysis shows that the upper and lower signs in (9.5) and (9.6) give rise to the first and second alternatives, respectively. Thus at each point of the Lemaître system there are free-falling (upper sign) or radially dispersing (lower sign) objects. In the first case all objects, and also the star's surface itself, fall freely toward the center (in the Schwarzschild reference system, of course), which means that there is an unchecked collapse of the celestial body. In the second case, all objects, including the surface, escape into infinity with a decelerated motion (i.e. along geodesic lines); this picture corresponds to a powerful explosion, which converts a superdense celestial body into diffuse matter.

In the last chapter we shall present a modified version of the theory of gravitation, which lacks the singularity inherent in the Schwarzschild solution.

At sufficiently large distances, with accuracy to terms of the order of $1/r$, the metric (9.4) reduces to

$$ds^2 = ds_0^2 - \frac{r_g}{r}(dx^{0^2} + dr^2), \tag{9.7}$$

where ds_0^2 is the square of an interval in the Minkowski world. When $r \to \infty$, $ds \to ds_0$. The above formula also expresses the metric of a celestial body of arbitrary shape, since at very large distances deviations from spherical symmetry decrease faster than r^{-n}, where $n \geqslant 3$.

10. EXPERIMENTAL FOUNDATIONS FOR RELATIVISTIC GRAVITATION THEORY

Before discussing the experimental confirmation of Einstein's theory of gravitation, one should check whether it agrees with Newton's gravitation theory. In the preceding section it was shown that in the case of sufficiently weak fields there is complete agreement. In strong fields there are considerable deviations from the above, resulting in a series of new effects. Sooner or later these effects will be compared with astronomical data and checked experimentally, but for the time being this is possible only for three effects, which play a fundamental role in general theory of relativity. These effects are the precession of a planet's orbit, the red shift, and the deflection of light rays in a gravitational field. Below we shall present a summary of these, without going into mathematical details.

Let us consider a finite motion of a test body around some massive celestial body with spherical shape. In nonrelativistic theory, the orbit of the test particle is an ellipse, therefore the value r of the radius should recur whenever φ changes by 2π. Here r is the distance of the test body from the right-hand focus, and φ is the angle measured from $r_{\min}$, i. e. from the perihelion. On the other hand, calculation based on general theory of relativity shows that the orbit of the material point is not a simple ellipse. It can, however, be represented as an ellipse performing a precession in the orbital plane, with its perihelion shifting in the direction of the particle's motion. In one full revolution, when φ increases by 2π, the perihelion moves through some angle $\Delta\varphi$, so that the value of $r_{\min}$ is now repeated when the angle φ increases by $2\pi + \Delta\varphi$, where

$$\Delta\varphi = \frac{6\pi k_0 M}{ac^2(1-e^2)}. \tag{10.1}$$

Here M is the mass of the celestial body, a is the semimajor axis of the ellipse, and e its eccentricity. For all planets of the solar system the numerical value of $\Delta\varphi$ is very small. Thus, over one century, it is only 3.8'' for the earth and 43'' for Mercury. This precession is many times smaller than the one appearing in Newtonian theory as a result of perturbations from other planets and deviation of their shapes from spherical. The accuracy of astronomical observations is such, however, that it was possible in the case of Mercury to confirm the prediction for $\Delta\varphi$ to within 3%.

As to the second effect, simple examination of light propagation in a gravitational field reveals that its rays are not straight lines. As an example, let us consider a ray of light which comes from infinity, passes by a clestial body, and continues into infinity. During the above it deviates from a straight line, turning toward the center through an angle

$$\Delta\varphi = \frac{2r_g}{r}, \tag{10.2}$$

where r is the impact parameter. In the nonrelativistic limit, when $r_g = 0$, this becomes $\Delta\varphi = 0$, therefore the above deviation is a relativistic effect. The largest effect is obtained when $r = R$, R being the body's radius. For the earth this is equal to 0.0057'', and for the sun it is 1.75''. Observations made during the solar eclipse in 1952 gave the result $\Delta\varphi = 1.70''$, which is in good agreement with the theoretical value.

Finally, let us discuss the red-shift effect. In a Minkowski world, a plane monochromatic wave is characterized by the four-dimensional wave vector $k^i\left(\frac{\omega}{c}, \mathbf{k}\right)$, whose square is by definition equal to zero:

$$g_{ik}^{(0)} k^i k^k = \frac{\omega^2}{c^2} - k^2 = 0. \tag{10.3}$$

The scalar product of k_i by x^i gives the phase of the wave, called the eikonal in the general case:

$$\psi(x^i) = \mathbf{kr} - \omega t = -g_{ik}^{(0)} k^i x^i,$$

from which we obtain

$$k_i = -\frac{\partial\psi}{\partial x^i}, \quad \omega = -c\frac{\partial\psi}{\partial x^0}. \tag{10.4}$$

When the wave is not plane the eikonal is no longer expressed by the simple formula $-k_i x^i$, but the relationship (10.4) still holds With the help of (10.4), (10.3) can be rewritten as

$$g^{(0)ik} \frac{\partial\psi}{\partial x^i} \cdot \frac{\partial\psi}{\partial x^k} = 0,$$

which is called the eikonal equation. When a gravitational field is present, this becomes

$$g^{ik} \frac{\partial\psi}{\partial x^i} \frac{\partial\psi}{\partial x^k} = 0. \qquad (10.5)$$

In a constant gravitational field the eikonal does not contain time in an explicit form; therefore the frequency $\omega_0 = -\frac{\partial\psi}{\partial t}$, which is measured with the world time $t = \frac{x^0}{c}$, is an integral of equation (10.5), i.e. it does not change along the ray. An observer, on the other hand, understands as frequency the one measured with the proper time $\tau = \sqrt{g_{00}}\, t$, namely

$$\omega = -\frac{\partial\psi}{\partial\tau} = -\frac{\partial\psi}{\partial t}\frac{\partial t}{\partial\tau} = \frac{\omega_0}{\sqrt{g_{00}}}. \qquad (10.6)$$

If light is emitted at point 1 and observed at point 2, then from (10.6) it follows that

$$\omega_2 = \sqrt{\frac{g_{00}(1)}{g_{00}(2)}}\,\omega_1, \qquad (10.7)$$

where $g_{00}(1)$ and $g_{00}(2)$ are the values of $g_{00}(x^i)$ at points 1 and 2, respectively. In the case of weak fields $g_{00} = 1 + \frac{2\varphi}{c^2}$ and

$$\omega_2 \approx \omega_1\left(1 + \frac{\varphi_1 - \varphi_2}{c_2}\right), \quad \Delta\omega \approx \omega_1 \frac{\varphi_1 - \varphi_2}{c^2}. \qquad (10.8)$$

Since always $\varphi < 0$, the frequency decreases when $|\varphi_1| > |\varphi_2|$ and increases when $|\varphi_1| < |\varphi_2|$. Thus, the light arriving at the earth from the sun's surface undergoes a red shift; and conversely, light traveling from the earth to the sun experiences a blue shift. This effect has also been confirmed experimentally. Using the Mössbauer effect, Pound and Rebka succeeded in an ingenious way in measuring the shift in light frequency at two different heights above the earth's surface.

Chapter II

THE PROPERTIES OF MATTER AT LOW TEMPERATURES AND HIGH DENSITIES

11. SYSTEM OF IDENTICAL PARTICLES

Let us consider a system containing a large number N of identical particles, which is either isolated or thermostatically controlled. The particles being identical here means that they all have exactly the same mass, charge, spin, and interaction law. The last requirement results from the first three. Because of the interaction between the particles, thermal equilibrium is established after a certain time, and is characterized by their distribution over all possible quantum states. Let us assume that the interaction is sufficiently weak, so that each particle moves independently of the others. Under these conditions the system can be approximated by single-particle states, which are obtained from the Schrödinger equation $H\psi_k = \varepsilon_k \psi_k$. The state of the whole ensemble is then determined by the number of particles in the different single-particle states, called *occupation numbers*. These occupation numbers undergo fluctuations but their mean value $\bar{n}_k$ remains constant during thermal equilibrium.

In nature there are two totally different types of distribution functions pertaining to two types of ensembles, whose particles have half-integral or integral spins. In the case of half-integral spin the wave function is antisymmetric: it changes sign whenever the coordinates and spin of two particles are interchanged. As a result, each quantum state can hold no more than one particle, namely $n_k = 0, 1$ (the Pauli principle). These particles are called *fermions*, and they obey the Fermi-Dirac statistics. In the case of integral spin the wave function is symmetric, and consequently each quantum state can hold any number of particles: $n_k = 0, 1, 2, 3, \ldots$ These particles are called *bosons* and obey the Bose-Einstein statistics.

In statistical physics it is shown that in statistical equilibrium

$$\bar{n}_k = \frac{1}{e^{\frac{\varepsilon_k - \mu}{kT}} \pm 1}. \tag{11.1}$$

Here $\bar{n}_k$ is the average occupation number of the k-th state with energy ε_k, T is temperature, k is the Boltzmann constant, and μ is a parameter called the chemical potential, which depends on temperature and the particle density N/V. The plus sign is for Fermi-Dirac statistics, the minus sign for Bose-Einstein statistics. The chemical potential is determined from the condition

$$\sum_k n_k = N, \tag{11.2}$$

where summation is over all single-particle states. The energy of the system is

$$\sum_k \varepsilon_k n_k = E. \tag{11.3}$$

When $\bar{n}_k \ll 1$, which happens when $\varepsilon_k - \mu \gg k$, equation (11.1) becomes

$$\bar{n}_k = e^{\frac{\varepsilon_k - \mu}{kT}}, \tag{11.4}$$

which is the classical *Maxwell-Boltzmann* distribution (Boltzmann statistics). In the next section it will be shown that the condition $\varepsilon_k - \mu \gg kT$ for a Boltzmann distribution is equivalent to

$$\frac{N}{V} \ll (2\pi mkT/h^2)^{3/2}, \tag{11.5}$$

where V is the volume of the gas and h is the Planck constant divided by 2π.

Henceforth, we shall consider only translational motion of the particles in the gas. In this case the energy assumes the continuous values

$$\varepsilon = c\,(m^2c^2 + p_x^2 + p_y^2 + p_z^2)^{1/2}, \tag{11.6}$$

and in the nonrelativistic limit of small velocities $v \ll c$

$$\varepsilon = mc^2 + \frac{1}{2m}(p_x^2 + p_y^2 + p_z^2). \tag{11.7}$$

We can now speak of the number of particles in the momentum interval $(P_k, P_k + dP_k)$, $k = x, y, z$. This is equal to $\bar{n}_p \Delta Z_p$, where ΔZ_p is the number of quantum states in this momentum interval. From quantum mechanics it is known that

$$\Delta Z_p = (2s + 1) \frac{dP_x \, dP_y \, dP_z}{(2\pi\hbar)^3} dV, \tag{11.8}$$

where dV is the three-dimensional volume element and s is spin. By changing to spherical coordinates in momentum space and integrating over the angles, we obtain the distribution over the absolute values of momentum

$$dN_p = \frac{(2s + 1) V p^2 dp}{2\pi^2 \hbar^3 \left(e^{\frac{\varepsilon - \mu}{kT}} \pm 1\right)}. \tag{11.9}$$

12. DEGENERATE ELECTRON GAS

Below we shall see that superdense stars of the white-dwarf type contain a free electron gas, which satisfies special conditions similar to those existing in metals. In both cases we can, without much error, assume $T = 0$ in explaining a number of important physical effects. It is true that the temperature of white dwarfs can be very high, but we shall see that the main statistical properties of the electron gas do not differ markedly from those obtained at absolute zero.

For the distribution of electrons with respect to the absolute value of momentum we obtain from (11.9) (for electrons $s = 1/2$)

$$dN_p = \frac{V}{\pi^2 \hbar^3} \frac{p^2 dp}{e^{\frac{\varepsilon - \mu}{kT}} + 1}. \tag{12.1}$$

From this we see that, when $T = 0$, $\varepsilon > \mu$ yields $dN_p = 0$ and $\varepsilon < \mu$ yields

$$dN_p = V \cdot \frac{p^2 dp}{\pi^2 h^3}. \qquad (12.2)$$

Here the left-hand side is the number of particles with momentum $(p, p+dp)$, and the right-hand side represents the number of quantum states (see equation (11.8)) in the same momentum interval.

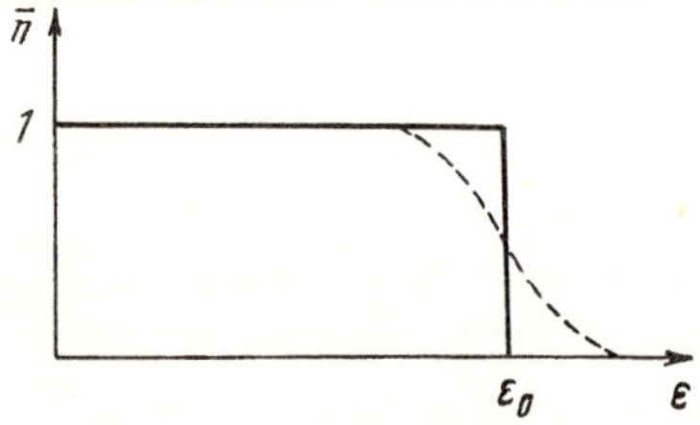

FIGURE 1. Average occupation numbers in a degenerate Fermi gas. The solid line gives the particle distribution at absolute zero, the dashed line at $T \ll T_0$.

Thus when $T=0$ all quantum states (the state of a free electron is determined by the projections p_x, p_y, p_z of the momentum and s_z of the spin, which takes the values $\pm h/2$) with energy $\varepsilon=\mu$ are filled and those with energy $\varepsilon>\mu$ are empty. At $\varepsilon=\mu$ there is a discontinuity in the distribution function — the occupation number drops abruptly from one to zero. The solid line in Figure 1 shows this distribution, which is referred to as degenerate (a Maxwell-Boltzmann momentum distribution is designated as normal).

The above result is obvious. At absolute zero the energy of the system should be the lowest possible. It would seem that all electrons should therefore be in a state of rest, i. e. be concentrated at the lowest level with the rest energy $m_e c^2$. On the other hand, they have to obey the Pauli principle (which was used in deriving the distribution function (11.1)), according to which there can be no more than one particle in each state. Therefore the electrons fill all states, one per state, from the one with the lowest energy $m_e c^2$ to some maximum energy ε_e, called the *Fermi energy*. From the previous discussion it follows that the Fermi energy is equal to the chemical potential of the gas

$$\varepsilon_0 = \mu. \qquad (12.3)$$

In a Bose gas there is no restriction on the occupation numbers, therefore at $T=0$ all particles are concentrated at the level $\varepsilon = mc^2$.

The total number of particles is obtained from (12.2) by integrating from zero to the so-called *Fermi momentum* p_0, which corresponds to the energy ε_0:

$$N = V \frac{p_0^3}{3\pi^2 h^3}, \quad \varepsilon_0 = c\,(m_e^2 c^2 + p_0^2)^{1/2}.$$

Consequently,

$$p_0 = (3\pi^2)^{1/3} \cdot h \cdot n^{1/3}, \tag{12.4}$$

where $n = N/V$ is the particle density.* We thus obtain for the kinetic Fermi energy

$$\varepsilon_0' = \varepsilon_0 - m_e c^2 = \frac{1}{2m_e} (3\pi^2)^{2/3} \cdot h^2 \cdot n^{2/3} \text{ for } p_0 \ll m_e c, \tag{12.5}$$

$$\varepsilon_0 = (3\pi^2)^{1/3} \cdot h \cdot c \cdot n^{1/3} \text{ for } p_0 \gg m_e c. \tag{12.6}$$

The total energy of the electron gas is

$$E = c\int_0^{p_0} (m_e^2 c^2 + p^2)^{1/2} dN_p = \frac{V \cdot c}{\pi^2 h^3} \int_0^{p_0} (m_e^2 c^2 + p^2)^{1/2} p^2 dp =$$

$$= V \frac{m_e^4 c^5}{8\pi^2 h^3} [x_e (1 + 2x_e^2)(1 + x_e^2)^{1/2} - \ln (x_e + \sqrt{1 + x_e^2})], \tag{12.7}$$

where $x_e = \frac{p_0}{m_e c}$ is the Fermi momentum, measured in units of $m_e \cdot c$.

The above equations are also applicable to cases when the temperature is not too high. Thus, if $kT \ll \varepsilon_0 - m_e c^2$, the thermal energy is not sufficient to disturb the main part of the degenerate distribution. Only the levels lying in a thin layer around the boundary $\varepsilon = \varepsilon_0$ become excited, but those in the interior of the distribution remain untouched. Since there are now electrons with energy $\varepsilon > \varepsilon_0$, the sharp edge of the distribution becomes somewhat smoothed out (dashed line in Figure 1). With increasing temperature more and more electrons enter the region $\varepsilon > \varepsilon_0$, and when finally $kT \gg \varepsilon_0 - m_e c^2$ a Maxwell distribution is established in the gas. The ratio of the kinetic Fermi energy to the Boltzmann constant, $(\varepsilon_0 - m_e c^2)/k = T_0$, is called the degeneracy temperature. At nonrelativistic energies, when $P_0 = (3\pi^2)^{1/3} h n^{1/3} \ll m_e c$, the degeneracy temperature is equal to

$$T_0 = \frac{(3\pi^2)^{2/3} h^2}{2m_e k} n^{2/3}, \tag{12.8}$$

and in the relativistic case

$$T_0 = \frac{c \cdot p_0}{k} = \frac{1}{k} (3\pi^2)^{1/3} hcn^{1/3}. \tag{12.9}$$

* [Throughout this book the author seems to use the terms "density," "number density," and "concentration" almost interchangeably. The translator has adhered to the original Russian to avoid additional confusion.]

A Fermi gas is thus degenerate at temperatures $T \ll T_0$, and has a Maxwell-Boltzmann distribution at $T \gg T_0$.

In the case of complete degeneracy, the entropy of a Fermi gas is zero. In fact, entropy is a measure of maximum [sic] disorder, and in degeneracy there is maximum ordering of the motion of particles and their distribution over states.

13. SMALL DEVIATIONS FROM COMPLETE DEGENERACY

Let us consider a fermion gas whose temperature is low compared to the degeneracy temperature. According to (12.1), the number of particles and the energy of the system are

$$N = \frac{V}{\pi^2 h^3} \int_0^\infty \frac{p^2 dp}{e^{\frac{\varepsilon-\mu}{kT}} + 1}, \tag{13.1}$$

$$E = \frac{c \cdot V}{\pi^2 h^3} \int_0^\infty \frac{\sqrt{m_e^2 c^2 + p^2}}{e^{\frac{\varepsilon-\mu}{kT}} + 1} p^2 dp. \tag{13.2}$$

The first equation determines the chemical potential as a function of temperature and particle density, $\mu(T, n)$. Substitution of the above value of the chemical potential in (13.2), followed by integration, gives the total energy. In the general case it is impossible to integrate (13.1) and (13.2) in terms of elementary functions. Such integration can, however, be carried out for very low ($T \ll T_0$) or very high ($T > T_0$) temperatures.

We shall now examine the case $T \ll T_0$. This restriction is equivalent to $kT \ll \mu'$, where $\mu' = \mu - m_e c^2$ (see equation (12.5)). In order to obtain analytical expressions, we shall treat separately the nonrelativistic and relativistic cases. The chemical potential and energy from (13.1) and (13.2) are expanded in power series of the small parameter $\pi kT/\varepsilon_0' = \pi T/T_0$. Omitting the procedure for calculating the integrals, which can be found in any textbook on statistical physics, we shall present only the final result.

In the nonrelativistic case $p \ll m_e c$ we obtain

$$\mu' = \mu_0' \left[1 - \frac{1}{12}\left(\frac{\pi T}{T_0}\right)^2 - \frac{1}{80}\left(\frac{\pi T}{T_0}\right)^4 + \dots\right]. \tag{13.3}$$

$$E = Nm_e c^2 + \frac{3}{5} N\varepsilon_0' \left[1 + \frac{5}{12}\left(\frac{\pi T}{T_0}\right)^2 - \frac{1}{16}\left(\frac{\pi T}{T_0}\right)^4 + \dots\right], \quad (13.4)$$

where $\mu_0' = \varepsilon_0' = kT_0$ is the chemical potential at absolute zero.

In the relativistic case the expressions are simpler, and their integration gives

$$\mu = \mu_0 \left[1 - \frac{1}{3}\left(\frac{\pi T}{T_0}\right)^2 + \dots\right], \quad (13.5)$$

$$E = \frac{3}{4} N\varepsilon_0 \left[1 + 2\left(\frac{\pi T}{T_0}\right)^2 + \frac{27}{15}\left(\frac{\pi T}{T_0}\right)^4 + \dots\right], \quad (13.6)$$

In (13.5) there is no $(\pi T/T_0)^4$ term.

These equations show that the results obtained in the preceding section for $T=0$ can be extended, without much loss of accuracy, to the case of temperatures which are quite small fractions of the degeneracy temperature.

A purely electronic gas forms an unstable system as a result of the Coulomb repulsion forces. In metals and white dwarfs, however, the electrons travel in the crystalline lattice formed by the positive ions, therefore the system as a whole is stable.

So far we have considered the electron gas to be ideal, neglecting the particles' interaction energy compared to their kinetic energy. We shall now check to see when this approximation is justified. The average Coulomb-interaction energies of an electron with nuclei or with other electrons are of the same order and approximately equal to Ze^2/l, where $Z \cdot e^2$ is the nuclear charge and $l = \left(\frac{Z}{n}\right)^{1/3}$ is the average distance between the electron and the nucleus (due to the screening effect, the mean interaction energy of an electron with charges outside a sphere of radius l is equal to zero). On the other hand, as can be easily proven, the average kinetic energy of a particle is of the order of the Fermi energy $\varepsilon_0' = \varepsilon_0 - m_e c^2$. Consequently, the ratio between the average interaction and kinetic energies is of the order of

$$\frac{Z \cdot e^2}{l\varepsilon_0'} \approx \frac{2m_e Z^{2/3} e^2}{(3\pi^2)^{2/3} h^2} n^{-1/3} \approx 5 \cdot 10^7 Z^{2/3} \cdot n^{-1/3}. \quad (13.7)$$

In the case of white dwarfs densities are quite high and the above ratio becomes very small compared to one, therefore the ideal-gas approximation is justified here. It should be noted that with increasing density the deviation of an electron gas from the state of an ideal gas becomes less and less significant. In white dwarfs the electron gas can also be relativistic, and in this case

$$\frac{Ze^2}{l\varepsilon_0'} \approx \frac{1}{3}\frac{e^2}{h\cdot c}Z^{2/3} \ll 1. \tag{13.8}$$

14. FORMATION OF A FREE ELECTRON GAS

Compared with stars like the sun, the radius of white dwarfs is smaller by one, two, or even three orders of magnitude, while the masses are approximately equal. As a result, their densities and pressures should be very high. When densities are higher than those of solids, the properties of matter undergo significant qualitative changes. High densities cause the appearance of a series of new phase states which do not exist under terrestrial conditions. One of these is the ionization of atoms and formation of a free electron gas.

The physics of the formation of a free electron gas is given in a monograph by Schatzmann /258/, which also includes references to original papers on the subject. Here we shall examine this subject qualitatively. When the density increases, i. e. when the distance between particles decreases, there is at first a broadening of the atomic energy levels, caused by the tunnel effect. The volume of each atom constitutes a potential well for the electrons in it. When the centers of the atoms approach each other the barrier between adjacent walls becomes thinner, permitting penetration of electrons from one atom into another. The probability of tunneling rises exponentially with decreasing barrier thickness and decreasing difference between the barrier height and particle energy. As a result of the latter, electrons in higher energy states tunnel through the barrier more easily than those in the ground state. The tunnel effect obviously leads to an additional broadening of the energy levels. At a certain density, an energy level becomes so broad that it merges with the continuous spectrum, and the electron in this state then becomes free. Naturally the first to break away from the atom are the optical electrons, followed by those in the K shell. Complete ionization of the atoms sets in when the lowest level merges with the continuous spectrum.

We shall now estimate the density required for full ionization of matter. The penetration of electrons into adjacent atoms becomes effective only when the average distance l between neighbors becomes comparable with the radius of the K shell, which is a_0/Z, where Z is the nuclear charge and $a_0 = \hbar^2/m_e \cdot e^2 = 0.529 \cdot 10^{-8}$ cm is the Bohr radius:

$$l \approx \alpha \cdot \frac{a_0}{Z}.$$

In the above, α is a number of the order of one. The density of atoms is $n \approx 1/l^3$. Therefore the density required for complete ionization should be

$$\rho = nAm_p \gtrsim \frac{m_p}{\alpha^3 a_0^3} Z^4 \approx \frac{10}{\alpha^3} Z^4, \qquad (14.1)$$

where A is the mass number of the nucleus and m_p is the mass of a proton. According to quantum-mechanical calculations carried out for hydrogen atoms $\alpha \approx 6$, i.e. a density of approximately $0.05\ \mathrm{g \cdot cm^{-3}}$ is necessary for ionization of a hydrogen gas. For heavier elements, complete ionization occurs at much higher densities.

The average density in white dwarfs exceeds the value (14.1) required for complete ionization, therefore they contain a free electron gas. In typical representatives of the above, the main part of the mass is in a state of complete ionization, in the envelope (whose thickness in these stars is relatively small) the ionization is partial, and at the surface itself there should also be neutral atoms (if temperature effects are neglected).

We shall now examine the state of the electron gas in white dwarfs, to see whether it is relativistic or nonrelativistic, degenerate or nondegenerate. The electron gas will be relativistic if $p_0 > m_e c$. This, together with (12.4), results in the condition for relativity

$$n_e \gtrsim \frac{1}{10} \lambda_e^{-3} \approx 2 \cdot 10^{30}\ \mathrm{cm}^{-3}, \qquad (14.2)$$

where n_e is the electron concentration and $\lambda_e = \hbar/m_e \cdot c = 3.862 \cdot 10^{-11}$ cm is the Compton wavelength of an electron (to obtain criterion (14.2) we used equation (12.4) for a degenerate gas, which is the more stringent condition; for the same density n_e, the effect of temperature would only strengthen the relativity of the gas, since the average energy of the particles would be larger than in a degenerate gas). The density in typical white dwarfs is higher than the above value, therefore the electrons in them are predominantly relativistic. At these densities the degeneracy temperature is

$$T_0 = \frac{cp_0}{k} \approx 0.7 \cdot n_e^{1/3} \gtrsim 10^{10}. \tag{14.3}$$

The centers of ordinary stars have temperatures of about 10^6—10^8 degrees. The temperature in white dwarfs cannot be higher than this. We therefore arrive at the conclusion that typical white dwarfs contain a relativistic-degenerate electron gas and fully or partially (depending on the chemical composition) ionized atoms. This is a neutral plasma, which is appropriately designated as the *electron-nuclear phase* of matter and will henceforth be denoted by *Ae*.

15. THE PHASE STATES OF MATTER

In the preceding section we arrived at the conclusion that in white dwarfs there is a free degenerate electron gas. We shall now examine the behavior of the atomic nuclei in this degenerate electron gas, to see whether they move freely in it or perform oscillations around certain fixed equilibrium points. In the first case there would be a mixture of electron and nuclear gases, in the second there would be a crystalline lattice with a free electron gas. The latter case is similar to that of a metal. The following discussion will show that under certain conditions, which apparently exist in some white dwarfs, a modified solid is formed /65, 192/.

Let us assume that we have a modified crystalline lattice, and find the range of temperatures and densities where this phase is stable (in this discussion we follow the work of D. A. Kirzhnits /65/). For the sake of simplicity we shall consider only one kind of nucleus. Let us divide the medium into neutral spherical cells with a single nucleus and Z electrons. The radius of a cell is then

$$R = \left(\frac{3}{4\pi n_A}\right)^{1/3} = \left(\frac{3\eta Z m_p}{4\pi\rho}\right)^{1/3}, \tag{15.1}$$

where n_A is the density of nuclei, A is the mass number, Z is the atomic number, $\eta = A/Z$, $\rho \approx n_A m_A \approx n_A m_p A$ is the density of mass, and m_A and m_p are the mass of the nucleus and of a proton, respectively.

If we assume that the charge of the electron cluster in a cell is uniformly distributed, then the potential energy of the nucleus is

$$U(r) = -\frac{3Z^2e^2}{2R} + \frac{Z^2e^2}{2R^3}\cdot r^2, \qquad (15.2)$$

where r is the displacement of the nucleus from the center of the cell, $r \leqslant R$. The nucleus is therefore in a potential well whose depth is $U_0 = 3Z^2e^2/2R$. The potential $U(r)$ is like that of an oscillator, and comparison with the corresponding expression $m_A\,\omega^2r^2/2$ for an isotropic harmonic oscillator gives the frequency

$$\omega = \left(\frac{Ze^2}{\eta m_p R^3}\right)^{1/2}. \qquad (15.3)$$

The energy of the oscillator is $\varepsilon_n = h\omega\left(n + \frac{3}{2}\right)$, where $n = 0, 1, 2 \ldots$ is the quantum number. At zero temperature $n = 0$, therefore the vibrational energy of the nucleus is

$$\varepsilon_0 = \frac{3}{2}h\omega = \frac{3}{2}h\left(\frac{Ze^2}{\eta m_p R^3}\right)^{1/2} = \frac{3eh}{2\eta m_p}\left(\frac{4\pi}{3}\rho\right)^{1/2}. \qquad (15.4)$$

The value of ε_0 is much smaller than the depth of the potential well:

$$\frac{\varepsilon_0}{U_0} = \frac{2h}{e\sqrt{\eta m_p R_A}}\cdot\frac{1}{Z}\left(\frac{R_A}{ZR}\right)^{1/2} \approx \frac{1}{Z}\left(\frac{R_A}{ZR}\right)^{1/2} \ll 1. \qquad (15.5)$$

Here R_A is the radius of the nucleus, and is obviously smaller than R. This indicates a strong coupling between the nucleus and its cell, and stability of the corresponding state.

According to (15.5), when R decreases ε_0 rises faster than $U(r)$, and when R becomes of the order of R_A the nucleus ceases to be held by the potential well. The present model, however, ceases to apply primarily due to a different reason. This is a change of phase, namely the appearance of free neutrons, which sets in before the above-mentioned phenomenon.

Let us now consider the effect of temperature, which amounts to an increase of the average nuclear energy. At temperatures above zero the behavior of the oscillator is determined by Bose-Einstein statistics, giving the following for the average energy of a nucleus:

$$\bar{\varepsilon} = \frac{3}{2}h\omega + \frac{3h\omega}{e^{\frac{3h\omega}{kT}} - 1} = \varepsilon_0 \operatorname{cth}\frac{\varepsilon_0}{kT}.$$

The first term represents the zero-vibration energy, and the second is the average thermal energy. Let us determine as before the ratio between the energy of a nucleus and the depth of the potential well:

$$\frac{\bar{\varepsilon}}{U_0} = \frac{\varepsilon_0}{U_0}\,\mathrm{cth}\,\frac{\varepsilon_0}{kT} = c_1 Z^{-5/3}\rho^{1/6}\,\mathrm{cth}\,\frac{c_2\rho^{1/2}}{T}, \qquad (15.6)$$

where

$$c_1 = \frac{h\,(4\pi/3)^{1/6}}{e\,(\eta m_p)^{2/3}} = 1.25\cdot 10^{-2}, \quad c_2 = \frac{\sqrt{3\pi}\,eh}{\eta m_p\,k} = 3.38\cdot 10^{3}.$$

Here we used $\eta=2$, and the derivation of (15.6) was based on (15.1) and (15.4). The condition for the existence of a solid-state phase is thus $\bar{\varepsilon}\ll U_0$, and the alternative case is that of a plasma phase. Since for $x\gg 1$, $\mathrm{cth}\,x\approx 1$, and for $x\ll 1$ $\mathrm{cth}\,x=\frac{1}{x}$, we obtain from (15.6) the following criterion for the formation of a solid-state phase:

$$\left.\begin{aligned} T\cdot 10^4 T_6^2 \lesssim \rho \lesssim 3\cdot 10^{11} Z^{10}, \\ 50\,Z^{-5}T_6^3 \ll \rho \ll 7\cdot 10^4 T_6^2, \end{aligned}\right\} \qquad (15.7)$$

where $T_6=10^{-6}T$ is the temperature, measured in units of 10^6 degrees. The first inequality applies to low temperatures, when the argument of the hyperbolic cotangent is large $\left(x=3380\,\frac{\sqrt{\rho}}{T}\gtrsim 1\right)$; the second is for high temperatures, when $x\ll 1$. The temperatures in white dwarfs are evidently such that the above inequalities are fulfilled, and there is sufficient justification for the belief that matter in them is in the solid state.

16. NEUTRONIZATION OF HYDROGEN GAS

We now proceed to the discussion of phase transitions occurring during a gradual increase of density. We shall start with the simplest case, namely that of a hydrogen gas or, to be exact, a proton-electron plasma. Other chemical compositions will be studied in the next section. White dwarfs cannot contain hydrogen in appreciable quantities (see Section 19), therefore a discussion of this case is somewhat academic. We shall start out from a hydrogen gas which is already ionized, and also assume that the electron gas is degenerate $(T<T_0)$.

As long as $\varepsilon_e<(m_n-m_p)c^2$, where ε_e is the electrons' Fermi energy and m_n is the neutron mass, no qualitative changes take place in the proton-electron gas. An increase of mass density is followed only by raising of the Fermi energy ε_e and of the average proton energy. When $\varepsilon_e\geqslant(m_n-m_p)c^2$, though, the proton-electron gas becomes unstable and creation of neutrons sets in. The inevitability of this process becomes evident if we consider the two possible states — the proton-electron "$p+e$" state and the neutron "n" state. Under ordinary conditions a free neutron is unstable, and disintegrates with an average lifetime $\tau=10^3$ sec into a proton, an electron, and an antineutrino. In the presence of a degenerate electron gas, however, a neutron can become a stable particle, and this occurs when the Fermi energy ε_e exceeds the maximum energy of the electron formed during disintegration. At the same time, transitions in the opposite direction also become possible. Neutrons can obviously be present in a medium in amounts comparable to the number of protons even in the absence of degeneracy if the temperature is high, $kT\gtrsim(m_n-m_p)c^2$. Thus, because of the two-way β-decay processes

$$n\rightarrow p+e+\bar{\nu}_e, \quad p+e\rightarrow n+\nu_e, \tag{16.1}$$

where ν_e and $\bar{\nu}_e$ are the symbols for a neutrino and an antineutrino, respectively, thermodynamic equilibrium is finally reached between the neutron- and proton-electron phases. There are two forms of neutrino, namely the electron neutrino ν_e, which is created as a decay product together with an electron, and the muon neutrino ν_μ, which is encountered during decay processes in association with a μ-meson (see Section 21). According to statistical physics, equilibrium between the phases is established when

$$\mu_p+\mu_e=\mu_n, \tag{16.2}$$

where μ_p, μ_e, and μ_n are the chemical potentials of a proton, an electron, and a neutrino, respectively. In the right-hand side of (16.2) we omitted the chemical potential of the neutrino, since it is equal to zero. The vanishing of the chemical potentials of the neutrino and antineutrino is a consequence of their absence in the medium. In fact, after their creation these particles leave the star, experiencing almost no interactions (the interaction cross-section between a neutrino and electrons is of the order of 10^{-44} cm^2). To be more precise, the main cause for the absence of neutrinos is not their weak interactions, but the fact that they have no rest mass. Had the rest mass been different from zero, a degenerate neutrino gas could have accumulated in a sufficiently cold star; in such a case, we would have had to attribute even to these particles a definite chemical potential.

We already know that the chemical potential is a function of temperature and particle density, therefore (16.2) provides a single equation for the values of T, n_p, n_e, and n_n, where n_k is the density of k-particles. The condition for electric neutrality

$$n_e = n_p \tag{16.3}$$

provides the second equation. No other restrictions exist. Therefore, in determining the state of the system, two of the above quantities have to be taken as independent variables, and we could choose temperature and mass density. It is, however, more convenient to choose proton density in place of the latter, since this yields a simpler relationship between the quantities. Thus (16.2) and (16.3) enable us to determine the neutron density $n_n = f(T, n_p)$. To find the explicit form of this function, though, we shall have to make further assumptions concerning the temperature.

Henceforth we shall use the notation T_k for the degeneracy temperature of k-type particles. At the temperatures which interest us, it is reasonable to assume that the electron gas is always degenerate, i. e. $T \ll T_e$. As to the nucleon gas, we are interested in examining both cases, when it is degenerate and when it is nondegenerate.

Let both proton and neutron gases be degenerate, $T \ll T_p$ and $T \ll T_n$. Then according to (12.3) the chemical potential is equal to the Fermi energy:

$$\left.\begin{aligned} \mu_e = \varepsilon_e = c\left[m_e^2c^2 + (3\pi^2)^{2/3}\hbar^2 n_e^{2/3}\right]^{1/2},\ n_p = n_e, \\ \mu_k = \varepsilon_k = m_k c^2 + \frac{(3\pi^2)^{2/3}\hbar^2}{2m_k} n_k^{2/3},\ k = n,\ p. \end{aligned}\right\} \tag{16.4}$$

Substituting (16.4) into (16.2) we obtain

$$n_n = \left\{\frac{m_n}{m_p} n_p^{2/3} + \frac{2m_n}{(3\pi^2)^{2/3}\hbar^2}\left[\varepsilon_e - (m_n - m_p)\cdot c^2\right]\right\}^{3/2}.$$

Introducing the notation

$$\alpha = \frac{m_n - m_p}{m_e} \approx 2.54,\ n_0 = \frac{m_e^3 c^3}{3\pi^2\hbar^3} = 5.87 \cdot 10^{29}\mathrm{cm}^{-3},$$

we can rewrite the above in the form

$$n_n = n_p\left\{1 + 2\frac{m_p}{m_e}\left(\frac{n_0}{n_p}\right)^{2/3}\left(\sqrt{1+\left(\frac{n_p}{n_0}\right)^{2/3}} - \alpha\right)\right\}^{3/2}. \tag{16.5}$$

Equating to zero the expression in curly brackets, we obtain the threshold for neutron creation:

$$n_p^{(0)} = \frac{\alpha^2 - 1}{3\pi^2}\left(\frac{m_e \cdot c}{h}\right)^3 \approx 7.5 \cdot 10^{30}\ \mathrm{cm}^{-3}. \tag{16.6}$$

Thus at densities $n_p \gtrsim 7.5 \cdot 10^{30}\ \mathrm{cm}^{-3}$ the neutrons become stable particles. The electron Fermi momentum corresponding to the threshold density (16.6) is $p_e \approx 2.2 m_e c$, so that even at the neutron-creation threshold the electrons are relativistic particles.

Directly above the threshold density, the number of neutrons rises much faster than the number of protons, and very soon it becomes predominant. In a degenerate material this is energetically more favorable. Thus for proton densities which obey the inequality

$$1 \ll \left(\frac{n_p}{n_0}\right)^{1/3} \ll \frac{m_p}{m_e},$$

we obtain from (16.5)

$$n_n \approx \left(2\frac{m_p}{m_e}\right)^{3/2}\left(\frac{n_0}{n_p}\right)^{1/3} \cdot n_p. \tag{16.7}$$

Since $(2m_p/m_e)^{3/2} \approx 2 \cdot 10^5$ and according to the above-mentioned inequality $10^{-3} \ll (n_0/n_p)^{1/3} \ll 1$, we note that n_n is three orders of magnitude larger than n_p.

Up till now our discussion referred only to the subnuclear density region, where the nucleons are nonrelativistic. In the supernuclear density region the disproportion between the concentrations of the two types of carriers begins to decrease, and in the ultra-relativistic energy region they become of the same order. In fact, if we introduce into the equilibrium equation (16.2) the relativistic expressions for the chemical potentials

$$\mu_e = \mu_p = (3\pi^2)^{1/3} h \cdot c \cdot n_p^{1/3}, \quad \mu_n = (3\pi^2)^{1/3} h \cdot c \cdot n_n^{1/3},$$

we obtain

$$n_n = 8 n_p. \tag{16.8}$$

Let us now examine the case when the nucleon gas is not degenerate, yet the temperature is not high enough to remove the degeneracy of the electron gas. We shall assume that the nucleons are nonrelativistic and the electrons relativistic. In the case of Boltzmann statistics (assuming $T > T_k$, $k = n, p$), the chemical potential is

$$\mu_k = m_k c^2 + kT\, ln\left[\frac{n_k}{2}\left(\frac{2\pi h^2}{m_k kT}\right)^{3/2}\right], \quad k = n, p \tag{16.9}$$

and from equation (16.2) we obtain

$$n_n = n_p \cdot \exp\left[\frac{(3\pi^2)^{1/3} hcn_p^{1/3} - am_e c^2}{kT}\right]. \tag{16.10}$$

Finally, let us examine the temperature range

$$\frac{m_e c^2}{k} \ll T \ll \frac{m_p c^2}{k}, \tag{16.11}$$

where $m_e c^2/k \approx 6 \cdot 10^9$ and $m_p c^2/k \approx 10^{13}$, and we shall assume that there is altogether no degeneracy. At such temperatures the thermal emission is rich in hard quanta, $\varepsilon_\gamma > m_e \cdot c^2$. Consequently there should also be a large number of positrons, which are created by photons paired with electrons. On the other hand, these temperatures are not high enough to create mesons, baryons heavier than nucleons, or pairs of these particles.

Let us examine the statistical equilibrium related to the reactions

$$\left.\begin{aligned} p + e^- \rightarrow n + \nu_e, \quad & n \rightarrow p + e^- + \bar{\nu}_e, \\ e^+ + e^- \rightleftarrows \gamma, \quad & n + e^+ \rightarrow p + \bar{\nu}_e. \end{aligned}\right\} \tag{16.12}$$

Annihilation of an electron-positron pair into one γ quantum and the inverse process of creating a pair from a quantum are possible in the field of a foreign particle. It should be noted that the single-photon processes $e^+ + e^- \rightleftarrows \gamma$ can also take place in a plasma with $n(\omega) < 1$ ($n(\omega)$ is the index of refraction) at high temperatures, when degeneracy is not complete. The quanta with which we are concerned have an energy

$$2m_e \cdot c^2 \lesssim h\omega \lesssim h \cdot c \cdot n^{1/3}.$$

The restriction on the maximum energy of quanta results from the requirement $\lambda > l$, namely that the medium be continuous (λ is the wavelength and $l \sim n^{-1/3}$ is the average distance between particles of the medium). The threshold densities for these processes are of the order of $1.5 \cdot 10^{32}\ cm^{-3}$. These processes were first studied in /87/, and they play an important role in energy-loss phenomena. The chemical potential of photons is zero, which results from non-conservation of their number. This also follows directly from the equilibrium condition for electron-electron scattering with emission or absorption of photons (this process requires the presence of a foreign particle so as not to violate the momentum-conservation law), $e+\gamma \rightleftarrows e$. At equilibrium $\mu_e + \mu_\gamma = \mu_e$, therefore, $\mu_\gamma = 0$. From (16.12) one obtains the equilibrium conditions

$$\left.\begin{aligned} &\mu_p + \mu_{e^-} = \mu_n, \quad n_p + n_{e^+} = n_{e^-}, \\ &\mu_{e^-} + \mu_{e^+} = 0. \end{aligned}\right\} \tag{16.13}$$

Because of the high temperatures, the number of electron-positron pairs is much larger than the number of protons, $n_p \ll n_{e^+}$. As a result, $n_{e^+} \approx n_{e^-}$ and $\mu_{e^+} \approx \mu_{e^-} \approx 0$, $\mu_p \approx \mu_n$. Thus in the temperature range (16.11) $n_{e^+} \approx n_{e^-}$ and $n_p \approx n_n$.

With respect to the material presented in this section, we should add a few remarks. At the densities considered here a hydrogen gas is unstable with respect to the formation of complex nuclei (see Section 19). According to the calculations made by Ya. B. Zel'dovich /43/, hydrogen H^1 turns into He^4 at densities $\rho \approx 5 \cdot 10^4\ g \cdot cm^{-3}$. Because of the tunnel effect, the nucleons can combine with each other even at absolute zero, forming α-particles and other nuclei. Neutronization of hydrogen gas and its applications in astrophysics were discussed in /228, 189, 128, 278, 243/ and elsewhere.

In the following section we shall see that the neutronization process also takes place in matter consisting of atomic nuclei, but then the laws and relationships are completely different. The outline given here is simplified in yet another respect, namely that at densities close to nuclear there appear in the medium, in addition to the nucleons, also hyperons and other elementary particles in comparatively large numbers. There are also additional details which were not included in the discussion of this section.

17. NEUTRONIZATION OF MATTER

When densities exceed a certain value, neutronization also takes place in matter consisting of complex atomic nuclei (the "*Ae*-phase"). In order to study this question, we need a clear picture of the state of nucleons in the nucleus. The underlying physics can be depicted graphically on the basis of the model of a degenerate nucleon gas. In this model, the nucleus is seen as a potential well with a width equal to its diameter and a depth of approximately 30 Mev, uniformly filled with neutrons and protons. The temperature of the nucleon gas should obviously be taken as zero, since the excitation energy of a nucleus is very high compared to its thermal energy. The nucleons in the nucleus thus form a completely degenerate Fermi gas.

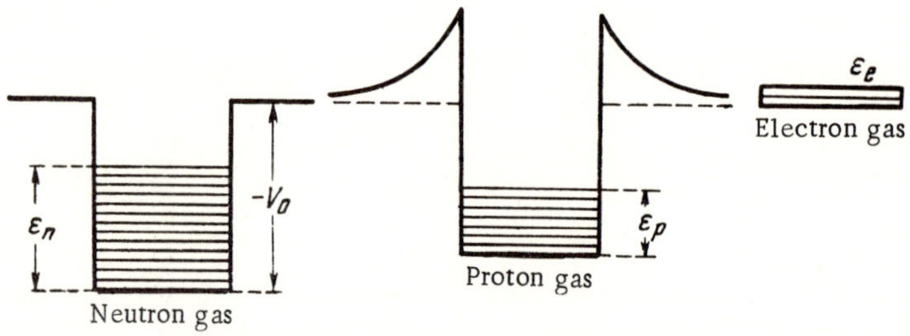

FIGURE 2. Model of an atomic nucleus with a degenerate nucleon gas. On the right, a degenerate free electron gas.

Under these physical conditons we can once again talk of two alternative states of matter, namely the "neutron," or "n," and the "proton-electron," or "$p+e$," state or phase (Figure 2). Transitions between these phases are possible due to the direct and inverse β-decay processes

$$n \to p + e + \bar{\nu}, \quad p + e \to n + \nu. \tag{17.1}$$

In the case of stars the evolutional periods are long enough to permit establishment of thermodynamic equilibrium. The state of equilibrium is determined by the well-known relationships between the chemical potentials in the presence of force fields:

$$\mu_n - V_0 = \mu_p - V_0 + V_c + \mu_e, \tag{17.2}$$

where V_0 is the depth of the nuclear potential well and V_c is the average Coulomb-interaction energy pertaining to one proton. We again omitted the chemical potential of the neutrino since, as explained in the last section, it is zero due to nonconservation of neutrinos. The depth V_0 of the potential well is assumed to be the same for protons and neutrons, since the nuclear forces are independent of the nucleon's charge state (protons and neutrons are considered as different charge states of a single particle, called nucleon).

We already know that in a completely degenerate Fermi gas the chemical potential coincides with the particles' Fermi energy, therefore equilibrium between the two states "n" and "$p+e$" is determined by the relationship

$$\varepsilon_p + \varepsilon_e + V_c = \varepsilon_n . \tag{17.3}$$

This equation can be given a visual interpretation. Let us imagine two connecting vessels, one (the left "well" in Figure 2) filled with a neutron "liquid" and the second (right "well") filled with a proton "liquid," with an electron "liquid" on top, and the Pauli principle in force. As a result of the Coulomb interaction among protons, the bottom of the proton well is raised by the amount V_c. Then, in accordance with (17.1), the "liquid" will flow from one well to another until the levels become equalized. This obviously corresponds to a state of minimum energy of the system.

The densities of protons and neutrons in the nucleus are

$$n_p = \frac{3}{4\pi r_0^3}\frac{Z}{A}, \qquad n_n = \frac{3}{4\pi r_0^3}\left(1 - \frac{Z}{A}\right),$$

where Z and $A - Z$ are the numbers of protons and neutrons in the nucleus, respectively, and $r_0 \approx 1.3 \cdot 10^{-13}$ cm is a constant entering into the expression $R = r_0 A^{1/3}$ for the radius of the nucleus. It is assumed here that the material consists of only one kind of nucleus and, moreover, that the nucleons in the nucleus are uniformly distributed. From (16.4) we obtain for the Fermi energies

$$\varepsilon_n = m_n c^2 + \frac{(3\pi^2)^{2/3} \cdot h^2}{2m_n} n_n^{2/3} \approx m_n c^2 + \frac{(3\pi^2)^{2/3} \cdot h^2}{2m_p} n_n^{2/3},$$

$$\varepsilon_p = m_p c^2 + \frac{(3\pi^2)^{2/3} \cdot h^2}{2m_p} n_p^{2/3}.$$

Substituting these expressions for ε_n and ε_p in (17.3) we obtain

$$\left(1 - \frac{Z}{A}\right)^{2/3} - \left(\frac{Z}{A}\right)^{2/3} = b\left(\varepsilon_e + V_c - a m_e \cdot c^2\right), \tag{17.4}$$

where

$$am_e c^2 = (m_n - m_p)\, c^2 \approx 1.3 \text{ Mev},$$

$$b = \frac{2m_p r_0^2}{h^2}\left(\frac{4}{9\pi}\right)^{2/3} \approx 1.38 \cdot 10^4 \text{ erg}^{-1}.$$

For the average nucleus V_c is of the order of 1 Mev, and at $n_e \lesssim 10^{31}$ cm^{-3} ($\rho \lesssim 10^7$ g·cm^{-3}) the Fermi energy of electrons is $\varepsilon_e \lesssim 1$ Mev. Therefore, when $\rho \lesssim 10^7$ g·cm^{-3}, the quantity $\varepsilon_e + V_c - am_e \cdot c^2$ is close to zero. In this density range the number of protons and neutrons is approximately equal, $Z/A \approx 0.5$. According to (17.4), the process of proton neutronization in the nucleus starts when the electron Fermi energy reaches the value

$$\varepsilon_e = am_e \cdot c^2 - V_c.$$

At high ε_e the ratio Z/A is a function of density, becoming smaller as ε_e increases. Finally, at high densities the neutronization process causes disruption of the nuclear stability, resulting in its disintegration. This leads to formation of a nucleon gas, in which the neutrons predominate numerically. A more complete discussion of Z/A as a function of density will be given in the next section.

18. NUCLEAR CHARGE AS A FUNCTION OF DENSITY

In the last two sections it was shown that, as the density is increased above $\rho = 10^7$ g·cm^{-3}, protons in the nucleus are converted into neutrons. Thus, under conditions of a degenerate electron gas with $\rho \gtrsim 10^7$ g·cm^{-3}, the nuclear charge (atomic number Z) is no longer an independent property, but dependent upon the density of matter (or, in other words, on the electron Fermi energy ε_e). This question was first studied in /124, 125/ in connection with the origin of elements, and later in more detail in /192, 101, 96/ and many other works.

Let us consider a nucleus with atomic number Z and mass number A and look at the behavior of A/Z with increasing density beyond the neutron-creation threshold. The earlier qualitative discussion was suitable for explaining the physical aspect of the problem. For a more accurate solution, however, it is preferable to resort to the following method.

At low densities (small ε_e) the numbers of protons and neutrons in the nuclei are either equal or negligibly different. When ε_e is increased, at a certain point the nucleus becomes unstable with respect to the inverse β-decay:

$$(A, Z) + e \rightarrow (A, Z-1) + \nu_e. \tag{18.1}$$

The electron Fermi energy which corresponds to this reaction is

$$\varepsilon_e = [M(A, Z-1) - M(A, Z)] \cdot c^2, \tag{18.2}$$

where $M(A, Z)$ is the mass of a nucleus with parameters A and Z. To calculate the threshold values of ε_e at which Z decreases by 1, it is best to consult the β-decay tables /39, 40/ if these contain the corresponding data; if the latter do not exist, it is possible to start out from the semiempirical equation for the nuclear mass. According to this equation the mass of a nucleus is /72/

$$\left.\begin{aligned} M(A, Z) &= (A-Z)m_n + Zm_p + \frac{B}{c^2}, \\ B &= -15.75 + 17.8A^{-1/3} + 0.71Z^2A^{-4/3} + \\ &\quad + 23.7\left(1 - \frac{2Z}{A}\right)^2 + 34A^{-7/4} \cdot \delta, \end{aligned}\right\} \tag{18.3}$$

where B is the binding energy of the nucleus in Mev and δ is a factor equal to 0, 1, or -1 for odd A, even A and odd Z, or even A and even Z, respectively. For large A the last term is insignificant and can be omitted. Strictly speaking, the above equation is valid only when $(A-2Z)/A \ll 1$. It is necessary to distinguish between the cases of odd Z and even Z. In the first case the transition (18.1) results in a stable nucleus $(A, Z-1)$, which turns into $(A, Z-2)$ only after the next threshold value of ε_e is reached. In the second case, the nucleus obtained in the transition from an even Z to an odd $Z-1$ is unstable and is therefore immediately converted into a stable nucleus with $Z-2$. This is due to the pairing term in the equation for the nuclear mass (the last term in (18.3)).

The right-hand side of (18.2) represents the energy $\varepsilon_{\beta-}$ for β-decay. In some cases ε_β is well known from direct measurements. As an example let us take three sequences of $\bar{\beta}$-transitions, in which the end product is oxygen in the first case, an iron isotope in the second case, and a plutonium isotope in the third case. These are /39, 40/

$$
\left.\begin{array}{l}
\text{I. } (16;3) \xrightarrow{61.2} (16;4) \xrightarrow{31.2} (16;5) \xrightarrow{36.3} (16;6) \xrightarrow{7.9} (16;7) \xrightarrow{11.5} (16;8), \\
\text{II. } (57;9) \xrightarrow{61.0} (57;10) \xrightarrow{57.3} (57;11) \rightarrow \ldots \rightarrow (57;15) \xrightarrow{38.8} \\
\xrightarrow{38.8} (57;16) \xrightarrow{35.1} (57;17) \rightarrow \ldots \rightarrow (57;24) \xrightarrow{5.56} (57;25) \xrightarrow{1.9} (57;26), \\
\text{III. } (240;25) \xrightarrow{71.2} (240;26) \xrightarrow{67.6} (240;27) \rightarrow \ldots \rightarrow (240;55) \xrightarrow{40.7} \\
\xrightarrow{40.7} (240;56) \xrightarrow{37.1} (240;57) \rightarrow \ldots \ (240;91) \xrightarrow{4.0} (240;92) \xrightarrow{0.4} \\
\xrightarrow{0.4} (240;93) \xrightarrow{2.0} (240;94).
\end{array}\right\} \quad (18.4)
$$

As before, the first number inside the parentheses is the mass number, and the second is the number of protons in the nucleus; the numbers over the arrows indicate the energies ε_β released during the decay.

At densities above the corresponding threshold energies these reactions obviously go in the reverse direction. In the second case, for example, when the electron Fermi energy reaches 1.9 Mev the iron isotope (57; 26) captures an electron (or emits a positron) and emits a neutrino, turning into the manganese isotope (57; 25). Later on, when ε_e reaches 5.56 Mev, there is a transition into Cr (57; 24). This continues until, after a series of β^+ decays, Ne (57; 10) is reached at $\varepsilon_e = 57.3$ Mev. The latter is stable up to $\varepsilon_e = 61$ Mev, after which it changes into the very heavy fluorine isotope F (57; 9). In accordance with our earlier remark about transitions from even to odd atomic number, during these an unstable nucleus can sometimes be formed, which immediately converts into one with even Z. Thus, for example, at $\varepsilon_e = 2$ Mev Pu (240; 94) becomes Np (240; 93), whose stability threshold is $\varepsilon_e = 0.4$ Mev and therefore lower than the above. As a result, the latter turns into U (240; 92) immediately after its creation, and this last nucleus is stable until $\varepsilon_e = 4$ Mev.

We thus see that the number of protons in the nucleus, and therefore also the quantity Z/A, decreases with rising electron Fermi energy (i. e. with increasing density), and isotopes which are unstable under ordinary conditions become stable at densities which exceed certain threshold values. The question is how long this process of enriching the nucleus with neutrons can go on. The smallest possible value of Z/A can be obtained by equating the binding energy of (18.3) to zero and using (18.2) and the neutrality condition $n_e = Zn_A$, where n_A is the density of nuclei (we assume the material to be chemically homogeneous). We shall return to this question in Section 20.

In calculating star models it will be necessary to know A_k/Z_k for different nuclei as a function of ε_e. These can be obtained from experimental data on β-decay, or, when these do not exist, by using equation (18.3) for the binding energy. The curves of A_k/Z_k as a function of ε_e are stepped lines. The A_k/Z_k lines for different nuclei (especially the middle ones) lie near each other and form a narrow band with increasing ordinate. One has to keep in mind that in white dwarfs and in the envelopes of baryon stars the chemical composition is usually heterogeneous. Assuming that every kind of nucleus is present, we can study their collective behavior instead of each one separately. In calculating white-dwarf models one can then use some sort of average line (Figure 3) instead of the series of A_k/Z_k lines. The error incurred will be small, and this approach may also be more reasonable in view of our earlier remark concerning the chemical composition involved. This averaged curve is a good approximation for medium nuclei, but contains appreciable errors in the case of very heavy or light nuclei.

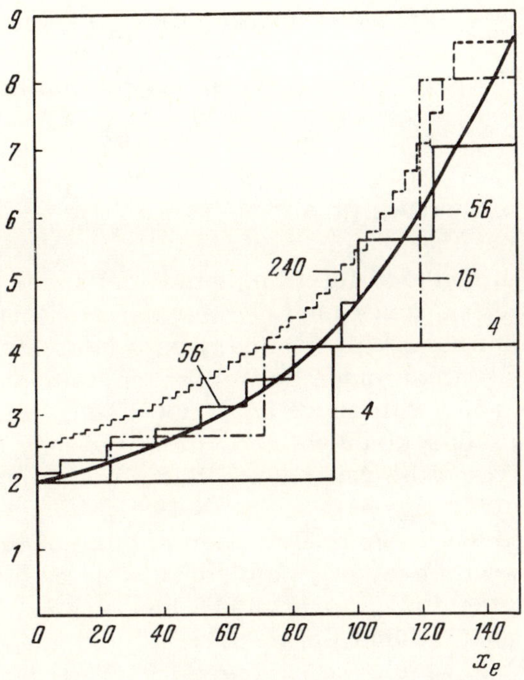

FIGURE 3. Average A/Z as a function of the parameter $x_e = \dfrac{p_e}{m_e c}$. The stepped lines are the true values of A/Z for nuclei with $A = 4, 16, 56$, and 240. The solid line shows the values $\overline{A/Z}$, which are averages over the different nuclei.

With increasing pressure the electron Fermi energy ε_e changes smoothly, but the density has jumps at those points where Z changes into $Z-1$. Generally, the mass density also has jumps at those points. These jumps can be quite large if much hydrogen is present, but this is ruled out in the case of white dwarfs (see Section 19). In the case of heavy elements or heterogeneous chemical compositions such jumps in the value of ρ are very small and of no significance.

In calculating star configurations it is inconvenient to use curves, and much preferable to have analytical expressions. The smooth curve in Figure 3 can be approximated with sufficient accuracy by the formula

$$\frac{A}{Z}=2+a_1\cdot x+a_2\cdot x^2+a_3\cdot x^3, \tag{18.5}$$

where $a_1=1.255\cdot10^{-2}$, $a_2=1.755\cdot10^{-5}$, $a_3=1.376\cdot10^{-6}$, and $x=p_e/m_ec$, p_e being the electron Fermi momentum. This equation is not correct for the envelope, which may contain a fair amount of hydrogen; furthermore, in (18.5) it is assumed that at the surface when $x\lesssim1$ the elements with $A/Z=2$ predominate.

19. NUCLEAR REACTIONS IN A COLD PLASMA

At sufficiently high densities (i.e. small distances between particles), the tunnel effect can cause nuclear synthesis reactions to occur even at absolute zero. Preliminary estimates of the probability of such effects were made by Zel'dovich, Cameron, and Salpeter /43, 146, 253/. Here we shall not discuss the energy aspects of these reactions, since we are interested only in problems of nuclear stability and the dependence of the mass numbers A_k on the electron Fermi energy /101/.

These reactions are essentially the combining of light nuclei with each other or with heavier nuclei, occurring as a result of the tunnel effect when the distances between them are comparatively small. Since we assume that the medium is cold, the only reactions to consider are those which are exothermic. The transmission coefficient through a potential barrier is very sensitive to a particle's mass, therefore on the astronomical time scale every type of nucleus must have a definite density threshold above which it cannot be a part of matter.

To calculate the probability, one has to choose a specific model for the potential barrier. Let us assume that the medium is divided into spherical cells, which are packed with maximum density. Each cell then consists of an electron cloud and an atomic nucleus (or even an incompletely ionized atom when the density is below that required for complete ionization), which performs zero-point vibrations around a fixed point in the electron cloud. The radius of the cell is determined from the neutrality conditions $4\pi R^3 n_e/3 = Z$:

$$R = \left(\frac{3Z}{4\pi}\right)^{1/3} \frac{a}{p_e}, \qquad (19.1)$$

where $a=(3\pi^2)^{1/3} h$. The cell radius is thus proportional to $Z^{1/3}$. The interaction energy between a nucleus and the electrons of its cell is, according to (15.2),

$$V(r) = -\frac{3}{2}\frac{Z^2e^2}{R} + \frac{Z^2e^2}{2R^3} r^2, \qquad (19.2)$$

where r is the distance between the nucleus and the center of the cell. The interaction with other cells can be neglected. From (19.2) it follows that a nucleus with mass m_A vibrates with the frequency

$$\omega = \frac{Z \cdot e}{m_A^{1/2} \cdot R^{3/2}}. \qquad (19.3)$$

The assumption that the nucleus is localized in a sphere of radius R is reasonable if the energy $h^2/m_A R^2$, dictated by the momentum uncertainty principle, is small compared to the energy of the oscillator. The ratio between these two energies is

$$\frac{6.6 \cdot 10^{-3}}{Z \cdot A^{2/3}} \rho^{1/6} \ll 1$$

for those densities at which the Ae-phase exists.

In (19.2) it was assumed that the electron cloud is distributed uniformly over the cell. Strictly speaking, this is not correct, but it can be shown that the deviation from uniform distribution is insignificant. The electron distribution in the cell is determined by

$$\varepsilon_e(r) - e\varphi(r) = \varepsilon_e(R), \qquad (19.4)$$

where $\varepsilon_e(r)$ is the electron Fermi energy at a distance r from the cell center, and $\varphi(r)$ is the electric potential of the self-consistent field. Equation (19.4) follows from the condition of constant total

Fermi energy along the cell radius. If one takes account of the fact that the nuclear vibrations cause its charge to be smeared out, then it is easy to see that $e\varphi(r) \ll \varepsilon_e(r)$. It then follows from (19.4) that the kinetic Fermi energy of the electrons in a cell is approximately constant.

Proceeding to the calculation of the tunnel-effect probability, we consider two adjacent cells with charges Z_1 and Z_2 (Figure 4). Let the first nucleus be lighter than the second, $Z_1 < Z_2$. We shall calculate the probability that the first nucleus passes through the barrier and combines with the second. The potential energy is

$$V(r) = \begin{cases} -\dfrac{3Z_1^2 \cdot e^2}{2R_1} + \dfrac{Z_1^2 \cdot e^2}{2R_1^3} \cdot r^2 + \dfrac{Z_1 \cdot Z_2 \cdot e^2}{s-r} - \\ \qquad -\dfrac{Z_1 \cdot Z_2 \cdot e^2}{|\mathbf{r}-\mathbf{l}|} \text{ for } r \leqslant \dfrac{R_1}{\cos\theta}, \\ -\dfrac{3Z_1 \cdot Z_2 \cdot e^2}{2R_2} + \dfrac{Z_1 \cdot Z_2}{2R_2^3} |\mathbf{r}-\mathbf{l}|^2 - \dfrac{Z_1^2 \cdot e^2}{r} + \\ \qquad + \dfrac{Z_1 \cdot Z_2 \cdot e^2}{s-r} \text{ for } r \geqslant \dfrac{R_1}{\cos\theta}. \end{cases} \tag{19.5}$$

The meaning of the symbols is explained in Figure 4. The first two terms in (19.5) are the interaction between the first nucleus and the electron cloud of its own cell, the third is the interaction between the two nuclei, and the last is the interaction between the first nucleus and the electron cloud of the second cell. The lower expression in (19.5) is the interaction energy for the case when the first nucleus is in the second cell; here the first two terms are the interaction energy with the electrons of the second cell, the third is the interaction with those of the first cell, and the fourth is the interaction between the two nuclei. The interaction with the remaining cells can be considered negligible.

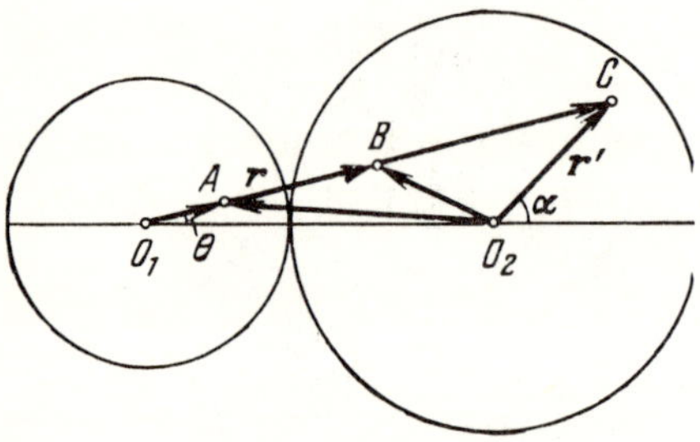

FIGURE 4. Calculation of the tunnel effect. $O_1O_2 = R_1 + R_2 = l$, $O_1C = S$, $O_2C = r'$. A and B are two points on the track of the first nucleus in the O_1C direction. C is the contact point between the two nuclei.

We have to introduce some simplifications in order to reach a full solution of the problem. First of all, we can neglect the interaction between the nucleus and the second cell as long as it remains in its own cell, i.e. the last two terms in the upper expression of

(19.5) can be dropped. It is also reasonable to replace $|\mathbf{r}-\mathbf{l}|=O_2B$ and r in the lower expression of (19.5) by their averages, since we shall later have to integrate over the volume of the second cell. The averaging is carried out only over those variables on which the transmission coefficient is not strongly dependent. We then have for the interaction energy

$$V(r)=\begin{cases} -\dfrac{3Z_1^2e^2}{2R_1}+\dfrac{Z_1^2e^2}{2R_1^3}r^2 & \text{for } r\leqslant\dfrac{R_1}{\cos\theta}, \\ V_0+\dfrac{Z_1\cdot Z_2\cdot e^2}{s-r} & \text{for } r\geqslant\dfrac{R_1}{\cos\theta}, \end{cases} \tag{19.6}$$

where

$$V_0=-\frac{3Z_1\cdot Z_2\cdot e^2}{2R_2}+\frac{Z_1\cdot Z_2\cdot e^2}{2R_2^3}|\mathbf{r}-\mathbf{l}|^2-\frac{Z_1^2\cdot e^2}{\bar{r}}.$$

The values used in calculations were $\bar{r}=R_1+0.5\,R_2$ and $|\mathbf{r}-\mathbf{l}|=0.5R_2$.

The probability of the first nucleus crossing into the second cell within a solid angle $\sin\theta\,d\theta$ is

$$\frac{1}{2}\sin\theta\,d\theta\,\frac{\omega_1}{2\pi}\cdot\exp\left[-\frac{2}{h}\sqrt{2m_1}\,I\,(s,\theta)\right].$$

Here m_1 and ω_1, are the mass and vibration frequency of the first nucleus, and I is determined by the expression

$$I(s,\theta)=\int_{r_1}^{s-r_0}\sqrt{V(r)-E_1}\,dr, \tag{19.7}$$

where $E_1=3h\omega_1/2-3Z_1e^2/2R_1$ is the ground-state energy of the first oscillator, measured not from the bottom of the potential well but from zero, r_1 is the distance from the center cell to the point where the energy level crosses the barrier,

$$r_1=\left(\frac{3h}{Z_1e}\right)^{1/2}\left(\frac{R_1^3}{m_1}\right)^{1/4},$$

and r_0 is a quantity of the order of the sum of the two nuclear radii,

$$r_0=1.2\cdot10^{-13}(A_1^{1/3}+A_2^{1/3}).$$

The probability of the second nucleus being at point C of the volume element dV is $\psi_2^2(r')\,dV$, where $\psi_2(r')=a_2^{-3/2}\pi^{-3/4}\exp(-r_2'^2/2a_2^2)$ is the wave function of the oscillator in the ground state. Consequently, the above probability is equal to

$$a_2^{-3}\pi^{-3/2}e^{-a_2^{-2}r'^2}\; r'^2 dr' \sin\alpha\, d\alpha,$$

where $a_2 = \sqrt{3h/m_2\omega_2}$ is the oscillation amplitude of the second oscillator, and m_2 and ω_2 are its mass and frequency, respectively. Finally, the probability of the two nuclei meeting at an arbitrary point C of the second cell and merging into a single nucleus can be written as

$$W = \frac{n_{\omega_1}W_1}{4\pi^{5/2}a_2^3}\int \exp\left[-\frac{r'^2}{a_2^2} - \frac{2\sqrt{2m_1}}{h}\, I(s,\theta)\right] \times$$

$$\times\, \delta\left(\theta - \arcsin\frac{r'\cdot\sin\alpha}{\sqrt{l^2 + r'^2 + 2lr'\cos 0}}\right)\cdot r'^2 dr' \sin\alpha \sin\theta\, d\alpha\, d\theta, \qquad (19.8)$$

where W_1 is the probability that after the collision a single complex nucleus is formed (it is obviously of the order of unity), and n is the number of nuclei adjacent to the first. A simple computation shows that $n \approx 27$ (with an equality sign in the case of chemical homogeneity). The δ-function was introduced so that the variables α and θ can be considered independent. Integration of (19.8) over θ yields

$$W = \frac{n\omega_1 W_1}{4\pi^{5/2}a_2^3}\int \frac{r'^3 dr' \sin^2\alpha\, d\alpha}{\sqrt{l^2 + r'^2 + 2lr'\cos\alpha}} \times$$

$$\times \exp\left[-\frac{r'^2}{a_1^2} - \frac{2}{h}\sqrt{2m_1}\, I(s,\theta_0)\right]. \qquad (19.9)$$

The integration over α is from 0 to π, and for r' it is from zero to R_2 (the upper limit can be changed to ∞).

Using (19.6), we obtain from (19.7) that

$$I(s,\theta_0) = \frac{Z_1 e}{(2R_1)^{3/2}}\left[\frac{R_1}{\cos^2\theta_0}\sqrt{R_1^2 - r_1^2\cos^2\theta_0} - \right.$$

$$\left. - r_1^2 \ln\frac{R_1 + \sqrt{R_1^2 - r_1^2\cos^2\theta_0}}{r_1\cdot\cos\theta_0}\right] +$$

$$+ \sqrt{\frac{Z_1 Z_2 e^2}{b_1}}\left\{\arcsin\left[b_1\left(s - \frac{R_1}{\cos\theta}\right)\right]^{1/2} + \right.$$

$$\left. + \left[b_1\left(s - \frac{R_1}{\cos\theta_0}\right) - b_1^2\left(s - \frac{R_1}{\cos\theta_0}\right)^2\right]^{1/2} - 2\sqrt{b_1 r_0}\right\}, \qquad (19.10)$$

where

$$b_1 = \frac{1}{Z_1 Z_2 e^2}\left(\frac{3hZ_1 e}{2R_1^{3/2} m_1^{1/2}} - V_0\right),\ \theta_0 = \arcsin\frac{r' \cdot \sin\alpha}{\sqrt{b^2 + r'^2 + 2l \cdot r' \cos\alpha}}.$$

Direct numerical calculations show that $I(s,\ \theta_0)$ is only very weakly dependent on the parameters α and r'. Substituting $\alpha = 0$ and $r' = 0$, we obtain $\theta_0 = 0$, $s = l$, and $I = I(l, 0)$. Further, since r' is small compared to l, we expand $(l^2 + r'^2 + 2lr' \cos\alpha)^{-1/2}$ in powers of r'/l and, keeping only the first two terms, we then obtain

$$W = \frac{n \cdot \omega_1 W_1 a_1}{16\pi^{3/2} l}\exp\left[-\frac{2\sqrt{2m_1}}{h} I_0\right], \tag{19.11}$$

where I_0 is determined from (19.10) by setting $\theta_0 = 0$ and $s = l$.

The results of computations for $n = 27$ and $W_1 = 1$ are shown in Figure 5. The quantity $x_e = p_e\ m_e c$ is plotted along the abscissa, and the ordinate gives the minimum value of the atomic number A for a nucleus which can exist in the medium at the given x_e (i.e. at the corresponding mass density). The lifetime of the nucleus was assumed to be of the order of $10^7 - 10^9$ years. For a certain reason, the threshold value of the mass number A is not very sensitive to changes in the lifetime. Thus, for example, this threshold changes very little when W (therefore also the lifetime, which is inversely proportional to W) is varied by one or two orders of magnitude.

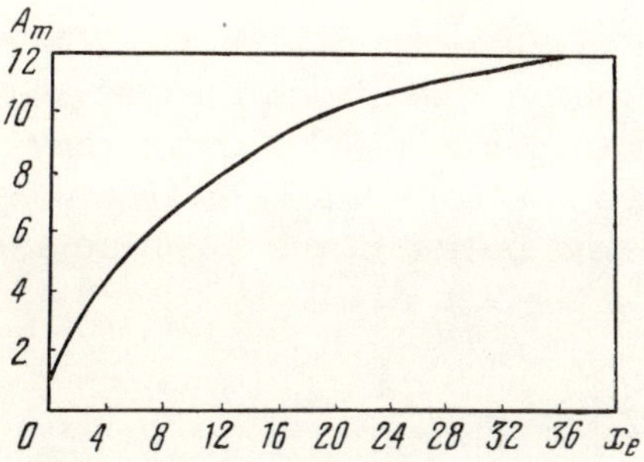

FIGURE 5. The smallest mass number A_m as a function of $x_e = p_e / m_e c$. The region above the $A_m(x_e)$ curve is that of stable (with respect to the tunnel effect) nuclei, and the region below is for unstable nuclei.

20. THE MOST STABLE STATE OF MATTER IN THE SUBNUCLEAR REGION

1. The Ae-phase. Up till now, our discussions of thermodynamic equilibrium were based on the assumption that the nuclear mass number remains constant. At the same time, we saw that the number of protons in the nucleus does not remain constant —

under conditions of a degenerate electron gas it decreases with increasing density. A state of equilibrium is reached through β-decay processes, therefore reaction times are very small and any deviations from equilibrium are quickly eliminated. On the other hand, it was shown in the last section that the mass numbers cannot be considered as constant or independent of density. Pycnonuclear reactions cause fusion of light nuclei. This process leads to the presence of many nuclei with different A, and the equilibrium composition (and thereby also the average of A) becomes dependent on density. Therefore the energy minimum obtained on the basis of β-processes alone is only relative, and the corresponding state of degenerate matter does not have the absolute minimum of energy. An absolutely stable state is one in which not only β-processes but also pycnonuclear reactions are forbidden. At temperatures above zero, the concept of forbiddenness should be replaced by that of statistical equilibrium between direct and inverse processes. The energy of such a state should be the lowest among those of all possible states.

We shall now discuss the state of absolute equilibrium, taking into account not only β-processes, but also nuclear transformations. For the binding energy of an atomic nucleus we shall use the semi-empirical formula (18.3). Although this formula is not always very accurate (e.g. for nuclei with many more neutrons than electrons), it is most convenient for an analytical examination of our problems.

Let us consider the energy density for a given A:

$$\rho = \frac{n}{A}[(A-Z)m_n + Zm_p - B(A,Z)] + \frac{3}{4}a\left(\frac{Z}{A}n\right)^{1/3}. \qquad (20.1)$$

The first term gives the energy of the atomic nuclei, and the second is the energy of the degenerate relativistic electron gas; n is the number of baryons per unit volume, n/A is the number density of atomic nuclei with parameters A and Z, and $B(A, Z)$ is the nuclear binding energy:

$$\left.\begin{aligned} B(A,Z) &= c_0A - c_1A^{2/3} - c_2\frac{Z^2}{A^{1/3}} - c_3A\left(1-\frac{2Z}{A}\right)^2, \\ c_0 &= 15.7, \quad c_1 = 17.8, \quad c_2 = 0.71, \\ c_3 &= 23.7, \; a = (3\pi^2)^{1/3}\, h\cdot c = 6.1145\cdot 10^{-11}\,\text{Mev}\cdot\text{cm}. \end{aligned}\right\} \qquad (20.2)$$

The last term of (18.3) was here omitted, since it contributes little in the case of medium and heavy nuclei. The last term in (20.1) represents the energy of the electron gas, which is assumed to be relativistic; the velocity of light is taken as unity, and energy is measured in Mev. Substituting (20.2) in (20.1) we find

$$(n, A, Z) = n\Big[(m_n + c_3 - c_0) - (m_n - m_p + 4c_3)\, ZA^{-1} +$$
$$+ c_1 A^{-1/3} + c_2 Z^2 A^{-4/3} + 4c_3 \frac{Z^2}{A^2}\Big] + \frac{3}{4} a \left(n \frac{Z}{A}\right)^{4/3}. \qquad (20.3)$$

Let us find the minimum of this expression for a given number of baryons:

$$\left(\frac{\partial \rho}{\partial Z}\right)_{n,A} = 0, \quad \left(\frac{\partial \rho}{\partial A}\right)_{n,Z} = 0.$$

The derivative with respect to Z gives

$$\varepsilon_e = a\,(ny)^{1/3} = (m_n - m_p + 4c_3) - 2y\,(4c_3 + c_2 A^{2/3}), \qquad (20.4)$$

where $y = Z/A$. It is easy to see that (20.4) is identical with (18.2) if one takes account of the relationship

$$\frac{\partial \rho}{\partial Z} = [M(A, Z) - M(A, Z-1)]\, \frac{n}{A}.$$

Equation (20.4) determines the dependence of Z/A on the baryon density n at a given A:

$$\frac{1}{a^3 y}\, [(m_n - m_p + 4c_3) - 2y\,(4c_3 + c_2 A^{2/3})]^3 = n. \qquad (20.5)$$

Using either the above equation or (20.4), expression (20.3) for the energy density can be rewritten in the form

$$\rho = n\Big[(m_n + c_3 - c_0) - \frac{1}{4}(m_n - m_p + 4c_3)\, y +$$
$$+ c_1 A^{-1/3} - \frac{1}{2}(4c_3 + c_2 A^{2/3})\, y^2\Big]. \qquad (20.6)$$

This is the minimum value of energy density $\rho\ (n, A)$ under the assumption that the medium consists of a single type of nucleus with a fixed mass number A. The main term in the above expression for the energy density is nm_n, the contribution of the remaining terms being quite small: of the order of, or less than, one percent of nm_n. These terms become extremely important, however, in such important cosmogonic problems as the sources of internal energy and the evolution of celestial bodies. This expression $\varepsilon = \rho + n(m_n - m_p)\, y - nm_n$ is obviously the density of the internal energy of a star.

Let us now determine the minimum energy when varying both Z and A. Equating to zero the derivative of (20.3) with respect to A, we obtain

$$(m_n - m_p + 4c_3)\, y - \frac{1}{3} c_1 A^{-1/3} - \frac{4}{3} c_2 A^{2/3} y^2 - \\ - 8c_3 y^2 - a n^{1/3} \cdot y^{4/3} = 0. \qquad (20.7)$$

We now use (20.4) and leave out the last term of this expression, obtaining

$$\frac{Z}{A} = \left(\frac{c_1}{2c_2}\right)^{1/2} \frac{1}{\sqrt{A}} = \frac{3.54}{\sqrt{A}}. \qquad (20.8)$$

Substitution of this $y = Z/A$ in (20.5) and (20.6) yields

$$\frac{1}{a^3}\left[(m_n - m_p + 4c_3)\left(\frac{2c_2}{c_1} A\right)^{1/6} - 2\left(\frac{c_1}{2c_2}\right)^{1/3}\left(c_2 A^{1/3} + \frac{4c_3}{A^{1/3}}\right)\right]^3 = n, \qquad (20.9)$$

$$\rho_{\min} = n\left[(m_n + c_3 - c_0) - \frac{m_n - m_p + 4c_3}{4}\left(\frac{c_1}{2c_2}\right)^{1/2} \frac{1}{A^{1/2}} + \right. \\ \left. + \frac{3c_1}{4A^{1/3}} - \frac{c_1 c_3}{c_2} \cdot \frac{1}{A}\right]. \qquad (20.10)$$

Expression (20.10) is thus the minimum value of the energy density for a degenerate electron-nuclear neutral plasma at a fixed n. Equations (20.8) and (20.9) determine the dependence of A and Z on the baryon density.

We have found the most stable state of the degenerate electron-nuclear gas. It is characterized by one independent parameter, namely the number of baryons, here taken per unit volume. As we shall see, in the most stable state of the Ae-phase each value of n has one corresponding nucleus with the parameters $A(n)$ and $Z(n)$. With rising density the mass number increases, but the ratio Z/A becomes smaller. This means that in the corresponding stellar configuration (a completely cold white dwarf, in which have been exhausted all the reserves of internal energy, which could in principle be emitted as thermal or radiant energy) A and Z are functions of the distance r from its center. With distance from the center $A(r)$ diminishes, tending toward a limiting value $A = 56$ at the surface, while the ratio Z/A approaches the limit 26 / 56.

The above results were obtained under the assumption that the electrons are relativistic. This is not a serious restriction, since the dependence of A and Z on density becomes significant only when $\varepsilon_e > m_e c^2$.

2. The Aen-phase. When the electron Fermi energy becomes

$$Z\varepsilon_e \geqslant [Am_n - M(A, Z)]\, c^2, \tag{20.11}$$

then the degenerate electron-nuclear plasma must also contain free neutrons. Here M is the mass of a nucleus with parameters A and Z. The equality sign corresponds to the threshold for the appearance of neutrons as a stable component of matter. Thermodynamic equilibrium between the different components can be established with the reactions

$$(A, Z) + Ze \rightleftarrows A \cdot n + Z\nu_e.$$

The corresponding relationship between the chemical potentials will then be

$$M(A, Z) + Z\varepsilon_e = A(m_n + T_n), \tag{20.12}$$

where T_n is the kinetic Fermi energy of neutrons (the neutron gas is also assumed to be fully degenerate). The appearance of free neutrons is related to the requirement of minimum energy. We shall use the term Aen-phase for the state of matter when free neutrons are present in addition to the atomic nuclei and electrons. It should be noted that in this phase the proton is an unstable particle, since now

$$m_p + \varepsilon_e > m_n + T_n.$$

Therefore an accidental free proton immediately captures an electron and turns into a neutron, $p + e \to n + \nu_e$.

The existence of free neutrons in the degenerate electron-nuclear plasma can be deduced by starting out from the principle of minimum internal energy. Let us write out the expression for energy density in the Aen-phase:

$$\rho = \frac{n - n_n}{A}[(A - Z)\, m_n + Zm_p - B(A, Z)] + n_m m_n + \\ + \frac{3a^2}{10m_n} n_n^{5/3} + \frac{3}{4} a \left[\frac{Z}{A}(n - n_n)\right]^{4/3}, \tag{20.13}$$

where n_n is the number density of neutrons. The next-to-last term represents the energy density of a degenerate nonrelativistic neutron gas, and the last term is the energy density of a relativistic electron gas. Account was taken of the condition of neutrality for the plasma

$$Z(n-n_n)=A\cdot n_e.$$

Using (20.2), the energy density can be rewritten in the form

$$\rho=nm_n+(n-n_n)\left[(c_3-c_0)-(m_n-m_p+4c_3)\frac{Z}{A}+\right.$$
$$\left.+\frac{c_1}{A^{1/3}}+c_2A^{2/3}\frac{Z^2}{A^2}+4c_3\frac{Z^2}{A^2}\right]+\frac{3a^2}{10m_n}n_n^{5/3}+$$
$$+\frac{3}{4}a\left[\frac{Z}{A}(n-n_n)\right]^{4/3}. \qquad (20.14)$$

This function depends on four arguments, $\rho\,(n,\ n_n,\ A,\ Z)$, and we have to find the energy minimum for a fixed n.

In analogy with the preceding case, we shall first consider a plasma consisting of a single type of atomic nucleus with a certain mass number A. During thermodynamic equilibrium the energy of the system should be a minimum:

$$\left(\frac{\partial\rho}{\partial n_n}\right)_{n,A,Z}=\frac{a^2}{2m_n}n_n^{2/3}-ya\,[y\,(n-n_n)]^{1/3}-[(c_3-c_0)-$$
$$-y\,(m_n-m_p+4c_3)+c_1A^{-1/3}+(c_2A^{2/3}+4c_3)\,y^2]=0,$$
$$\left(\frac{\partial\rho}{\partial Z}\right)_{n,n_n,A}=\frac{n-n_n}{A}\{a\,[y\,(n-n_n)]^{1/3}-(m_n-m_p+4c_3)+$$
$$+2\,(c_2A^{2/3}+4c_3)\,y\}=0.$$

As before, $y=Z/A$, and $a\,[y\,(n-n_n)]^{1/3}$ is the electron Fermi energy. The first equation leads to the relationship (20.12), and the second equation gives

$$\varepsilon_e=a\,[y\,(n-n_n)]^{1/3}=(m_n-m_p+4c_3)-2y\,(4c_3+c_2A^{2/3}). \qquad (20.15)$$

Considering (20.15), we find from the derivative of ρ with respect to n_n that

$$T_n=\frac{a^2}{2m_n}n_n^{2/3}=(c_3-c_0)+c_1A^{-1/3}-y^2\,(4c_3+c_2A^{2/3}). \qquad (20.16)$$

Equations (20.15) and (20.16) yield the functions $n_n\,(n)$ and $y\,(n)$.

Thus, in the case of specified A we have

$$\left.\begin{aligned}\left\{\frac{2m_n}{a^2}[(c_3-c_0)+c_1A^{-1/3}-y^2(4c_3+c_2A^{2/3})]\right\}^{3/2}&=n_n,\\ \frac{1}{a^3}\left\{\frac{m_n-m_p+4c_3}{y^{1/3}}-2y^{2/3}(4c_3+c_2A^{2/3})\right\}^3+n_n&=n.\end{aligned}\right\} \quad (20.17)$$

Using (20.15) and (20.16), the energy density (20.14) can be written as

$$\begin{aligned}\rho=nm_n+n\Big[(c_3-c_0)+c_1A^{-1/3}-\frac{m_n-m_p+4c_3}{4}-y-\\ -\frac{1}{2}y^2(4c_3+c_2A^{2/3})+n_n\Big[-\frac{2}{5}(c_3-c_0)-\frac{2}{5}c_1A^{-1/3}+\\ +\frac{m_n-m_p+4c_3}{4}y-\frac{1}{10}y^2(4c_3+c_2A^{2/3})\Big].\end{aligned} \quad (20.18)$$

We now proceed to determine the most stable state of matter in the Aen-phase. To this end, (20.15) and (20.16) should be supplemented by the equation

$$\begin{aligned}\left(\frac{\partial\rho}{\partial A}\right)_{n,n_n,Z}=\frac{n-m}{A}\Big[-y\varepsilon_e+(m_n-m_p+4c_3)\,y-\\ -\frac{1}{3}c_1A^{-1/3}-\frac{4}{3}c_2A^{2/3}y^2-8c_3y^2\Big]=0.\end{aligned} \quad (20.7)$$

Substituting in this equation the expression for ε_e from (20.15), we arrive at the result obtained for the previous case:

$$y^2=\frac{c_1}{2c_2}\frac{1}{A}.$$

Keeping in mind (20.7), we find from (20.15) and (20.16) that

$$\left.\begin{aligned}\left(\frac{2m_n}{a^2}\right)^{3/2}\left[(c_3-c_0)+\frac{1}{2}c_1A^{-1/3}-\frac{2c_1c_3}{c_2}\frac{1}{A}\right]^{3/2}&=n_n,\\ \frac{1}{a^3}\left(\frac{2c_2}{c_1}\right)^{1/2}\Big[(m_n-m_p+4c_3)\,A^{1/6}-\sqrt{2c_1c_2}\,A^{1/3}-&\\ -8c_3\sqrt{\frac{c_1}{2c_2}}\,A^{-1/3}\Big]^3+n_n&=n.\end{aligned}\right\} \quad (20.19)$$

These equations enable us to calculate A and n_n for a given value of n.

It remains to write the final equation for the energy density. This is obtained through simple changes in (20.18), with the help of (20.15),

(20.16), and (20.17):

$$\rho_{\min} = n \cdot m_n + n \Big[(c_3 - c_0) + \frac{3}{4} c_1 A^{-1/3} - \frac{c_1 c_3}{c_2} \frac{1}{A} -$$

$$- \frac{m_n - m_p + 4c_3}{4} \sqrt{\frac{c_1}{2c_2}} A^{-1/2} \Big] + n_n \Big[\frac{m_n - m_p + 4c_3}{4} \times$$

$$\times \sqrt{\frac{c_1}{2c_2}} A^{-1/2} - \frac{2}{5}(c_3 - c_0) - \frac{c_1 c_3}{5c_2} \frac{1}{A} - \frac{9}{10} c_1 A^{-1/3} \Big]. \qquad (20.20)$$

This is the smallest energy density value which, in principle, a material in the *Aen*-phase can have for a given number of baryons n.

According to (20.11), at the *Aen*-phase threshold the electron Fermi energy is

$$\varepsilon_e^{(n)} = \frac{Am_n - M}{Z} = \frac{A}{Z} b(A, Z), \qquad (20.21)$$

where $b(A, Z)$ is the binding energy of one nucleon in a nucleus with parameters A and Z. The numerical values of $\varepsilon_e^{(n)}$ can be determined from the schemes presented in (18.4), or by starting out from the binding-energy equation (18.3) and including the dependence of the ratio A/Z on ε_e (see Figure 3). For nuclei with mass number $A = 4, 16, 40, 57, 80, 197, 240$, the values of ε_e obeying equation (20.21) are respectively $\varepsilon_e^{(n)} = 14.6, 19.6, 22.6, 23.3, 23.7, 23.75$, and 23.36 Mev. It may seem that for $\varepsilon_e^{(n)}$ one should take its value for a helium atom. It should be stressed, though, that at the corresponding densities no light nuclei can exist in the plasma since, as we saw in Section 19, the tunnel effect causes them to fuse into heavier nuclei. It is therefore reasonable to assume that at the threshold for the appearance of free neutrons the electron Fermi energy is $\varepsilon_e^{(n)} \approx 23$ Mev. From Figure 3 we find that the corresponding average value of A/Z is $A/Z = 2.8$. The corresponding threshold value of the mass density is found from

$$\rho \approx \sum_i m_p A_i n_i \approx n_p n_e \left(\frac{\overline{A}}{Z}\right) \approx$$

$$\approx \frac{m_p}{3\pi^2} \left(\frac{m_p c}{\hbar}\right)^3 \left(\frac{\overline{A}}{Z}\right) x_e^3 \approx 10^6 \left(\frac{\overline{A}}{Z}\right) x_e^3,$$

where the summation is understood to be over all the types of nucleus, n_i is the density of the i-th type of nucleus, and $x_e = \varepsilon_e / m_e c$.

Setting $x_e = 46$ and $\overline{A/Z} = 2.8$, we obtain $\rho^{(n)} \approx 2.7 \cdot 10^{11}$ g·cm^{-3}. Thus the threshold for the existence of free neutrons is

$$\varepsilon_e^{(n)} \approx 23 \text{ Mev}, \quad \frac{A}{Z} \approx 2.8, \quad \rho^{(n)} \approx 2.7 \cdot 10^{11} \text{g} \cdot \text{cm}^{-3}. \tag{20.22}$$

If the medium is in an absolutely stable state at the $Ae \longrightarrow Aen$ phase-transition threshold,

$$A = 122, \quad y = 0.32, \quad n = 1.83 \cdot 10^{35} \text{cm}^{-3}. \tag{20.23}$$

For obvious reasons there is no phase in which free protons could exist simultaneously with the nuclei and neutrons. It is energetically more advantageous for protons to combine with a corresponding number of neutrons to form an atomic nucleus than it is to remain outside the nucleus. This simply means that in the Aen-phase the neutron Fermi energy T_n never reaches the value of the electron Fermi energy; or, to be more precise, $T_n < \varepsilon_e - (m_n - m_p)$. It is not difficult to find a rigorous mathematical proof for this statement.

The Aen-phase ends with the complete disintegration of atomic nuclei and the formation of an electron-nucleon (enp) phase. The disintegration of nuclei is brought about by an excessive rise in the electron Fermi energy. Accurate determination of the threshold for the $Aen \longrightarrow enp$ phase transition is not possible because of inadequate knowledge concerning the nuclear potential in the corresponding density range. It can, however, be estimated from the requirement that the binding energy of a nucleon be zero:

$$b(A, Z) = -(c_3 - c_0) - c_1 A^{-1/3} + 4c_3 y - (4c_3 + c_2 A^{2/3}) y^2 = 0.$$

From this

$$y = \frac{2c_3 \pm \sqrt{4c_3^2 - (4c_3 + c_2 A^{2/3})(c_3 - c_0 + c_1 A^{-1/3})}}{4c_3 + c_2 A^{2/3}}. \tag{20.24}$$

The upper sign does not correspond to reality, therefore one has to consider the lower sign only. For the most probable nucleus $y = \sqrt{\frac{c_1}{2c_2}} A^{-1/2}$, and as a result

$$\frac{2c_3 - \sqrt{4c_3^2 - (4c_3 + c_2 A^{2/3})(c_3 - c_0 + c_1 A^{-1/3})}}{4c_3 + c_2 A^{2/3}} = \sqrt{\frac{c_1}{2c_2}} \cdot A^{-1/2}. \tag{20.25}$$

The root of this equation is $A \approx 702$, which means that, before complete nuclear disintegration, the value of A corresponding to an absolutely minimum internal energy (the most probable A) is approximately 700. Substituting this A in (20.19), we find that at the enp-phase threshold

$$\left.\begin{array}{l} n \approx 1.25 \cdot 10^{37}\text{cm}^{-3}, \\ n_n \approx 7 \cdot 10^{36}\text{cm}^{-3}. \end{array}\right\} \tag{20.26}$$

Figure 6 plots $y=Z/A$ against the logarithm of baryon density for different mass numbers and also for the most stable state in the Ae- and Aen-phase. Figure 7 shows the dependence of the mass number of the most stable nucleus on the common logarithm of the number density of baryons. Finally, Figure 8 gives the internal-energy difference

$$\Delta\varepsilon = \rho_A - \rho_{\min},$$

calculated for one baryon, as a function of $\log n$.

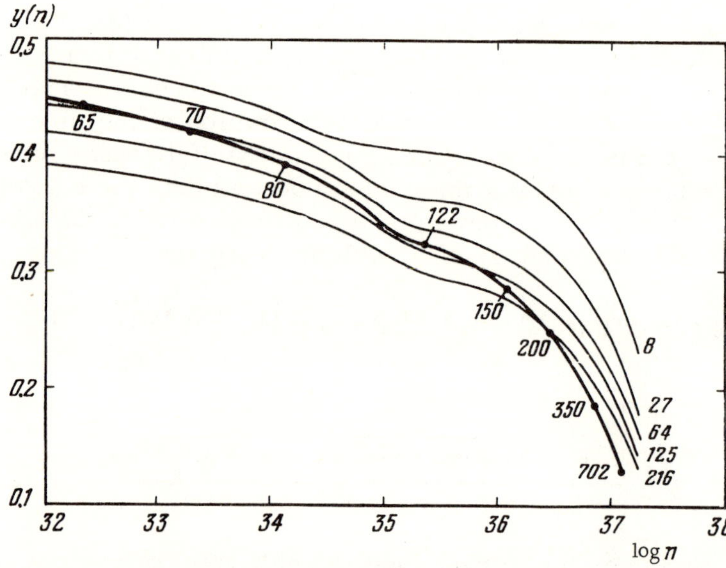

FIGURE 6. Plots of $y(n)$ (n is the number of baryons per unit volume, and $y=Z/A$) for nuclei with mass number $A=8, 27, 64, 125, 216$. The decreasing nuclear charge in a degenerate medium is caused by neutronization. The heavy curve gives the most stable state in the Ae- and Aen-phases, the changing parameters along it being not only y but also A. The value of A is indicated at some points on this line. The flattening of the curves in the vicinity of $\log n \approx 35$ is due to the $Ae \to Aen$ phase transition.

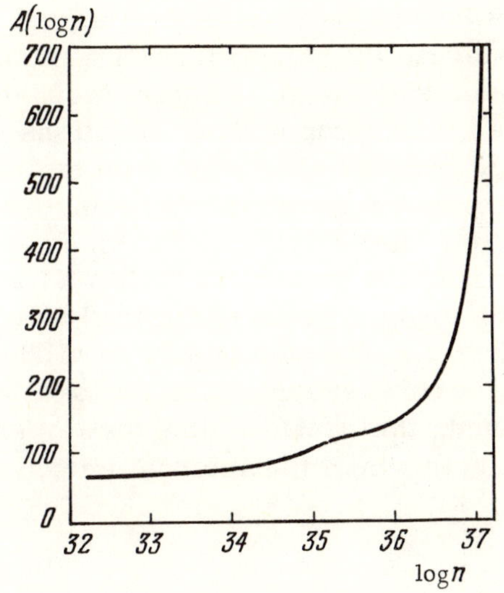

FIGURE 7. Mass number of the most stable nucleus in the *Ae*- and *Aen*-phase as a function of baryon density. The small bulge at log $n \approx 35$ is due to the $Ae \rightarrow Aen$ phase transition. Just before formation of a continuous nuclear material, i.e. the *enp*-phase, the most stable nucleus is $A \approx 700$.

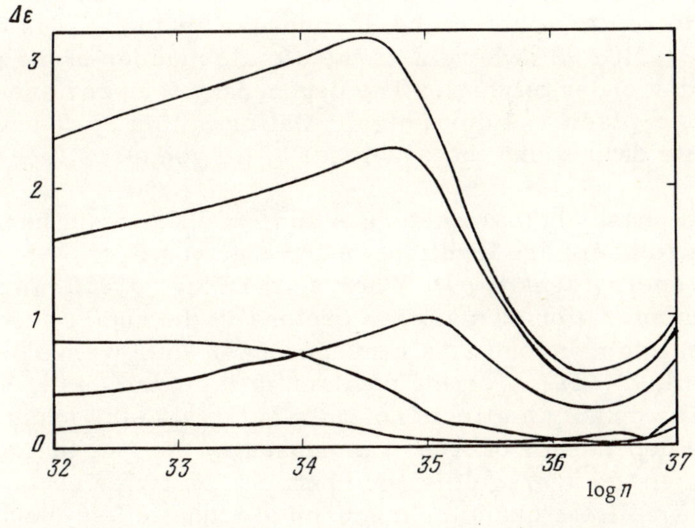

FIGURE 8. The variable along the ordinate is $\Delta\varepsilon = (\rho_A - \rho_{min})/n$, where n is the number density of baryons, ρ_{min} is the energy density of the most stable state in the *Ae*- or *Aen* phase and ρ_A is the energy density for given mass numbers $A = 8$, 12, 27, 125, 216. The first extrema in the $\Delta\varepsilon(n)$ lines occur at the $Ae \rightarrow Aen$ phase transformation points. Obviously, $\Delta\varepsilon$ is the internal-energy difference calculated for one baryon. The energy is measured in Mev.

The analysis given in this section, as well as the results presented in Figure 8, indicate that superdense stars should not be seen as extinct celestial bodies, depleted of their internal-energy reserves. It is expected that pycnonuclear reactions will continue in white dwarfs and in the envelopes of baryon stars, leading to fusion of light nuclei into heavy ones and to the inverse processes of splitting of heavy nuclei. In contrast to β-processes, which take place very rapidly, these nuclear reactions, after a first stage in which the light nuclei disappear, proceed fairly slowly. As a result, in white dwarfs and baryon-star envelopes there will be a slow transition from internal to thermal energy. As the available energy reserves are consumed, the state of the matter gradually tends toward the most stable, in which the internal energy has an absolute minimum.

SUMMARY

In this chapter we discussed the properties of a neutral plasma, consisting of free electrons and "bare" atomic nuclei, at temperatures below the degeneracy temperature of the electron gas. It is convenient to refer to this as the Ae-phase of matter. It is formed at densities $\rho \gtrsim 10\alpha^{-3}Z^4$, where Z is the atomic number of the nucleus, and α is of the order of unity. The degeneracy temperature of electrons in an Ae-plasma is quite high. Matter in this state can be found in white dwarfs and the envelopes of baryon stars (see Chapter VIII).

In the Ae-phase, the parameters A and Z of the atomic nuclei (A is the mass number) are functions of the electron Fermi energy (or of the total-energy density ρ). When $\varepsilon_e \gtrsim 1$ Mev ($\rho \gtrsim 10^7$ g·cm^{-3}) there is neutronization of matter — protons in the nuclei become transformed into neutrons as a consequence of the inverse β-decay $(A, Z)+e \rightarrow (A, Z-1)+\nu_e$ or the positron decay $(A, Z) \rightarrow (A, Z-1)+$ $+e^+ + \nu_e$ The density ρ affects not only Z, but also the mass number A. The dependence of A on ρ is caused by the fact that at high densities, when the average distance between particles becomes sufficiently small, the quantum-mechanical tunnel effect causes the light nuclei to fuse into heavy nuclei and thus disappear.

When the electron Fermi energy reaches the value

$$Z\varepsilon_e = (Am_n - M)c^2,$$

where M is the mass of a nucleus with parameters A and Z, free neutrons appear in the plasma. An Aen-phase is then formed, in which, alongside the electrons and atomic nuclei, there are also free neutrons. At the threshold for the appearance of free neutrons

$$\varepsilon_e \approx 23, \quad A/Z \approx 2.8, \quad \rho \approx 2.7 \cdot 10^{11} \mathrm{g} \cdot \mathrm{cm}^{-3}.$$

Here a free proton is an unstable particle. With further increase in the density, the ratio Z/A continues to diminish and A increases. This phase ends when $n \approx 10^{37}$ cm^{-3} and $n_n \approx 7 \cdot 10^{36}$ cm^{-3} (n_n is the number density of free neutrons), after which a continuous nuclear material is formed – the *enp*-phase. The dependence of the ratio Z/A on the electron Fermi energy ε_e and the number density of baryons n is shown in Figures 3 and 6.

In the *Ae*- and *Aen*-phases, the mass number of the most stable nucleus, whose internal energy is the smallest (and the nuclear binding energy is the largest), is dependent on the baryon density n. With increasing n it changes from $A = 56$ to $A \approx 122$ at the *Aen*-phase threshold and to $A \approx 700$ before the full disintegration of atomic nuclei and the subsequent formation of continuous nuclear matter. The curve for $A(n)$ is shown in Figure 7.

Chapter III

IDEAL DEGENERATE BARYON GAS

21. LIST OF ELEMENTARY PARTICLES

In this section we present some information on elementary particles, which will be needed for the following chapters. Thirty years ago the term "elementary" was meant literally, i. e. it was assumed that the particles referred to as such had no structure. Let us remember that in those days only six elementary particles were known, not counting the antiproton, the antineutron, and the neutrino. Nowadays this term is used in a nominal sense only — no one doubts that if not all, then at least a fair majority of these particles should have a complex, as yet unknown, internal structure. There are strong arguments in favor of the above statement.

In Tables 1—3 are listed the most important properties of those elementary particles which are known at present /133, 249/. We omit only meson resonances, which will be of no concern to us. Elementary particles fall into three classes, namely baryons, mesons, and leptons. Nucleons and heavier particles, whose decay products include a proton or a neutron, are called *baryons*. Baryons which are heavier than a nucleon are called *hyperons*. There are also the so-called *resonance* particles, which belong to the baryon class, and these will be discussed later on. Pions (π mesons), and all particles ranging in mass between these and nucleons, are called *mesons*. The ones which are heavier than pions are K mesons. The lepton group includes electrons, μ^- mesons (negative muons), and electron and muon neutrinos. Positrons and μ^+ mesons are antiparticles. Photons constitute a separate class.

In 1952 Fermi found a sharp maximum in the cross section for scattering of a pion (with total energy of 338 Mev) by a proton, resembling a resonance effect. Since then, many similar resonances have been discovered in the scattering of π and K mesons by nucleons. These effects are today interpreted as evidence for the existence of a large number of baryon and meson excited states, referred to as *resonance* particles. Mesonic resonances are

TABLE 1. Elementary particles with lifetime above 10^{-16} sec

Class	Particle	Mass, Mev	Spin	Isotopic spin, τ, τ_3	Mean lifetime, sec	Major decay scheme
Photon	γ	0	1	—	∞	
Leptons	$\nu_e, \bar{\nu}_e$	0	1/2	—	∞	
	$\nu_\mu, \bar{\nu}_\mu$	0	1/2	—	∞	
	$e^\pm$	$0.511006 \pm 2 \cdot 10^{-6}$	1/2	—	∞	
	$\mu^\pm$	105.659 ± 0.002	1/2	—	$(2.1983 \pm 0.0008) \cdot 10^{-6}$	$\mu \to e + \nu_\mu + \bar{\nu}_e$
Mesons	$\pi^\pm$	139.578 ± 0.013	0	± 1 1	$(2.604 \pm 0.007) \cdot 10^{-8}$	$\pi \to \mu + \bar{\nu}_\mu$
	π^0	134.974 ± 0.013		0	$(0.89 \pm 0.18) \cdot 10^{-16}$	$\pi^0 \to 2\gamma$
	$K^\pm$	493.82 ± 0.11	0	1/2, $\pm 1/2$	$(1.23 \pm 0.01) \cdot 10^{-8}$	$K \to \begin{cases} 2\pi \\ 3\pi \\ \pi + \mu + \bar{\nu}_\mu \\ \pi + e + \bar{\nu}_e \\ \mu + \bar{\nu}_\mu \end{cases}$
	$K^0 \bar{K}^0$			1/2, $\pm 1/2$	—	
	K^0_S	$497.76 + 0.16$		1/2,	$(0.862 \pm 0.006) \cdot 10^{-10}$	
	K^0_L			1/2,	$(5.29 \pm 0.15) \cdot 10^{-8}$	
	η		0	0	$\Gamma = (2.3 \pm 0.5)$ Kev	$\gamma\gamma$; $3\pi^0$; $\pi^+ \pi^- \pi^0$

TABLE 1 (continued)

Class	Particle	Mass, Mev	Spin	Isotopic spin, τ, τ_3	Mean lifetime, sec	Major decay scheme
Baryons	p	938.256 ± 0.005	1/2	$+1/2$	∞	
	n	939.550 ± 0.005	1/2	$-1/2$	$(0.96 \pm 0.04) \cdot 10^{3}$	$n \to p + e + \bar{\nu}_e$
	Λ	1115.57 ± 0.07	1/2	0	$(2.52 \pm 0.03) \cdot 10^{-10}$	$\Lambda \to \begin{cases} p + \pi^- \\ n + \pi^0 \end{cases}$
	Σ^+	1189.43 ± 0.17		$+1$	$(0.810 \pm 0.013) \cdot 10^{-10}$	$\Sigma^+ \to n + \pi^+,\ p + \pi^0$
	Σ^0	1192.55 ± 0.11	1/2	1 0	$< 1.0 \cdot 10^{-14}$	$\Sigma^0 \to \Lambda + \gamma$
	Σ^-	1197.42 ± 0.09		-1	$(1.64 \pm 0.06) \cdot 10^{-10}$	$\Sigma^- \to n + \pi^-$
	Ξ^-	1321.25 ± 0.18	1/2?	1/2 $-1/2$	$(1.66 \pm 0.04) \cdot 10^{-10}$	$\Xi \to \Lambda + \pi$
	Ξ^0	1314.7 ± 0.7		1/2	$(3.03 \pm 0.18) \cdot 10^{-10}$	
	Ω^-	1672.4 ± 0.6	3/2?	0	$(1.3 \pm 0.35) \cdot 10^{-10}$	$\Omega^- \to \Xi + \pi;\ \Lambda + K^-$

observed in processes during which several particles are created. Thus, for example, when in proton-antiproton annihilations five pions are created, in most cases the energy of the π^+ meson and a π^0 meson in their center-of-mass system is 763 Mev. From this it was concluded that at first a meson resonance ρ^+ with mass $m_\rho = 763$ Mev is created, which later decays into two mesons: $p+\bar{p}\rightarrow\rho^+ + 2\pi^- + \pi^+,\ \rho^+\rightarrow\pi^+ + \pi^0$. From the width of the energy band (half-width of the energy distribution curve) it was found that the lifetime is 10^{-22} sec. Resonance particles decay through nuclear interactions, so their mean lifetime is only of the order of 10^{-23} sec. Therefore, due to their very brief existence, these excited baryon states can practically not be found in ready form.

TABLE 2. Baryons

	Particle or resonance	Mass, Mev	Isotopic spin	Spin
N	p	938.3	1/2	1/2
	n	939.6	1/2	1/2
	N′(1470)	1470	1/2	1/2
	N(1518)	1525	1/2	3/2
	N(1550)	1550	1/2	1/2
	N(1680)	1680	1/2	5/2
	N(1688)	1690	1/2	5/2
	N′(1710)	1710	1/2	1/2
	N(2190)	2200	1/2	7/2
	N(2650)	2650	1/2	?
	N(3030)	3030	1/2	?
Δ	Δ(1236)	1236	3/2	3/2
	Δ(1640)	1640	3/2	1/2
	Δ(1950)	1950	3/2	7/2
	Δ(2420)	2420	3/2	11/2
	Δ(2850)	2850	3/2	15/2
	Δ(3230)	3230	3/2	?
Z	Z_0(1865)	1865	0	?
Λ	Λ	1115.6	0	1/2
	Λ(1405)	1405	0	1/2
	Λ(1520)	1518.8	0	3/2
	Λ′(1670)	1670	0	1/2
	Λ′(1690)	1695	0	3/2

	Particle or resonance	Mass, Mev	Isotopic spin	Spin
Λ	Λ(1815)	1815	0	5/2
	Λ(1830)	1830	0	5/2
	Λ(2100)	2100	0	7/2
	Λ(2350)	2350	0	?
Σ	Σ	(+)1189.4 (0)1192.6 (−)1197.4	1	1/2
	Σ(1385)	1382.2	1	3/2
	Σ(1660)	1660	1	3/2
	Σ(1700)	1700	1	?
	Σ(1770)	1765	1	5/2
	Σ(1910)	1910	1	5/2
	Σ(2030)	2030	1	7/2
	Σ(2250)	2250	1	9/2
	Σ(2455)	2455	1	11/2
	Σ(2595)	2595	1	?
Ξ	Ξ	(0)1314.7 (−)1321.3	1/2	1/2
	Ξ(1530)	(0)1528.9 (−)1533.8	1/2	3/2
	Ξ(1815)	1815	1/2	?
	Ξ(1930)	1915	1/2	?
Ω^-	Ω^-	1672.4	0	3/2

Resonances are denoted by letters with the mass value in parentheses. Table 2 includes all the resonances which were known by the middle of 1963 /133/, but their number continues to grow.

An important property of particles is their spin, i.e. their intrinsic mechanical moment in the absence of an orbital moment.

TABLE 3. Decay channels of fundamental baryons

Particle	Decay channels and their relative probabilities, %
n	$pe^-\nu$
Λ	$p\pi^-$ (65.3); $n\pi^0$ (34.7); $pe\nu$ ($8.8\cdot10^{-4}$); $p\mu\nu$ ($1.35\cdot10^{-4}$)
Σ^+	$p\pi^0$ (52.8); $n\pi^+$ (47.2); $p\gamma$ ($1.9\cdot10^{-3}$); $n\pi^+\gamma$ ($\approx10^{-3}$); $\Lambda e^+\nu$ ($2.2\cdot10^{-5}$); $n\mu^+\nu$ ($<7\cdot10^{-5}$); $ne^+\nu$ ($<2\cdot10^{-5}$)
Σ^0	$\Lambda\gamma$ (100); Λe^+e^- ($5.45\cdot10^{-3}$)
Σ^-	$n\pi^-$ (100); $ne^-\nu$ ($1.14\cdot10^{-3}$); $n\mu^-\nu$ ($6.2\cdot10^{-4}$); $\Lambda e^-\nu$ ($6.6\cdot10^{-5}$); $n\pi^-\gamma$ ($\approx10^{-3}$)
Ξ^0	$\Lambda\pi^0$ (100); $p\pi^-$ ($<0.9\cdot10^{-3}$); $pe^-\nu$ ($<1.3\cdot10^{-3}$); $\Sigma^+e^-\nu$ ($<1.5\cdot10^{-3}$); $\Sigma^-e^+\nu$ ($<1.5\cdot10^{-3}$); $\Sigma^+\mu^-\nu$ ($<1.5\cdot10^{-3}$); $\Sigma^-\mu^+\nu$ ($<1.5\cdot10^{-3}$); $p\mu^-\nu$ ($<1.3\cdot10^{-3}$)
Ξ^-	$\Lambda\pi^-$ (100); $\Lambda e^-\nu$ ($9\cdot10^{-4}$); $n\pi^-$ ($<1.1\cdot10^{-3}$); $\Lambda\mu^-\nu$ ($<1.3\cdot10^{-3}$); $\Sigma^0e^-\nu$ ($<0.5\cdot10^{-3}$); $\Sigma^0\mu^-\nu$ (<0.5); $ne^-\nu$ (<1)
Ω^-	$\Xi^0\pi^-$ (?); $\Xi^-\pi^0$ (?); ΛK^- (?)

Another quantum number, which has a formal similarity to the spin, is the *isotopic spin*. This concept was introduced in low-energy nuclear physics after it was found that the nuclear forces are independent of the electric charge of the proton or neutron. As a result, protons and neutrons were then regarded as two different charge states of the same particle, named nucleon. The nucleon is thus assigned a new intrinsic degree of freedom (similar to the spin), which can assume two values, corresponding to a proton and a neutron. This degree of freedom resembles the spin and is referred to as an isotopic spin. The quantum number describing the absolute value (more precisely, the square) of the isotopic spin is denoted by the letter τ, and the z-component of this spin is denoted by τ_3. In the case of a nucleon with $\tau=1/2$, the value $\tau_3=+1/2$ of the second quantum number corresponds to a proton, and the value $\tau_3=-1/2$ corresponds to a neutron, i.e. these particles constitute an isotopic doublet. This picture applies formally also to the case of an electron, which can be in two different states corresponding to two possible directions of the spin. The three pions form an isotopic triplet with quantum number $\tau = 1$, with the three possibilities $\tau_3 = -1, 0, +1$ corresponding to the π^-, π^0, and π^+ mesons, respectively.

Since nuclear forces do not depend on the charge state of particles (i. e. on their quantum number τ_3), the masses of the individual components of isotopic multiplets should be identical. The small difference in their mass values is due only to electromagnetic interactions. Analogously to the splitting of states belonging to different values of the projection of the orbital moment (removal of l-degeneracy) in the presence of a magnetic field, here we have splitting into isotopic multiplets due to electromagnetic interactions.

It was found that in all processes known to date the total number of baryons, and also that of leptons, is strictly conserved; thus each baryon (lepton) is assigned a number +1, and an antibaryon (antilepton) the number −1. These are sometimes referred to as the baryon or lepton charges.

In the phenomenological classification of particles it was found very useful to introduce a new quantum number, called *strangeness*. It is defined as follows:

$$S = 2Q - 2\tau_3 - B, \tag{21.1}$$

where Q is the particle's charge (in units of the elementary charge e), τ_3 is the z-component of the isotopic spin, and B is the baryon charge. In strong interactions the strangeness is conserved. The sum $S+B$ is called hypercharge.

We shall not discuss such important particle characteristics as the space, time, and charge parity. These quantum numbers arise during space inversion (changing $\mathbf{r}$ into $-\mathbf{r}$), time reversal (changing t into $-t$), and the operation of charge conjugation (changing a particle into its antiparticle). These concepts do not concern us here.

Particles interact with each other through four types of fields. These fields, in order of ascending strength, are: *gravitational*, *weak*, *electromagnetic*, and *strong*. Gravitational interactions are characteristic of all particles without exception, but they obviously play no role in the world of elementary particles. At least, until now no effect has been discovered which can be attributed to gravitation. Weak interactions affect all leptons, mesons, and baryons. Electromagnetic interactions exist between all particles which have an electric charge or a magnetic moment. Strong interactions operate between baryons and mesons. These forces bind the nucleons in the nucleus. They are also responsible for such processes as the scattering of baryons by baryons and mesons by baryons, the creation of mesons in baryon collisions, and others. Of the listed interactions, an adequate theory exists only for the electromagnetic case, and to a certain degree also for gravitational forces. For weak and especially strong interactions there is no logically consistent theory, and the phenomenological approach predominates. If we regard strong interactions as having a force of the order of unity, then the electromagnetic forces are of the order of 10^{-2}, weak forces of the order of 10^{-5}, and gravitational forces of the order of 10^{-38}.

The world of elementary particles has the following conservation laws. First of all, the conservation laws for energy, momentum, and angular momentum are observed. These follow, respectively, from the invariance of physical phenomena with respect to time translation, translation in space (parallel displacement), and rotation in space. The conservation laws for electric, baryon, and lepton charge are also absolute. These conservation laws are apparently based on certain symmetry properties of particles.

Strong and electromagnetic interactions are invariant with respect to $\hat{C}$, $\hat{P}$, and $\hat{T}$ operations (charge conjugation, space inversion, and time reversal, respectively), separately, while weak interactions are invariant only with respect to the $\hat{T}$ operation, i.e. with respect to time reversal. The violation of $\hat{C}$ invariance leads to nonconservation of parity in weak interactions. The strangeness $\hat{S}$ is conserved in strong and electromagnetic interactions but not in weak interactions. All decay processes presented in the last column of Tables 1 and 3 are caused by weak interactions.

We now proceed to the definition of antiparticles. Two particles whose mass and spin are completely identical, but whose electric charge, baryon (lepton in the case of leptons) charge, parity, and strangeness (for leptons $S=0$) are of opposite sign, are called antiparticles. In the tables are listed not only leptons, but also antileptons. All baryons have their antiparticles, but these were not listed in Tables 1—3.

Nucleons obviously have a structure, which is unknown to us. As evidence for this is the fact that, in nuclear collisions involving high-energy particles, a proton and a neutron generate hyperons and baryon resonances, which then go through different decays and finally end up as nucleons once again. From this it is evident that hyperons and resonances are excited states of nucleons.

In recent years there has been much talk of hypothetical strongly interacting particles with fractional electric and baryon charges, called *quarks*. Table 4 shows the properties ascribed to these particles /57/.

TABLE 4. Quantum numbers of quarks and antiquarks

Symbol	Quantum numbers					
	charge Q	isotopic spin		strange-ness, S	baryon charge B	spin, s
		τ	τ_3			
Q_p	2/3	1/2	+1/2	0	1/3	1/2
Q_n	−1/3	1/2	−1/2	0	1/3	1/2
Q_λ	−1/3	0	0	−1	1/3	1/2
$\overline{Q}_p$	−2/3	1/2	−1/2	0	−1/3	1/2
$\overline{Q}_n$	+1/3	1/2	+1/2	0	−1/3	1/2
$\overline{Q}_\lambda$	+1/3	0	0	+1	−1/3	1/2

Baryons and mesons appear to be compound particles consisting of quarks. Thus, for example, a proton, a neutron, and a Λ hyperon are described, respectively, as the combinations $Q_pQ_pQ_n$, $Q_pQ_nQ_n$, and $Q_pQ_nQ_\lambda$, and a meson as a combination of quark and antiquark, e. g. a π^- meson as the combination $Q_p\overline{Q}_p$. The mass of quarks should be large, at least larger than that of the heaviest baryon. Attempts have been made to discover these particles experimentally, but all results have been negative.

Apparently, quarks are only theoretical constructions, which express the symmetry properties of baryons and mesons as belonging to certain groups. They have been very helpful in the phenomenological classification of particles according to groups. Since baryons and mesons are represented as compound particles consisting of quarks, they have received the collective name *hadrons*.

22. THE BARYON GAS

Our discussion at the end of Chapter II was developed up to densities of the order of nuclear densities. At densities $\rho \approx 10^{13}$ g · cm^{-3} the atomic nuclei vanish completely and a continuous nuclear material is formed, consisting mainly of neutrons and a relatively small number of protons and electrons (a nuclear plasma or, more precisely, the *nep*-phase of matter). Thermodynamic equilibrium is established through the reactions

$$n \to p + e + \bar{\nu}_e, \quad p + e \to n + \nu_e, \quad p \to n + e^+ + \nu_e.$$

In the ideal-gas approximation the particle concentrations are determined from the relationships

$$\mu_n = \mu_p + \mu_e, \qquad n_e = n_p. \tag{22.1}$$

The first of these follow directly from consideration of the above reactions at thermodynamic equilibrium, and the second expresses the condition of neutrality for the plasma.

The degeneracy temperatures of particles in a nuclear plasma are very high, namely $T_n \approx 5\cdot 10^{11}$, $T_p \approx 10^{10}$, and $T_e \approx 10^{12}$ degrees. On the other hand, due to the enormous neutron energy losses the temperature inside the corresponding superdense stars should be considerably lower than these. Consequently, matter in the *nep*-phase can be considered strongly degenerate. One can therefore write

$$\left.\begin{aligned} \mu_k &= \varepsilon_k = m_k c^2 + \frac{a^2}{2m_k c^2} n_k^{2/3}, \quad k = n;\ p, \\ \mu_e &= \varepsilon_e = a n_e^{1/3}, \end{aligned}\right\} \tag{22.2}$$

where $a = (3\pi^2)^{1/3}\hbar c$. The nucleon gas is here assumed to be non-relativistic, and the electron gas relativistic, as is in fact the case in the nuclear density range.

From (22.1) and (22.2) we obtain

$$\begin{aligned} n_e = n_p &= \left[\sqrt{\frac{m_p}{m_n} n_n^{2/3} + \frac{m_p c^4 (2m_n - m_p)}{a^2}} - \frac{m_p c^2}{a}\right]^3 \approx \\ &\approx \left[\sqrt{n_n^{2/3} + \frac{m_p^2 c^4}{a^2}} - \frac{m_p c^2}{a}\right]^3. \end{aligned} \tag{22.3}$$

For $n_n = 10^{38}\,\mathrm{cm}^{-3}$ this yields $n_e = n_p = 1.23\cdot 10^{36}\,\mathrm{cm}^{-3}$, which means that in the nuclear density region the concentration of neutrons is approximately a hundred times that of protons (or electrons). This fact was the reason for naming these configurations neutron stars. In Section 23 we shall see that when the interaction energy is taken into account, the disproportion between the concentrations of protons and neutrons diminishes (see equation (28.8)).

Let us now consider supernuclear densities /14, 10, 92, 17/. With increasing density, ε_e rises in proportion to $n_e^{1/3}$, and the Fermi kinetic energy of neutrons and protons increases as $n_n^{2/3}$ and $n_p^{2/3}$, respectively. At sufficiently high values of $n = n_n + n_p$, when the particles' Fermi energies exceed the rest energies of unstable baryons such as hyperons and resonances, the nucleon-electron phase loses its stability. In this situation the creation of hyperons and baryon resonances is energetically favored. This is due to the fact that, to achieve minimum energy for the system, it is preferable for some nucleons to be in the hyperon or resonance states of lower energy rather than in the upper levels of the Fermi distribution. This physical aspect of the effect is similar to the one encountered in the conversion of protons into neutrons. But in following this principle of minimum energy we also have to show that it can in fact be realized. The process of neutronization occurs through $\beta^{\pm}$-decay. In the present case, the transition from nucleons to hyperon or resonance states can take place through one of two processes, namely the emission of π mesons or low-probability lepton processes. Transitions through either way are caused by weak interactions only. The decay processes we have in mind are inverse to those given in Table 3. For the Λ hyperon, for example, we would have

$$n \rightarrow \Lambda + \pi^0, \qquad p \rightarrow \Lambda + \pi^+,$$
$$p \rightarrow \Lambda + e^+ + \nu_e, \quad p \rightarrow \Lambda + \mu^+ + \nu_\mu.$$

Under the conditions existing in a degenerate nucleon-electron plasma, however, the formation process of a Λ hyperon does not follow these channels, but rather the capture of electrons by protons $p+e^- \rightarrow \Lambda + \nu_e$, since in the latter case the threshold density is lower. In fact, for the former decays a neutron Fermi energy $\varepsilon_n \gtrsim 1251$ Mev is required, whereas in the last case it is $\varepsilon_n = \varepsilon_p + \varepsilon_e \gtrsim 1116$ Mev.

We shall now list those reactions which, under the given conditions, bring about the appearance of hyperons and the establishment of thermodynamic equilibrium. Thus, in view of the remark made in connection with the creation of Λ hyperons, of the decay schemes presented in Table 3 we should obviously consider the following:

$$\left.\begin{array}{ll}
n \rightarrow p + e + \bar{\nu}_e, & p + e \rightarrow n + \nu_e, \\
\Lambda \rightarrow p + e + \bar{\nu}_e, & p + e \rightarrow \Lambda + \nu_e, \\
\Sigma^0 \rightarrow p + e + \bar{\nu}_e, & p + e \rightarrow \Sigma^0 + \nu_e, \\
\Sigma^- \rightarrow n + e + \bar{\nu}_e, & n + e \rightarrow \Sigma^- + \nu_e, \\
\Sigma^+ + e \rightarrow n + \nu_e, & n \rightarrow \Sigma^+ + e + \bar{\nu}_e, \\
\Xi^- \rightarrow n + e + \bar{\nu}_e, & n + e \rightarrow \Xi^- + \nu_e, \\
\Xi^0 \rightarrow p + e + \bar{\nu}_e, & p + e \rightarrow \Xi^0 + \nu_e, \\
\Omega^- \rightarrow n + e + \bar{\nu}_e, & n + e \rightarrow \Omega^- + \nu_e.
\end{array}\right\} \qquad (22.4)$$

We have given only part of the possible transitions, namely those involving nucleons and electrons; there are additional lepton channels, in which only hyperons take part.

The thermodynamic equilibrium corresponding to reactions (22.4) between the different components of matter is determined by equations

$$\mu_k^0 = \mu_n, \quad \mu_k^+ = \mu_n - \mu_e, \quad \mu_k^- = \mu_n + \mu_e. \tag{22.5}$$

Here μ_n and μ_e are the chemical potentials of neutrons and electrons, and μ_h^0, μ_h^+ and μ_k^- are the chemical potentials for neutral, positively charged, and negatively charged baryons, respectively. Earlier we did not mention baryon resonances, but equations (22.5) apply to these particles as well.

From now on we shall assume that the temperature of the medium is zero or considerably lower than the degeneracy temperature T_k of any particle. As will be shown below, in the supernuclear density range, where there is a baryon gas with a rich "chemical" composition, the concentrations of all particles are so high that the degeneracy temperatures T_k are also very elevated. Therefore the chemical potentials of the particles coincide with their Fermi energies, $\mu_k = \varepsilon_k$, and the latter are dependent only upon the densities n_k of the corresponding particles. Thus equations (22.5) establish relationships between the concentrations of particles during thermodynamic equilibrium.

23. THE COMPLETE SYSTEM OF EQUATIONS

Starting out from some general physical principles, we shall study the "chemical" composition of matter in the supernuclear density range, but without appealing to specific physical processes to justify the appearance of some particle or another. We shall assume that the mass density is so high that the material already contains all those elementary particles which can become stable in a superdense medium. We shall now check which particles belong to this category.

Under the physical conditions which are of interest to us, possible components of matter can be all baryons, from the lepton group — electrons and $\mu^\pm$ mesons, and from the meson group only π^- mesons. The stability of hyperons and baryon resonances is ensured through the presence of a degenerate nucleon gas, and the stability of μ^- mesons by the presence of a degenerate electron gas. This is because, as a result of the Pauli principle, the decay products of these particles cannot find a vacant place in phase space, since it is solidly filled (with one particle in each quantum state) up to the Fermi momentum level by the same particles (protons, neutrons, and electrons). All hyperons and baryon resonances have a half-integral spin, therefore the gas formed by them will also be strongly degenerate (it is assumed that $T \ll T_k$); thus transitions between these particles, which are unstable under laboratory conditions, will also be forbidden. Negative muons also form a degenerate gas, thus ensuring the stability of the π^- meson, which disintegrates into a μ^- meson and a neutrino. As to the other known elementary particles, namely photons, positrons, neutrinos, μ^+, π^0, π^+, and different kinds of K mesons and their resonances, at absolute zero they should all be absent from the medium, since there is nothing to prevent their disintegration, annihilation, or escape (in the case of neutrinos or photons) from inside the star.

For every kind of unstable particle there is a threshold density above which it becomes stable. Under the physical conditions being considered, a special place is occupied by the π^- mesons. Since these particles obey Bose statistics, they all concentrate at the level of lowest energy $\varepsilon_\pi = m_\pi c^2$. They thus form a highly degenerate Bose gas.

As we see, all baryons, the negative muon, and from among the bosons the negative pion, which are unstable under terrestrial conditions, become completely stable at supernuclear densities. The appropriate physical conditions can occur in certain special celestial bodies (see Chapters VI and VII). On the other hand, under the special conditions existing in a degenerate proton-electron gas the proton, which is a typically stable particle, becomes unstable. The property of particle stability is thus a relative concept, dictated by the physical conditions in the medium.

If we want to find out which particles can exist in the medium and what their concentrations are for a given total number density of baryons, we obviously have to apply the principle of minimum total energy of the system. Nature is economical, choosing from among all possible states the one for which the energy of the system is smallest (degenerate matter is an ideally ordered system whose entropy is zero, therefore it cannot serve as a criterion for the choice of a most probable state).

The energy per unit volume is

$$\rho = \frac{c}{2\pi^2 h^3} \sum_k a_k \int_0^{p_k} \sqrt{m_k^2 c^2 + p^2}\, p^2 dp + m_\pi c^2 n_\pi , \tag{23.2}$$

where the summation is assumed to be over all forms of baryons as well as an electron and a μ meson, p_k is the Fermi momentum for the k-th type of particle, n_π is the number density of π^- mesons, and $a_k = 2s_k + 1$, where s_k is the spin of the k-th particle, a_k is a constant which allows for the number of possible spin states for the particle. For leptons and all basic baryons (perhaps to the exclusion of the Ω^- hyperon, whose spin has not yet been accurately determined, but is assumed to be $s_k = 3/2$) $s_k = 1/2$ and $a_k = 2$; but among resonance particles one also encounters other spin values, such as $s_k = 3/2, 5/2, 7/2$, and in the case of the $\Delta(2420)$ resonance even $s_k = 11/2$. As has already been pointed out, at $T = 0$ the pions have no kinetic energy, therefore (23.1) contains only their rest energy $m_\pi c^2$. Furthermore, in (23.1) it is assumed that the density of matter is so high that all the above-mentioned particles are present in it and even constitute an ideal gas. The Fermi momentum and particle density are related through

$$p_k = \left(\frac{6\pi^2}{a_k}\right)^{1/3} \hbar n_k^{1/3}, \qquad n_k = \frac{a_k}{6\pi^2\hbar^3} p_k^3. \tag{23.2}$$

From the condition that the energy (23.1) be a minimum we now have to find the equations which determine the particle concentrations. Before doing this, though, we should find the restrictions to be imposed on these concentrations. There are two additional conditions, namely the expressions for the conservation of baryon charge and the conservation of electric charge. The problem is as follows: given the total number of baryons n, one has to find out how it is divided among the different components, i.e. to determine the concentrations n_k. Accordingly, the first restriction reduces to the equation

$$n = \frac{1}{6\pi^2\hbar^3} \sum_k a_k p_k^3 = \text{const}, \tag{23.3}$$

where the summation is over baryons only. In real celestial bodies n depends on the coordinates, but during thermodynamic equilibrium the number n is of course constant at each fixed point of a star. Actually, (23.3) means that during any processes leading to a redistribution of the number of particles, the total number of baryons (baryon charge) in the star remains unchanged.

During elementary processes, of course, the number of leptons (lepton charge) is also conserved. On the other hand, when conditions are such that the newly created neutrinos and antineutrinos can leave the body of the star relatively easily, the number of leptons in it cannot be considered fixed; during the establishment of thermodynamic equilibrium the total number of leptons in a celestial body is not conserved. The number of leptons (electrons and μ^- mesons) becomes fixed only after thermodynamic equilibrium is established, since the parameter which determines the total number of leptons is then once again the number of baryons.

Gravitational forces are extremely weak compared to electromagnetic forces, therefore even a very small disturbance (about 10^{-18} excess charged particle for one particle with balanced charge) of neutrality leads to instability. Consequently, electric neutrality has to apply not only to a celestial body as a whole, but also to each volume element in it, i. e.

$$\frac{1}{6\pi^2h^3}\Big(\sum_{(++)} 2a_kp_k^3 + \sum_{(+)} a_kp_k^3 - \sum_{(-)} a_kp_k^3 - \sum_{e,\mu} a_kp_k^3\Big) - n_\pi = 0, \qquad (23.4)$$

where summation in the second term is over all positively charged baryons, in the third over the negatively charged baryons, in the fourth over the leptons, and in the first over the positive particles with double electric charge. The latter are found among the Δ resonances, whose isotopic spin is $I = 3/2$. Table 2 lists 6 such particles. In connection with this we should note that all the resonance states of baryons shown in Table 2 form isotopic multiplets. The exception is Λ resonances, whose isotopic spin is zero. Thus the Δ resonances form an isotopic quadruplet (consisting of four particles with charges $-e$, 0, $+e$, $+2e$), the Σ resonances form triplets (three particles with charges $-e$, 0, $+e$), the N resonances form doublets (two particles with charges 0, $+e$), and the Λ resonances are singlets (one particle with charge 0).

We thus have to find the minimum of expression (23.1) under the conditions (23.3) and (23.4). As is well known, in this case the problem is reduced to finding the minimum of the function

$$\Phi = \frac{1}{2\pi^2h^3}\Big[\sum ca_k \int_0^{p_k} \sqrt{m_k^2c^2 + p^2}\, p^2dp + 2\pi^2h^3m_\pi c^2 n_\pi + \frac{\alpha}{3}\sum_k a_kp_k^3 + \frac{\beta}{3}\Big(\sum_{(++)} 2a_kp_k^3 + \sum_{(+)} a_kp_k^3 - \sum_{(-)} a_kp_k^3 - \sum_{e,\mu} a_kp_k^3 - 6\pi^2h^3n_\pi\Big)\Big], \qquad (23.5)$$

in which the parameters α and β are still undetermined.

Equating to zero the first derivatives of the function Φ with respect to the particle concentrations n_k (or, for convenience, in the case of fermions one can differentiate with respect to the Fermi momentum p_k), we find the necessary conditions for minimum system energy. Here attention should be drawn to one mathematical detail, which has an important physical consequence. The point is that the vanishing of the derivatives of Φ is a necessary condition for a minimum only if the minimum is attained inside the allowed interval of each variable. If, however, this minimum is attained at an extreme value of one of the variables, then the derivative with respect to that variable does not necessarily have to be zero. Thus, for example, the baryon resonance Δ(3230) is heavier than all baryons listed in Table 2; therefore it should appear only at very high densities, later than all the others. At baryon densities below the Δ(3230) threshold, the function Φ should have its lowest value when $N_\Delta(3230) = 0$; but this condition is not obtained by setting the derivatives of Φ equal to zero. We should therefore be careful, and use the following equations only after making sure that the concentration with respect to which differentiation led to the equation in question does not vanish. This will correspond to the division, to be introduced below, of the whole interval of baryon densities (or mass densities) into separate subintervals, differing from each other in their elementary-particle composition, which can be referred to as different phases of matter. We can of course construct the Φ function for each phase and later find its minimum. It seems preferable, though, to consider first the highest phase, when the medium contains all the enumerated particles, i. e. to write equation (23.5). Later, in the equations expressing the equilibrium condition among the components, one can omit all parameters pertaining to particles which are not present in the phase at hand, thus automatically obtaining all necessary functions for determining the particle concentrations in that particular phase.

Whenever differentiation with respect to n_k occurs, it can be replaced by differentiation with respect to the corresponding Fermi momentum p_k, except in the case of n_π. Equating the derivatives of Φ to zero, we obtain

$$\left.\begin{aligned} a_k\varepsilon_k p_k^2 + a_k(\alpha + 2\beta)\,p_k^2 &= 0 \quad \text{for} \quad B^{++}, \\ a_k\varepsilon_k p_k^2 + a_k(\alpha + \beta)\,p_k^2 &= 0 \quad \text{for} \quad B^{+}, \\ a_k\varepsilon_k p_k^2 + a_k(\alpha - \beta)\,p_k^2 &= 0 \quad \text{for} \quad B^{-}, \\ a_k\varepsilon_k p_k^2 + a_k\alpha p_k^2 &= 0 \quad \text{for} \quad B^{0}, \\ a_k\varepsilon_k p_k^2 - a_k\beta p_k^2 &= 0 \quad \text{for} \quad k = e \text{ and } \mu^-, \\ m_\pi c^2 - \beta &= 0\,. \end{aligned}\right\} \tag{23.6}$$

Here $\varepsilon_k = c\left(m_k^2 c^2 + p_k^2\right)^{1/2}$ is the Fermi energy, B signifies a baryon, and the upper sign indicates the electric charge of the particle.

Equation (23.6) together with (23.3) and (23.4) completely determine the values of the Fermi momenta p_k and the parameters α and β. Eliminating α and β from (23.6), we obtain the following set of equations:

$$\varepsilon_k^{(0)} = \varepsilon_n, \tag{23.7}$$

$$\varepsilon_k^{(+)} + \varepsilon_e = \varepsilon_n, \tag{23.8}$$

$$\varepsilon_k^{(-)} - \varepsilon_e = \varepsilon_n, \tag{23.9}$$

$$\varepsilon_k^{(++)} + 2\varepsilon_e = \varepsilon_k^0, \tag{23.10}$$

$$\varepsilon_\mu = \varepsilon_e, \tag{23.11}$$

$$\varepsilon_e = m_\pi c^2, \tag{23.12}$$

where $\varepsilon_k^{(0)}$, $\varepsilon_k^{(+)}$, and $\varepsilon_k^{(-)}$ are, respectively, the Fermi energies of neutral, positively charged, and negatively charged baryons.

This system of equations applies to the most general case, when the medium contains all particles which can become stable.

In accordance with our earlier remarks, the last equation is valid only at densities above the pion-formation threshold. This, of course, also applies to all baryons heavier than a neutron, and to muons. The above system of equations coincides with the one given in (22.5).

This is not surprising, since the derivation of the well-known relationships between the chemical potentials during equilibrium is also based on the principle of minimum energy and on the laws of conservation of electric and baryon charge.

Equations (23.7)—(23.12) together with (23.3)—(23.4) determine the particle concentrations n_k in the most general case, when the gas of elementary particles is of such high density that it contains all baryons, negative leptons, and π^- mesons. As we can see, in a state of thermodynamic equilibrium the Fermi energies of baryons with the same electric charge are equal. This also holds for leptons.

In the treatment given in this section it was not necessary that $T = 0$. It is obvious that the results obtained above remain in force even at temperatures different from zero, as long as the baryon gas can be considered strongly degenerate. Moreover, since the material is very dense, the permitted temperatures can attain high values.

24. PARTICLE STABILITY THRESHOLDS

In a gas consisting of elementary particles, the independent variable is logically assumed to be the total number density n of baryons.

If n is known, equations (23.3), (23.4), and (23.7)—(23.11) enable one to determine the concentrations of all types of particles. Knowledge of n is evidently equivalent to knowing the density of the total energy or the mass. The chemical composition of the material is thus a unique function of the baryon density n, and our problem is to find out which particles exist in the medium and how their concentrations n_k change with increasing n.

In the preceding chapter we led our discussion up to the point of nuclear densities, where the matter consists of a nucleon-elec... gas. Suppose we now raise the density. We shall then have to deal with a succession of phases. A separate investigation of all these phases is not possible, and meaningless besides. Special consideration will be given only to the formation of the first two phases beyond the nucleon-electron phase.

As will be shown below, when ascending in the supernuclear region, the first to become stable is the muon, immediately afterward the Σ^- hyperon, followed by the other particles in a certain order. Let the density be such that the Fermi energy of electrons exceeds the rest energy of μ^- mesons, and the sum of the neutron and electron Fermi energies is smaller than the energy of the Σ^- particles (the *nucleon-lepton* phase). In this case the set of equations (23.3), (23.4), and (23.7)—(23.12) reduces to the following:

$$\left.\begin{aligned} \sqrt{m_p^2c^2+p_p^2} &= \sqrt{m_n^2c^2+p_n^2} - p_e, \\ \sqrt{m_\mu^2c^2+p_\mu^2} &= p_e, \\ p_e^3+p_\mu^3 &= p_p^3. \end{aligned}\right\} \tag{24.1}$$

We omitted equation (23.3) for the sum of the baryon concentrations $n_n+n_p=n$, since in this region $n_p \ll n_n$ and $n \approx n_n$. As in the preceding phase, the electrons here are also very relativistic: $\varepsilon_e \approx cp_e$. The number of muons here is very small compared to the number of electrons; therefore in calculating the concentrations of protons and electrons we can, as before, use equation (20.11). From (24.1) we obtain for the density of μ^- mesons

$$n_\mu = n_e\left[1-\left(\frac{A_\mu}{n_e}\right)^{2/3}\right]^{3/2}, \tag{24.2}$$

where $A_\mu = (1/3\pi^2)(m_\mu c/h)^3 = 5.24\cdot 10^{36}$ cm^{-3} is the threshold for formation of muons. This electron density corresponds to a neutron density of approximately $5\cdot 10^{38}$ cm^{-3}, which is higher by a factor of about 2.5 than the density of particles in ordinary nuclear material.

The next phase sets in at a density when the sum of the neutron and electron Fermi energies becomes equal to the rest energy of Σ^- hyperons. The first Σ^- hyperons appear at this density. From equations (23.3), (23.4), and (23.7)—(23.12), omitting the terms pertaining to those particles which are not yet present in the medium, we obtain

$$\left.\begin{aligned} &\varepsilon_p + \varepsilon_c = \varepsilon_n, \quad \varepsilon_n + \varepsilon_e = \varepsilon_\Sigma, \quad \varepsilon_\mu = \varepsilon_e, \\ &p_p^3 + p_n^3 + p_\Sigma^3 = 3\pi^2 h^3 n, \\ &p_p^3 - p_e^3 - p_\mu^3 = 0. \end{aligned}\right\} \tag{24.3}$$

These equations are solved numerically. Substituting in these $p_\Sigma = 0$ and $\varepsilon_\Sigma = m_{\Sigma^-} c^2$, we obtain the following equation for the Σ^- hyperon threshold:

$$\left\{\left[\left(m_\Sigma c^2 - \varepsilon_n'\right)^2 - m_\mu^2 c^4\right]^{3/2} + \left(m_\Sigma c^2 - \varepsilon_n'\right)^3\right\}^{2/3} - \\ - \left(m_\Sigma c^2 - 2\varepsilon_n'\right)^2 + m_p^2 c^4 = 0.$$

Here ε_n' is the value of the neutron Fermi energy at the Σ^- hyperon threshold. Graphic solution of this equation yields the result

$$\left.\begin{aligned} &cp_n = 517\,\text{Mev}, \\ &n_n = 6.1\cdot 10^{38}\text{cm}^{-3}, \quad n = 6.4\cdot 10^{38}\text{cm}^{-3}, \\ &\rho = 1.12\cdot 10^{15}\text{g}\cdot\text{cm}^{-3}, \; t_n = 4\,\text{arsh}\frac{p_n}{m_n c} = 2.1. \end{aligned}\right\} \tag{24.4}$$

This means that Σ^- hyperons appear in a degenerate baryon-lepton gas when the density is higher by a factor of about three than nuclear. In the next chapter we shall see that inclusion of the nuclear interaction energy considerably lowers the stability threshold of this particle, bringing it down to the nuclear density.

We shall now consider the most general case, when the medium contains all baryons, the leptons e and μ^-, and the π^- mesons. The concentrations of particles with identical baryon and electric charges are related through a simple analytical expression. Thus, from equations (23.7)—(23.9) it follows that

$$\left.\begin{aligned} n_k^{(0)} &= \frac{1}{2} a_k n_n \left[1 - \left(\frac{A_k}{n_k}\right)^{2/3}\right]^{3/2}, \\ A_k &= \frac{1}{3\pi^2}\left(\frac{m_k c}{h}\right)^3 \left(1 - \frac{m_p^2}{m_k^2}\right)^{3/2}, \end{aligned}\right\} \qquad (24.5)$$

$$\left.\begin{aligned} n_k^{(+)} &= \frac{1}{2} a_k n_p \left[1 - \left(\frac{B_k}{n_p}\right)^{2/3}\right]^{3/2}, \\ B_k &= \frac{1}{3\pi^2}\left(\frac{m_k c}{h}\right)^3 \left(1 - \frac{m_p^2}{m_k^2}\right)^{3/2}, \end{aligned}\right\} \qquad (24.6)$$

$$\left.\begin{aligned} n_k^{(-)} &= \frac{1}{2} a_k n_{\Sigma^-} \left[1 - \left(\frac{c_k}{m_{\Sigma^-}}\right)^{2/3}\right]^{3/2}, \\ C_k &= \frac{1}{3\pi^2}\left(\frac{m_k c}{h}\right)^3 \left(1 - \frac{m_{\Sigma^-}^2}{m_k^2}\right)^{3/2}, \end{aligned}\right\} \qquad (24.7)$$

$$\left.\begin{aligned} n_k^{(++)} &= \frac{1}{4} a_k n_\Delta \left[1 - \left(\frac{D_k}{n_\Delta}\right)^{2/3}\right]^{3/2}, \\ D_k &= \frac{2}{3\pi^2}\left(\frac{m_k c}{h}\right)^3 \left(1 - \frac{m_\Delta^2}{m_k^2}\right)^{3/2}, \end{aligned}\right\} \qquad (24.8)$$

where $n_k^{(0)}$, $n_k^{(+)}$, $n_k^{(-)}$, and $n_k^{(++)}$ are the densities of neutral, positive, and negative baryons, and of resonances with double electric charge, respectively, n_Δ is the density of a Δ (1236) resonance with the charge $Q=+2e$, and $a_k = 2s_k + 1$ is the earlier introduced static weight of the k-th particle. As we can see, the constants A_k, B_k, C_k, and D_{kk} act as threshold densities for the corresponding particles. The neutral k-th particle becomes stable in a degenerate gas of elementary particles only at neutron densities $n_k > A_k$. Similarly, the k-th positive and negative baryons attain stability only at densities $n_p > B_k$, $n_{\Sigma^-} > C_k$, and $n_\Delta > D_k$.

From equations (24.5)—(24.8) it follows that after gaining stability, just above the threshold, the concentration of a particle is not much different from the concentrations of baryons already present.

The phase which contains π^- mesons has an interesting peculiarity. Pions appear in the medium when the Fermi energy of electrons and muons becomes equal to the rest energy of these particles. With the appearance of π^- mesons the concentrations of e and μ^- leptons stop rising — they remain constant, but the concentration of π^- mesons rises very rapidly with increasing baryon density, and soon becomes of the same order of magnitude as the concentrations of the different types of baryons. The reason for this lies in the principle of minimum energy: it is energetically more favorable to have pions in a state of rest than to have leptons with the Fermi energy $\varepsilon_e = \varepsilon_\mu > m_\pi c^2$.

The threshold for π^- meson formation is obtained directly from equations (23.11) and (23.12), i. e. from the relation $\varepsilon_e = \varepsilon_\mu = m_\pi c^2$. We thus have

$$\left.\begin{aligned} n_e &= \frac{1}{3\pi^2}\left(\frac{m_\pi c}{h}\right)^3 = 1.20\cdot 10^{27}\,\mathrm{cm}^{-3}, \\ n_\mu &= n_e\left(1-\frac{m_\mu^2}{m_\pi^2}\right)^{3/2} = 3.35\cdot 10^{36}\,\mathrm{cm}^{-3} \\ &\text{for } n_n > 8.5\cdot 10^{39}\,\mathrm{cm}^{-3}. \end{aligned}\right\} \tag{24.9}$$

Equations (24.5)—(24.8) enable one to express the densities of neutral, positive, negative, and doubly positively charged baryons through the densities of neutrons, protons, Σ^- hyperons, and $\Delta^{++}(1236)$ resonances, respectively. Thus, to obtain the concentrations of all particles at a given density of the material, it only remains to express the densities of electrons, Σ^- hyperons, and $\Delta^{++}(1236)$ resonances through the density of neutrons. Henceforth, instead of the total density ρ of the material or baryon density n, we shall specify the neutron density n_n, and this is in practice equivalent to specification of the constant in equation (23.3).

We still have at our disposal four unused equations, namely the condition of neutrality (23.4), and two equations of the system (23.8)—(23.10), linking the Fermi energies of protons, Σ^- hyperons, and Δ (1236) resonances to the Fermi energy of neutrons:

$$\left.\begin{aligned} &\sum_{(++)} 2a_k p_k^3 + \sum_{(+)} a_k p_k^3 - \sum_{(-)} a_k p_k^3 - 2p_e^3 - \\ &\qquad\qquad - 2p_\mu^3 - 6\pi^2 h^3 n_\pi = 0, \\ &\varepsilon_p + \varepsilon_e = \varepsilon_n, \quad \varepsilon_\Sigma - \varepsilon_e = \varepsilon_n, \quad \varepsilon_\Delta^{++} + 2\varepsilon_e = \varepsilon_n. \end{aligned}\right\} \tag{24.10}$$

Using (23.2), we now rewrite equations (24.6)—(24.8) in terms of the Fermi momenta:

$$\left.\begin{aligned} p_k &= p_\alpha\left(1-\frac{b_{k\alpha}^2}{p_\alpha^2}\right)^{1/2}, \\ b_{k\alpha}^2 &= \left(m_k^2 - m_\alpha^2\right)c^2, \end{aligned}\right\} \tag{24.11}$$

where $\alpha = p$, Σ^-, or Δ (1236), respectively, for positive and negative baryons, and Δ^{++} resonances, and the index k covers all particles belonging to the charge groups of the above-mentioned baryons.

Substituting (24.11) in the first of equations (24.10), we obtain

$$p_\Delta^3\left[4+\sum_{(++)}a_k\left(1-\frac{b_{k\Delta}^2}{p_\Delta^2}\right)^{3/2}\right]+p_p^3\left[1+\sum_{(+)}\frac{1}{2}a_k\times\right.$$
$$\left.\times\left(1-\frac{b_{kp}^2}{p_p^2}\right)^{3/2}\right]-p_\Sigma^3\left[1+\sum_{(-)}\frac{1}{2}a_k\left(1-\frac{b_{k\Sigma}^2}{p_\Sigma^2}\right)^{3/2}\right]-$$
$$-p_e^3\left[1+\left(1-\frac{m_\mu^2c^2}{p_e^2}\right)^{3/2}\right]-3\pi^2h^3n_\pi=0, \qquad (24.12)$$

where according to the last three equations in (24.10)

$$\left.\begin{aligned} cp_p &= \left[(\varepsilon_\iota-\varepsilon_e)^2-m_p^2c^4\right]^{1/2},\\ cp_\Sigma &= \left[(\varepsilon_n+\varepsilon_e)^2-m_{\Sigma^-}^2c^4\right]^{1/2},\\ cp_\Delta &= \left[(\varepsilon_n-2\varepsilon_e)^2-m_\Delta^2c^4\right]^{1/2}.\end{aligned}\right\} \qquad (24.13)$$

After substituting these values of p_p, p_Σ, and p_Δ in (24.12), one obtains a single equation for the three quantities n_n, n_e, and n_π. It may seem that some equations have been left out, but that is not the case. In fact one has to differentiate between two separate phases, namely the one in which there are no pions and the one with pions. In the first case one should set $n_\pi=0$ in equation (24.12), leaving the two quantities n_n and n_e. If we now regard n_n as the independent variable and specify its value, we can obtain the electron density n_e from equation (24.12). After this the Fermi energies ε_n and ε_e are calculated, and then, from (23.7)—(23.11), also the Fermi energies $\varepsilon_k^{(+)}$ $\varepsilon_k^{(++)}$, and $\varepsilon_k^{(-)}$ of all baryon states. Once we have the Fermi energies, it is easy to calculate the Fermi momenta, and then, from the second equation (23.2), the densities n_k of the corresponding particles. After all this it is easy to find the total baryon density $n=\Sigma n_k$. In the second case, when the medium contains pions, the electron density is constant $n_e=1.2\cdot10^{27}$ cm^{-3} and $\varepsilon_e=m_\pi c^2$ (see (24.9)). Using (24.13) to eliminate from (24.12) the Fermi momenta of protons, Σ^- hyperons, and Δ^{++} (1236) resonances, we obtain a single equation for n_n and n_π. After defining n_n, we find n_π, and following the previous method, we determine the concentrations of all particles and the total baryon density n. In both cases equation (24.12) is solved numerically and not analytically.

Below the threshold for the appearance of the k-th particle, the momentum of a basic particle obeys $p_\alpha<b_{k\alpha}$ (α is the index for a basic particle, which we here define as a proton, a Σ^- hyperon, a

Δ^{++}(1236) resonance, and an electron). At these densities the corresponding term in (24.12) becomes imaginary and drops out of the calculation. The stability thresholds of particles, expressed through baryon densities, are given in Table 5.

TABLE 5. Particle stability thresholds

Particle	Baryon density	Particle	Baryon density	Particle	Baryon density
Σ^-(1197)	$6.4\cdot10^{38}$	Σ^-(1382)	$1.5\cdot10^{40}$	N'^+(1470)	$7.2\cdot10^{40}$
Δ^-(1236)	$\sim10^{39}$	Δ^{++}(1236)	$1.64\cdot10^{40}$	N^0(1550)	$7.76\cdot10^{40}$
Λ(1115.6)	$1.2\cdot10^{39}$	Σ^0(1382.2)	$2.5\cdot10^{40}$	N^+(1518)	$9.5\cdot10^{40}$
Σ^0(1192.6)	$4\cdot10^{39}$	Λ(1405)	$3.28\cdot10^{40}$	Δ^-(1640)	$9.6\cdot10^{40}$
Δ^0(1236)	$6\cdot10^{39}$	Σ^+(1382)	$3.6\cdot10^{40}$	N^+(1550)	$1.18\cdot10^{41}$
Σ^+(1189)	$8\cdot10^{39}$	Ξ^-(1534)	$5\cdot10^{40}$	Δ^0(1640)	$1.18\cdot10^{41}$
Ξ^-(1321)	$9\cdot10^{39}$	N'^0(1470)	$5.2\cdot10^{40}$	Σ^-(1660)	$1.2\cdot10^{41}$
Δ^+(1236)	$1.1\cdot10^{40}$	N^0(1518)	$6.4\cdot10^{40}$	Ω^-(1672)	$1.25\cdot10^{41}$
Ξ^0(1314.7)	$1.48\cdot10^{40}$	Λ(1518.8)	$6.46\cdot10^{40}$	Σ'^-(1700)	—
		Ξ^0(1528.9)	$6.84\cdot10^{40}$		

Note. The mass values are indicated in parentheses.

Figure 9 shows the function ε_e (ε_n), and Figure 10 presents the dependence of the concentrations of electrons, neutrons, protons Σ^--hyperons, and Δ^{++} (1236) resonances on the total number density of baryons. The $n_e(n)$ curve for electrons has a maximum at $n\approx 1.5\cdot10^{30}$ cm^{-3} and a minimum at $n\approx 2.5\cdot10^{40}$ cm^{-3}, after which it rises monotonically up to the point where $\varepsilon_e = m_\pi c^2$. Once ε_e has reached this value, the densities n_e and n_μ stop rising. The behavior of the n_μ (n) curve for μ^- mesons is similar to that of $n_e(n)$.

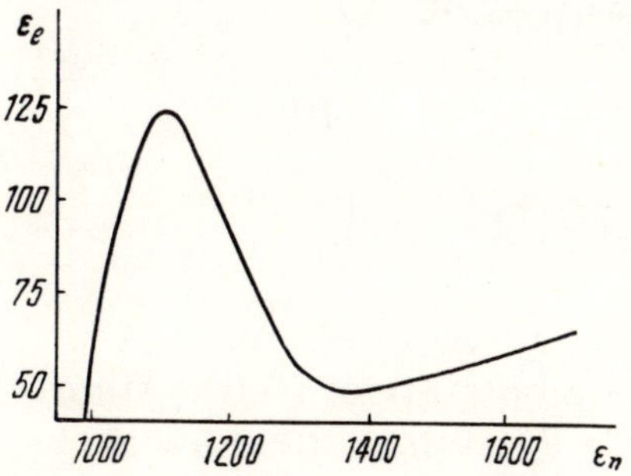

FIGURE 9. The Fermi energy of electrons as a function of the Fermi energy of neutrons.

At sufficiently high densities, when $p_k > m_k c$, the concentrations of the different baryons are quite similar: $n_k \approx n_n$. In this relativistic gas the presence of π^- mesons is possible only when there are more types of positive baryons than of negative baryons. In the presence of a π^- meson gas the system is more stable, corresponding to a state of minimum internal energy. In the absence of π^- mesons, the charge of the positive baryons has to be compensated by the charge of negative baryons and leptons. In this case the internal energy is naturally higher, because of the relatively high values of the Fermi energies of the negative particles. Earlier we saw that after the appearance of π^- mesons the lepton Fermi energies stop rising. This situation was studied in /14/. When there are as many types of positive as there are of negative baryons, the conditions necessary for the existence of π^- mesons are never fulfilled.

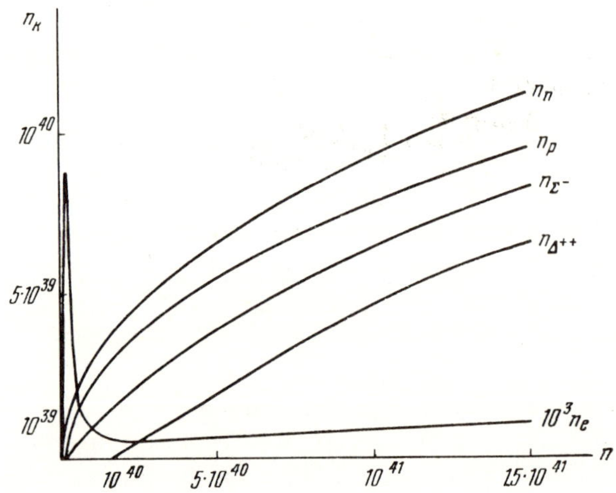

FIGURE 10. Dependence of the density of basic particles (neutrons, protons, Σ^--hyperons, Δ^{++} (1236) resonances, and electrons) on the density of baryons.

25. CLOSED SYSTEM IN EQUILIBRIUM

It is reasonable also to investigate a completely closed system /91/. The only ideal example for such a system is the universe, taken as a whole. In such a system, in addition to energy and to electric and baryon charge, lepton charge is also conserved. From our previous discussion we already know that, because of the escape of neutrinos, the lepton charge in individual celestial bodies is not conserved.

It has by now become established that the radius of curvature of space changes (increases) with time, i. e. the universe is in a nonstationary state. However, below we shall assume that this change is quite slow, and that matter is in a state of quasi-equilibrium. We shall also assume that space is uniform and isotropic. Needless to say, the assumption of thermodynamic equilibrium is not always justified. Equilibrium apparently exists at sufficiently high densities. We are interested in densities of the order of nuclear and higher, when the material consists only of elementary particles. Under these conditions any perturbation will propagate with a speed close to the velocity of light, therefore deviations from equilibrium, caused by fluctuations and by changes in space curvature, should quickly disappear. We shall impose no limitations on the temperature.

Let us introduce the concept of entropy for the whole universe. In accordance with our assumption that the state of the system is always close to equilibrium, the entropy, as a function of time, must have a maximum value, corresponding of course to the specific situation. In finding this maximum entropy, one has to take into account the laws of conservation of energy and of electric, baryon, and lepton charge. In fact, at any particular moment the process of establishing an equilibrium takes place without violation of these four conservation laws. Let us remember that every particle (baryon or lepton) has assigned to it the number +1, and an antiparticle the number −1; therefore the term baryon (lepton) charge means the difference between the number of baryons (leptons) and antibaryons (antileptons).

The requirement of maximum entropy leads to the following system of equations /91/:

for neutral baryons

$$\mu_k = \mu_n , \tag{25.1}$$

for positively charged baryons

$$\mu_k + \mu_e = \mu_n + \mu_\nu , \tag{25.2}$$

for negatively charged baryons

$$\mu_k + \mu_\nu = \mu_n + \mu_e , \tag{25.3}$$

for leptons

$$\mu_\mu = \mu_e , \tag{25.4}$$

for particles and their antiparticles

$$\tilde{\mu}_k + \mu_k = 0, \tag{25.5}$$

for neutral bosons

$$\mu_k = \mu_\gamma = 0, \tag{25.6}$$

and, finally, for negatively charged bosons

$$\mu_k = \mu_{\pi^-} = \mu_e - \mu_\nu. \tag{25.7}$$

The positively charged bosons are antiparticles of the negatively charged bosons. In the above equations, the symbols n, e, μ, γ, π^- and ν denote neutrons, electrons, muons, γ quanta, pions, and neutrinos, respectively. The chemical potential μ_k of a particle is a function of temperature and the concentrations of baryons and leptons. If several types of leptons are present, (25.4) should be replaced by $\mu_k = \mu_v$ or $\mu_k = \mu_e$ for neutral and negatively charged leptons, respectively.

Equations (25.1)—(25.7) can also be derived from the requirement of thermodynamic equilibrium between the direct and inverse reactions which convert the particles into each other. They are not difficult to obtain if one remembers that the total lepton, baryon, and electric charges on the right-hand and left-hand sides of the equalities have to be the same. Setting $T=0$ and $\mu_\nu=0$ in (25.1)—(25.7), one obtains equations (23.7)—(23.12).

One should keep in mind that, at a given temperature, for every particle there is a corresponding threshold value of the density of matter. The particle will be present in the medium only when the density is above its threshold value. Therefore, if densities are below the threshold for a given particle, equations (25.1)—(25.7) will yield an absurd result.

For a given energy and electric, baryon, and lepton charge (the electric charge in a closed universe is equal to zero), the relations (25.1)—(25.7) between the chemical potentials fully determine the concentrations of all particles and antiparticles. This is easily verified, if one recalls that the chemical potential of a particle is a function of temperature and of its density. Thus it follows from (25.5) that, at a given T, the concentration of an antiparticle is uniquely determined by the concentration of the corresponding particle. Furthermore, from (25.1)—(25.3) it is seen that the concentrations of all baryons having the same electric charge are determined by the concentration of one of them. According to (25.4) and (25.7), the same can be said about leptons and bosons. According to (25.6), the density of neutral bosons is determined by the temperature, while that of negative bosons is determined by temperature and the concentration of leptons. Finally, one has to take into account equations (25.2) and (25.3), from which it is seen that there are two relations between the concentrations of positive, negative, and neutral baryons. It is thus apparent that, in the final analysis, the concentrations of all particles can be expressed through the concentrations of neutrons, electrons, and neutrinos, and the temperature (n_n, n_e, n_ν, T), which are in their turn obtained from the aforementioned four conditions (for the energy and three types of charge).

Thus the electric, baryon, and lepton charges and the energy are the basic parameters which characterize the universe with all its properties. The term energy here refers only to that part which is related to matter. The total mass (energy) also includes the gravitational part; it has, of course, a definite value, yet the mass of matter does not remain constant but changes with time. The gravitational and mass-related parts of energy are functions of the radius of curvature (isotropic model); therefore, all is ultimately reduced to the value of the radius of curvature, which depends only on time. Consequently, the physical state of the universe is determined precisely by its age.

If the expansion of the universe is sufficiently fast, equilibrium radiation (electromagnetic and neutrino), produced at high densities, could have been preserved up to the present as independent and almost noninteracting with matter (this is quite accurate for neutrino radiation). During the course of expansion, the gravitational red shift causes the equilibrium temperature of blackbody radiation to decrease. As a matter of fact, precisely such a cosmological radiation, with temperature $T=3°$, called relic radiation, was discovered by astronomers in 1966. For certain reasons, relic neutrino radiation has not yet been observed. The discovery of such a cosmologic radiation could shed some light on the physical state of the universe during earlier times of its history.

From statistical physics it is known that, for the chemical potential of bosons, $\mu_k - m_k c^2$ has to be negative /67/. Comparing this with (25.5), one arrives at the following:

$$-m_k c^2 < \mu_k < m_k c^2, \tag{25.8}$$

where μ_k is the chemical potential of charged bosons. Thus the chemical potential of bosons can assume only those values which are inside the interval $(-m_k c^2, m_k c^2)$.

SUMMARY

The properties of a degenerate baryon gas at absolute zero were studied in the ideal-gas approximation.

At densities of the order of 10^{13} $g \cdot cm^{-3}$ disintegration of all atomic nuclei takes place, resulting in the formation of a phase consisting of free nucleons (neutrons and protons) and electrons (the nucleon-electron or *npe*-phase). Here, in the ideal-gas approximation, the concentration of neutrons is larger by two to three orders of magnitude than the concentration of protons (electrons).

When the density is higher by a factor of 3 than nuclear density, hyperons and baryon resonances appear in the medium. Generally, speaking, the former should attain the stability of the lighter baryons. On the other hand, the neutrality condition introduces some changes. Thus, for example, in spite of the fact that the rest mass of Λ, Σ^+, and Σ^0 particles is smaller than that of Σ^- particles, the former are Σ^- hyperons. The threshold densities at which the various particles become stable are given in Table 5. The Σ^- particles appear at a density $\rho = 6.4 \cdot 10^{38}$ cm^{-3}. The stability of hyperons and baryon resonances is ensured by the presence in the medium of a degenerate gas of electrons and nucleons. The appearance of these particles in the nuclear plasma is dictated by the requirement of minimum system energy for a given total number of baryons. Beside electrons, the leptons in a nuclear plasma with $\rho \geqslant 8.4 \cdot 10^{14}$ cm^{-3} also include μ^- mesons. Soon after the appearance of hyperons, the concentrations of all baryon forms become approximately equal, whereas the concentration of electrons, and consequently also of negative muons, is lower by three or more orders of magnitude.

Chapter IV

REAL BARYON GAS

26. THE SUPERNUCLEAR DENSITY RANGE

In the preceding chapter we discussed the theory of degenerate matter under the assumption that the particles form an ideal gas at all densities. In the below-nuclear density range this approximation can be considered justified. In the supernuclear density range this approximation of an ideal gas is apparently unsound. The development of an accurate theory for a real baryon gas is not an easy problem. To be sure, we know that all baryons are strongly interacting particles, but we do not know the exact value of the interaction energy of baryon pairs of any type or its dependence on particle velocity, nor the collective interaction between the particles. Starting out from the data at our disposal, we shall here discuss a theory for a real baryon gas /96, 250/. Our chief aim is to find the qualitative changes arising from the introduction of the interaction among particles.

We shall here examine a hypothetical (not existing in reality) equilibrium celestial body with an infinitely high central pressure. This will enable us to cover all possible phases and give a logical discussion of their sequence. In the central region of such a stellar configuration, where the density of matter is above nuclear, we are dealing with a gas of elementary particles. When moving away from the center toward the surface, one encounters a succession of phases, i. e. a change in the "chemical composition" of matter. In the region of the *baryon sphere* (as we call the region where matter consists only of elementary particles), the change in chemical composition with distance from the star's center is determined by

$$[\mu_k(n) + V_k(n)]\sqrt{g_{00}(r)} = \text{const}, \tag{26.1}$$

which applies to every type of neutral baryon. Here $n(r)$ is the total baryon density, and μ_k and V_k are, respectively, the chemical potential and average potential energy of the k-th type of neutral baryon in the plasma. It is convenient to write equation (26.1) for neutrons. The concentration $n_n(r)$ of neutrons uniquely determines the concentrations of all baryons, leptons, and bosons which can exist in the medium at a given density. An analogous equation for the chemical potentials can be written for that part of the star where the composition of matter also includes atomic nuclei.

At the change-of-phase surfaces $r=r_k$, beside the continuity in the chemical potentials of the neutral components of matter, there must also be continuity of the total pressure $P(r)$ and the pressure gradient. In the case of positively and negatively charged baryons, the quantities $\mu_k^+(r)+\mu_e(r)$ and $\mu_k^-(r)-\mu_l(r)$, respectively, must remain continuous. This is evident from the local equilibrium conditions between differently charged baryons. These requirements become clear if we remember that we are dealing with configurations which are in a state of thermodynamic and hydrodynamic equilibrium.

Below we shall divide the whole physical density interval into three arbitrary parts, namely: the region of supernuclear densities, the region of densities of the order of nuclear, and the region of below-nuclear densities.

We assume the supernuclear region to cover the baryon densities $n>10^{39}$ cm^{-3}. The most important characteristic of this region is the presence of a relatively large number of hyperons and baryon resonances as stable components of matter.

Thus, let us consider a stellar configuration in whose center the energy density is infinite. Now, let us trace the changes occurring in the chemical composition between the center and surface of this configuration. To carry this out, though, we have to know which particles can exist in the medium, their most important characteristics, e.g. spin, mass, and electric, baryon, and lepton charges, and their laws of interaction.

In Chapter III, which dealt with an ideal gas, consideration was given to all particles known by the end of 1968. Here we shall consider a simplified model of a real gas, in which we take into account: the proton p, neutron n, and Λ, Σ^0, Σ^-, Σ^+, Ξ^0, and Ξ^- hyperons from among the baryons; the electron and μ^- meson from among the leptons; and the π^- meson from among the bosons. The stability of baryons is ensured by the presence of a degenerate proton gas, and that of negative muons and pions by the degenerate electron gas. For all phases, as before, it is convenient to take the electrons and nucleons as the basic particles. The inclusion of baryon resonances does not introduce significant changes in the values of such important parameters of the configuration as mass, radius, and mass distribution. On this comparatively simple model of an elementary-particle gas we can then base a correct representation of superdense stellar configurations consisting of a degenerate baryon gas. Henceforth, these will be called ***baryon configurations***.

First let us assume that all baryon forms interact in the same manner. Then, in the general case and at high densities, when the medium contains all baryon types, the particle concentrations are determined by equations (23.3), (23.4), and (23.7)—(23.12) of the preceding chapter:

$$\varepsilon_b^{(+)} + \varepsilon_e = \varepsilon_b^{(-)} - \varepsilon_e = \varepsilon_b^{(0)}, \tag{26.2}$$

$$\varepsilon_l^{(-)} = \varepsilon_e, \tag{26.3}$$

$$\sum_b n_b = n, \tag{26.4}$$

$$\sum_b n_b^{(+)} - \sum_b n_b^{(-)} - \sum_l n_l = 0, \tag{26.5}$$

where $\varepsilon_b^{(+)}$, $\varepsilon_b^{(-)}$, $\varepsilon_b^{(0)}$, $\varepsilon_l^{(-)}$, ε_e are the Fermi energies for positive, negative, and neutral baryons, negative leptons, and electrons, respectively. In (26.4) the summation is over all baryons, in the first term of (26.5) it is over the positive baryons, in the second term over negative baryons, and in the third term over negative leptons. The first two equations are the conditions for thermodynamic equilibrium between the various components, and the last two are the conservation laws for baryon and electric charge. The above equations do not include π^- mesons. It will be shown below that at the given baryon composition densities are never such as to create the necessary conditions for the stability of these particles. The particles' Fermi energies ε_k are related to their number n_k per unit volume through

$$\varepsilon_k = \left(m_k^2 c^4 + a^2 n_k^{2/3}\right)^{1/2}, \tag{26.6}$$

where $a = (3\pi^2)^{1/3} \hbar c$ for particles with spin 1/2.

Let us find the chemical composition at the center of the hypothetical configuration with $\rho(0) = \infty$. Using (26.6), we obtain from (26.2) for $n \to \infty$:

$$n_b^{(+)1/3} + n_e^{1/3} = n_b^{(-)1/3} - n_e^{1/3} = n_b^{(0)1/3}.$$

Comparison between these equations and (26.5) leads to the result:

$$\left.\begin{aligned} n_b^{(+)}(0) = n_b^{(-)}(0) = n_b^{(0)}(0), \\ n_e(0) = n_{\mu}(0) = 0, \end{aligned}\right\} \tag{26.7}$$

which means that at the center all baryons have the same concentrations, but the leptons do not. It should be remembered that the result of (26.7) is valid only for the special composition of charged baryons chosen for this case, namely two positive and two negative baryons. For leptons to be conserved at $\rho(0)=\infty$, there should have been more types of positive baryons than of negative baryons. But at no particle composition it is possible to achieve an infinite increase in lepton concentration when $n \to \infty$. This situation is due to the existence of π^- mesons. In fact, as was shown in Section 24, the Fermi energies of electrons and μ^- mesons can only increase as far as the limit

$$\varepsilon_e = \varepsilon_\mu = m_\pi c^2, \tag{26.8}$$

after which their increase breaks off, because it is then energetically more favorable for the material to be in the π^- meson phase than in the lepton phase.

In the case of a configuration with finite central pressure, leptons remain even at $r = 0$.

Strictly speaking, equations (26.2) are correct only if the same interaction law applies to all baryon types. In the case of a macroscopic medium consisting of elementary particles, it is obviously reasonable to consider only the average potential energy of the particles, representing the interaction of one particle with the rest, i. e. with the medium. This one-particle potential energy depends on the baryon density, the chemical composition, and the type of particles. If the different types of baryons interact differently with the medium, then instead of (26.2) we have

$$\begin{aligned} \varepsilon_b^{(+)}(n) + V_b^{(+)}(n) + \varepsilon_e(n) = \varepsilon_b^{(-)}(n) + V_b^{(-)}(n) - \varepsilon_e(n) = \\ = \varepsilon_b^{(0)}(n) + V_b^{(0)}(n), \end{aligned} \tag{26.9}$$

where the $V(n)$ with indices are the potential energies of the different baryons in the medium. The present state of the theory of elementary particles does not yet enable us to solve our problem in a rigorous manner. Below we shall assume that in the supernuclear density region all baryon forms interact with the medium in the same way. For the nuclear density region, there are definite possibilities for a more or less correct treatment of the problem of particle concentrations.

It should be noted that the mass and radius of the configuration are insensitive to inaccuracies in the relative particle concentrations. The integral parameters of a stellar configuration depend significantly only on the average rest mass of the particles inside the baryon sphere. We shall return to this question later on, after the calculations of stellar configurations.

27. PARTICLE CONCENTRATIONS IN THE HYPERON REGION OF A STAR

Let us follow the changes occurring in the state of matter between the center of the extreme configuration (i. e. $\rho(0)=\infty$) and its exterior. The extreme phase, in which there are no electrons, is realized from a mathematical point of view only at the star's center, which is a geometrical point. With distance from the center, the baryon density diminishes monotonically, and the electron density increases. The dependence of particle density on distance from the center can be found only after integrating the differential equations of the gravitational field. On the other hand, the behavior of matter composition with decreasing baryon density can be determined from equations presented earlier.

At some distance $r=r_{\Xi^0}$, where the baryon density reaches a value of $9.93\cdot10^{39}$ cm^{-3}, the Ξ^0 hyperons disappear. This means that at a distance r_{Ξ^0} the Fermi energy of these particles becomes equal to their rest energy. With further distance from the star's center, the particles Σ^+, Ξ^-, Σ^0, Λ, and Σ^- disappear at the distances $r=r_{\Sigma^+}$, r_{Ξ^-}, r_{Σ^0}, r_Λ, and r_{Σ^-} where the baryon density has decreased to the values $n(r)=8.12\cdot10^{39}$, $3.83\cdot10^{39}$, $3.32\cdot10^{39}$, $1.24\cdot10^{39}$, and $6.1\cdot10^{38}$ cm^{-3}, respectively. In a sense, one can say that at $n(r_{\Sigma^-})=6.07\cdot10^{38}$ cm^{-3} the protons also disappear, since below this density the concentration of protons decreases sharply, becoming two orders of magnitude lower than that of neutrons.

At the threshold for the disappearance of the last hyperon, the electron Fermi energy is 124 Mev, which corresponds to a concentration $n_e=8.36\cdot10^{36}$ cm^{-3}. The maximum value of the electron Fermi energy, which is approximately 125 Mev, is reached just before that. With further increase in r (i. e. decrease of ρ), ε_e drops monotonically. Hence it follows that, at the chosen baryon composition, the conditions required for the appearance of negative pions never arise. In fact, stability of these particles would require an electron Fermi energy of $m_\pi c^2\approx 140$ Mev.

The above threshold densities, above which the corresponding particles become stable, were calculated from equations (26.2) — (26.5), i. e. under the assumption that the potential energies at the Fermi energy levels are the same for all baryon forms. This certainly involves considerable errors. According to (26.9), these errors are related not to the values of the potentials $V_b^{(k)}(\varepsilon_k)$ themselves, where ε_k is the Fermi energy of the corresponding particle, but to the difference between these potential energies for neutral and charged baryons. As a result, it can be expected that the errors for a certain particle concentration are not large, when the potential energies are omitted (or their differences set), and they naturally tend to zero with increasing density n.

To complete this description of phase transitions along the radius of the extreme configuration, one should also mention μ^- mesons. At a distance $r = r_\mu$ where the baryon density decreases to $6.0 \cdot 10^{39}$ cm^{-3}, the Fermi electron energy becomes equal to the rest energy of muons: $\varepsilon_e = m_\mu c^2 = 106$ Mev. From this point on there are negative muons, whose concentration is determined from the electron concentration through equation (24.2).

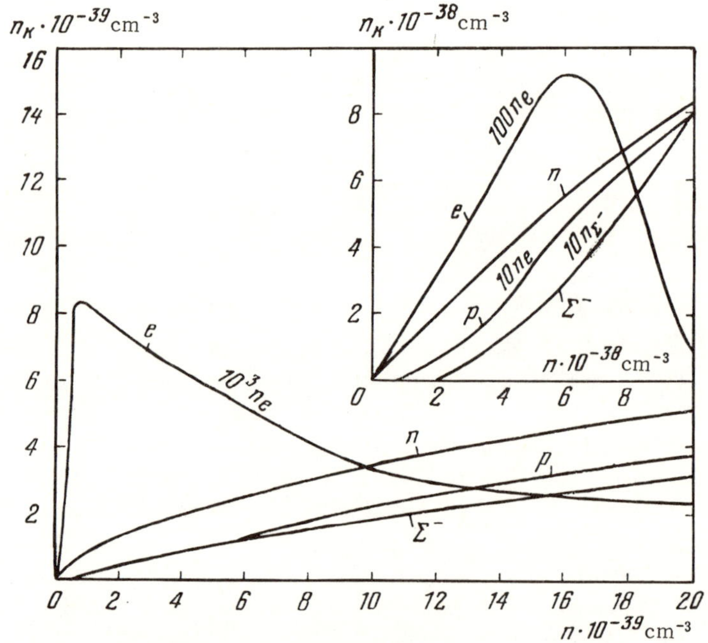

FIGURE 11. The densities of neutrons n, protons p, Σ^- hyperons, and electrons e, as a function of baryon density. The calculations were based on the assumption that the potential energies are the same for all varieties. In the upper right-hand corner are shown the same functions for the case when the real nuclear potential is included.

Between the concentrations of baryons with identical electric charge there is a simple analytical relationship. Thus, for example, according to (24.5)—(24.7) the densities of neutral, positive, and negative baryons can be expressed correspondingly through the densities of neutrons, protons, and Σ^- hyperons. Finding the densities of all particles requires a numerical solution of the algebraic equations (26.2)—(26.5), for which the procedure is described in Section 24. In these numerical solutions the electron, neutron, proton, and Σ^- hyperon play the role of basic particles, in the sense that the concentrations of the remaining particles can be calculated from their concentrations through equations (24.5)—(24.7). The densities n_e, n_n, n_p, and n_{Σ^-} of the basic particles as a function of the total baryon density n can only be calculated numerically. The results of these calculations are shown in Figure 11.

From Figure 11 it is seen that, when moving from the surface of the extreme configuration toward its center, the electron density n_e first increases monotonically, reaches its maximum in the vicinity of the hyperon-sphere surface, and then diminishes monotonically, tending toward zero at $r=0$. At a first glance it may seem that such a configuration is unstable, since in the gravitational potential well the Fermi level is uneven; therefore to even it out the electrons would have to move toward the center. This, however, is not so; the equilibrium condition (26.1) along the radius applies only to neutral particles, i.e. the energy surface is flat only for them. In the case of charged particles, a flat energy surface is obtained from a combination of levels, namely

$$\left.\begin{aligned}(\varepsilon_p+\varepsilon_e)\sqrt{g_{00}(r)} &= \text{const},\\ (\varepsilon_{\Sigma^-}-\varepsilon_e)\sqrt{g_{00}(r)} &= \text{const}.\end{aligned}\right\} \qquad (27.1)$$

Any displacement, however small, of the charged particles from their equilibrium positions leads to a disturbance of neutrality and is therefore accompanied by strong electric forces, which tend to keep the particles in their places.

28. THE NUCLEAR DENSITY RANGE

The nuclear range covers the baryon density interval $10^{37} \lesssim n \lesssim 10^{39}$ cm^{-3}. In the corresponding layer of a star, matter consists of neutrons, protons, electrons, Σ^- hyperons, and negative muons, and the outer boundary of the region also includes atomic nuclei. Among all these particles, neutrons are numerically predominant. Hyperons can be found only near the inner boundary of the region. With distance from the center, Σ^- hyperons are the last hyperons to disappear. In Sections 24 and 27 we found $n_{\Sigma^-}=6.4\cdot 10^{38}$ cm^{-3} for the stability threshold of these particles. However, those stability-threshold calculations were made with the assumption that the single-particle potential energies $V_k(\varepsilon_k)$ at the Fermi levels were identical for all baryon forms. The functions $V_k(\varepsilon_k)$ are unknown for the density range $n>10^{39}$ cm^{-3}, and we had no other choice. The situation in the nonrelativistic density range is different. Here there are many references /31 – 33/ in which a theory was developed for a hypothetical nuclear material having equal numbers of protons and neutrons. Although in our case we have $n_n \ll n_p$ in the nuclear density range, we nevertheless hope to obtain some results through the use of these references.

First, let us try to find more accurate results for the threshold density for Σ^- particles. From nuclear theory it is known that the potential energy of a nucleon depends on its momentum:

$$V_k(p) = V_k(0) + T_k(p), \tag{28.1}$$

where $V_k(0)$ is the fixed part of the potential, which depends only on n, and T_k is the kinetic energy of the k-th particle. The threshold for the appearance of Σ^- hyperons is obtained from (26.9) by setting $n_{\Sigma^-}=0$ and then using expression (28.1). For the concentrations of the other particles one then obtains the following set of equations:

$$\left.\begin{aligned}
&m_n c^2 + 2T_n + V_n(0) = m_p c^2 + 2T_p + V_p(0) + \varepsilon_e,\\
&m_n c^2 + 2T_n + V_n(0) = m_{\Sigma^-} c^2 + V_{\Sigma^-}(0) - \varepsilon_e,\\
&p_e\left[1 + \left(1 - m_\mu^2 c^2/p_e^2\right)^{3/2}\right]^{1/3} = p_p,\\
&p_n^3 + p_p^3 = \frac{a^3 n}{c^3}, \quad a = (3\pi^2)^{1/3}\hbar c.
\end{aligned}\right\} \tag{28.2}$$

For a nuclear material (with $n_p = n_n$), it was found that for nucleons $V(0) \approx -100$ Mev /274, 72/. It is to be expected that, in the case of Σ^- particles, $V_{\Sigma^-}(0)$ is not much different from $V(0)$. From the first of equations (28.2) it is seen that the quantity $|V_n(0) - V_p(0)|$ should be compared with $2T_n$ or $2T_p + \varepsilon_e$, and from the second that $|V_n(0) - V_{\Sigma^-}(0)|$ should be compared with $2T_n + \varepsilon_e$ or with $(m_{\Sigma^-} - m_n)c^2 = 258$ Mev. For $n \gtrsim 2\cdot 10^{38}$ cm^{-3} we have $2T_n \approx 2T_p + \varepsilon_e \gtrsim 150$ Mev, and on the other hand it is reasonable to assume that $|V_n(0) - V_k(0)| \ll 100$ Mev for $k=p$ and Σ^-. We would then have a good approximation if we neglect the potential differences compared to the other terms in the equations. It then follows from the first three equations in (28.2) that

$$p_e\left[1+\left(1-m_\mu^2c^2/p_e^2\right)^{3/2}\right]^{1/3}= \\ =\{m_p\left[(m_{\Sigma^-}-m_p)\,c^2-2cp_e\right]\}^{1/2}. \qquad (28.3)$$

Solving this equation, we find $cp_e=120.9$ Mev, $T_p=8.1$ Mev, and $T_n=$ $=67.9$ Mev. The Σ^- threshold particle concentrations which correspond to these numbers are

$$\left.\begin{aligned} n_n=1.994\cdot10^{38}\,\mathrm{cm}^{-3},\quad n_p=8.206\cdot10^{36}\,\mathrm{cm}^{-3},\\ n_e=7.727\cdot10^{36}\,\mathrm{cm}^{-3}.\end{aligned}\right\} \qquad (28.4)$$

The threshold for the formation of Σ^- hyperons is thus lower by a factor of approximately 3 than the one obtained in the ideal-gas approximation. At this threshold the medium also contains μ^- mesons; therefore the number of protons is larger than the number of electrons. Just below the threshold densities (28.4), μ^- mesons also disappear from the medium. Thus negative muons and Σ^- hyperons appear in the medium, starting at the regular nuclear density. One therefore wonders why there are no μ^- mesons and Σ^- hyperons in atomic nuclei. The point is that in this problem an important role is played not only by the density, but also by the degenerate electron gas. The absence of muons and hyperons in atomic nuclei is caused precisely by the absence in them of a degenerate electron gas at the corresponding density.

Attention should also be paid to the fact that in the nuclear density range, as seen from (28.4), the number of neutrons is higher by a factor of only 25 than that of protons. In the scientific literature there is a prevailing mistaken concept of *neutron* stars; these are considered to be superdense configurations in which the concentration of neutrons is higher by three orders of magnitude than the concentration of protons and electrons. These configurations are obtained from the study of a degenerate ideal nucleon-electron gas which, as we have seen in this and the preceding chapters, is unstable and can therefore not appear in nature.

Following the disappearance of the last hyperons, the star has a relatively narrow layer in which matter consists of neutrons, protons, and electrons. Let us find the particle concentrations in this density range. Here it is convenient to make use of the concept of the effective mass of nucleons in the nuclear material. Let us write out the relationship between the chemical potentials of the particles:

$$m_pc^2+\frac{p_p^2}{2m_p^*}+V_p(0)+\varepsilon_e=m_nc^2+\frac{p_n^2}{2m_n^*}+V_n(0), \qquad (28.5)$$

where m_p^* and m_n^* are the effective masses of protons and neutrons. The electrons are highly relativistic: $\varepsilon_e = cp_e$. From the neutrality condition it follows that

$$p_p = p_e \equiv p. \tag{28.6}$$

From the last two equations we find that

$$p = m_p^* c\left\{\left[1 + \frac{2}{m_p^* c^2}\left(\frac{p_n^2}{2m_n^*} + V_n - V_p + m_n c^2 - m_p c^2\right)\right]^{1/2} - 1\right\}, \tag{28.7}$$

and it is not difficult to show that

$$|V_n - V_p + m_n c^2 - m_p c^2| \ll \frac{p_n^2}{2m_n^*}.$$

Considering the latter, we find from (28.7) that $p \approx p_n^2/2m_n^* c$, i. e.

$$n_e = n_p \approx \left(a/2m_n^* c^2\right)^3 n_n^2. \tag{28.8}$$

According to /138, 274, 72/, in the nuclear density range we have $m_n^*(p_n) \approx 0.6\, m_n$ (the value of the effective mass depends on the momentum at the low levels, $m_n^*(0) \approx 0.5 m_n$), therefore $\left(a/2m_n^* c^2\right)^3 \approx$ $\approx 1.595 \cdot 10^{-40}$ cm^3. As an illustration we shall now present some figures. Thus, for $n_n = 2 \cdot 10^{38}$; 10^{38}; $8 \cdot 10^{37}$; $6 \cdot 10^{37}$ cm^{-3} the concentration ratios between protons and neutrons are, respectively, $n_p/n_n \approx 0.032$; 0.016; 0.013; 0.010. The plots of $n_k(n)$ for neutrons, protons, electrons, and Σ^- hyperons shown in the upper right-hand corner of Figure 11 were calculated with the nuclear potential, i. e. from equation (28.8).

With further approach toward the surface of the extreme configuration, i. e. in the below-nuclear density range, the *Aen*- and *Ae*-phases appear and disappear successively. These phases were studied in Sections 18 and 20. In these cases the ideal-gas approximation is apparently quite good.

SUMMARY

In studying the state of matter at densities $\rho \gtrsim 10^{13}$ g$\cdot$cm^{-3}, the ideal-gas approximation used in the preceding chapter is not appropriate. At densities of about nuclear and above, an important role is played by nuclear attraction forces. For example, their inclusion leads to the result that Σ^- hyperons appear in the medium not at a density $\rho \approx 6.1 \cdot 10^{14}$ g·cm^{-3}, but at a density which is lower by a factor of three; this means that these particles already appear at densities which exist in regular nuclei.

Beside electrons, negative muons are the only other leptons which can exist in the baryon plasma, and they appear at the nuclear density immediately before the appearance of hyperons.

At sufficiently high densities, when almost all hyperons and many resonances are represented in the medium, the only bosons which can attain stability are the negative pions, and that only if the algebraic sum of the charges of all baryon varieties is a positive number. If this condition is not satisfied, then free π^- mesons cannot exist in the degenerate baryon gas under any circumstances. The transition to the phase which includes negative pions takes place when the electron Fermi energy reaches the value $\varepsilon_e = m_\pi c^2$. In this phase the negative pions are concentrated in the lowest energy level, i. e. they are all in a state of rest. Right above the threshold for this phase the π^- mesons in the medium become abundant, and their concentration very quickly becomes equal to that of the baryons. With the appearance of π^- mesons, the densities of the leptons freeze at their values. As has already been mentioned, the reason behind this is that it is energetically more favorable for the system to have π^-mesons in their state of rest than to have electrons with energies $\varepsilon_e > m_\pi c^2$.

When the average distance between particles becomes of the order of $l \lesssim 2 \cdot 10^{-14}$ cm (to which corresponds a density $\rho \gtrsim 10^{40}$ g·cm^{-3}), very strong repulsion forces arise between the particles. Thus, in this density range it is incorrect not only to consider the medium as an ideal gas, but even to treat it as a gas at all. It is here appropriate to speak of a liquid, which consists of baryons.

At extremely high densities, when the average distance between particles in the baryon plasma becomes similar to the assumed "dimensions" of the particles themselves, it is entirely unrealistic to see the medium as a combination of such a large number of baryon forms. There is no doubt that of the baryons known to date, if not all, then at least an overwhelming majority of them are not at all elementary but have complex structures. It is not impossible that hyperons and resonances are excited states of certain basic baryons, say nucleons. If this is the case, then at densities $n \lesssim 1/4l^3$ (where l is approximately equal to the diameter of a compound particle), particles having a structure undergo a process similar to what atoms and atomic nuclei are subjected to at certain density values.

Needless to say, they decompose into their components. It may thus occur that, after formation of a plasma with the rich "chemical" composition described in the last two chapters, there is a new simplification at densities above a certain value: all baryons now disappear from the medium, resulting in the formation of a new phase consisting of more compact "elementary" (for which the term "primary" may be more suitable) particles. There have been similar discussions concerning quarks /61/. At one time, the prevailing fashion in the theory of elementary particles was the Sakata model, in which the basic particles in the baryon and boson groups were considered to be the proton, the neutron, and the Λ-hyperon.

Chapter V

THE EQUATION OF STATE

29. THE EQUATION OF STATE FOR THE *Ae*- AND *Aen*-PHASES

In the following chapters we shall deal with models of stellar configurations which consist of degenerate gaseous masses. These are the configurations of white dwarfs and baryon stars. The latter are understood to be celestial bodies consisting of a degenerate baryon gas. For calculating the parameters of these stellar configurations we need the equation of state, and we are interested only in the degenerate states of matter.

We shall start with the *Ae*-phase, which consists of bare atomic nuclei and a free degenerate electron gas. At low temperatures, the motion of the nuclei is reduced to zero-point vibrations around fixed equilibrium points. As a result, they contribute nothing to the pressure. The pressure is entirely due to electrons, but the energy density is determined by the atomic nuclei.

The energy density is

$$\rho = (m_n c^2 + b) \sum_k A_k n_k + \rho_e, \tag{29.1}$$

where b is the average binding energy of a nucleon in the nucleus (there is no point here in distinguishing between the mass of a proton and a neutron), n_k is the number of nuclei of a given type (with parameters A_k and Z_k) in unit volume, and ρ_e is the energy density of the electron gas. When the electron gas is degenerate, b is a function of ε_e (see Section 18). According to (12.7),

$$\rho_e = 4K_e \left[x_e (1 + 2x_e^2) \sqrt{1 + x_e^2} - \ln \left(x_e + \sqrt{1 + x_e^2} \right) \right], \tag{29.2}$$

where $x_e = p_e/m_e c = (3\pi^2)^{1/3} \hbar n_e^{1/3}/m_e c$ is the electron Fermi momentum in units of $m_e c$ (for $p_e \gg m_e c$, $x_e = \varepsilon_e/m_e c^2$), and

$$K_e = \frac{m_e^4 c^5}{32\pi^2 h^3}. \tag{29.3}$$

It is sometimes convenient to use another parameter t_e in place of x_e, where

$$t_e = 4 \operatorname{arsh} x_e. \tag{29.4}$$

With the help of this parameter, the electron energy density assumes the following compact form:

$$\rho_e = K_e (\operatorname{sh} t_e - t_e). \tag{29.5}$$

A number of simplifications can be introduced into the energy expression (29.1). Thus

$$\sum_k A_k n_k = \frac{A}{Z} \sum_k Z_k n_k = \frac{A}{Z} n_e,$$

where A/Z is the average value of the ratio A_k/Z_k (the averaging is carried out over all nucleus types present in the medium). Using the above and neglecting the small values of b and ρ_e, we obtain

$$\rho = \frac{32}{3} \cdot \frac{m_n}{m_e} \cdot \frac{A}{Z} \cdot K_e x_e^3. \tag{29.6}$$

It should be remembered that, owing to neutronization, the ratio A/Z is a function of x_e, which is approximated by the polynomial (18.5).

Now let us calculate the pressure. It is equal to the negative derivative of energy with respect to volume, at constant number of particles and entropy (in our case the entropy is zero). Since the partial pressure of the nuclei can be neglected, we have

$$P = -\left(\frac{\partial (V\rho)}{\partial V}\right)_{N_e} = -\left(\frac{\partial V \rho_e}{\partial V}\right)_{N_e},$$

where $N_e = V n_e$ is the number of electrons in a volume V. In differentiating ρ_e one should remember that x_e depends on the volume V. Keeping in mind (29.2), we find for the pressure

$$P = \frac{4}{3} K_e \left[x_e (2x_e^2 - 3) \sqrt{1 + x^2} + 3 \ln\left(x_e + \sqrt{1 + x_e^2}\right)\right]. \tag{29.7}$$

Using (18.5), the equation of state for matter in the Ae-phase can be written in the following parametric form:

$$\left.\begin{aligned}\rho &= \frac{32}{3}\left(\frac{m_e}{m_n}\right)^3 K_n x_e^3\left(2+a_1x_e+a_2x_e^2+a_3x_e^3\right),\\ P &= \frac{4}{3}\left(\frac{m_e}{m_n}\right)^4 K_n\left[x_e\left(2x_e^2-3\right)\sqrt{1+x_e^2}+\right.\\ &\qquad\left.+3\ln\left(x_e+\sqrt{1+x_e^2}\right)\right],\end{aligned}\right\}\qquad(29.8)$$

where a_1, a_2, a_3 are the constants from equation (18.5): $a_1=1.255\cdot10^{-2}$, $a_2=1.755\cdot10^{-5}$, $a_3=1.376\cdot10^{-6}$. We also introduced the new notation

$$K_n=\frac{m_n^4c^5}{32\pi^2h^3}=5.11\cdot10^{35}\,\mathrm{erg\cdot cm^{-3}},\qquad(29.9)$$

which will be used from now on.

Let us examine two limiting cases of the equation of state (29.8). In the nonrelativistic case, the parameter x_e is small compared to unity. Let us expand P in powers of x_e and discard the small terms in the expressions for ρ and P; eliminating the parameter x, we obtain

$$P=A\rho^{5/3},\qquad(29.10)$$

where

$$A=\frac{3^{2/3}}{160}\left(\frac{2}{\eta}\right)^{5/3}\cdot\frac{m_n}{m_e}\cdot\frac{1}{K_n^{2/3}}=3.74\cdot10^{-23}\left(\frac{2}{\eta}\right)^{5/3}.$$

We have $\eta=\frac{A}{Z}\approx 2$ for all nuclei, with the exception of hydrogen.

In the relativistic case $x\gg1$. If one neglects the effect of the dependence of η on the electron Fermi energy, then $P\sim x_e^4$ and $\rho\sim x_e^3$, and consequently

$$P=B\rho^{4/3},\qquad(29.11)$$

where

$$B=\frac{3^{1/3}}{32}\left(\frac{2}{\eta}\right)^{4/3}K_n^{-1/3}=5.64\cdot10^{-14}\left(\frac{2}{\eta}\right)^{4/3}.$$

In the expression for the energy density we omitted b and ρ_e.

The binding energy of a nucleon in the nucleus lies in the interval $0<b\lesssim 8$ Mev. At the threshold for the disappearance of the Ae-phase $P\approx5\cdot10^{29}$ erg·cm^{-3}, and the ratio between the partial energy

densities of electrons and nuclei is of the order of

$$\frac{\rho_e}{\rho_A} \approx \frac{3}{4} \cdot \frac{Z}{A} \cdot \frac{\varepsilon_e}{m_n c^2} = 0.0066.$$

Thus b and ρ_e are indeed small, and can play no important role in calculations of stellar configurations.

In the above equation of state, no account was taken of the interaction between particles, which consists in this case only of Coulomb forces. At the end of Section 13 it was shown that the potential energy of electrons, which is due to electric forces, is small compared to their kinetic energy, and the ratio between the two diminishes with increasing density. Therefore the ideal-gas approximation is here fully justified. A number of corrections to the pressure expression (29.8), resulting from Coulomb interactions, were discussed in /253/. There is no necessity to discuss these here, since, as pointed out before, they are quite small. The corrections of P can have some significance only for large Z and $x<1$. The changes due to temperature are also negligible. Of importance here is only the dependence of A/Z on the electron Fermi energy.

The equation of state (29.8) is valid up to $x=46$, which corresponds to a density $\rho \approx 2.4 \cdot 10^{32}$ erg·cm^{-3}. At higher densities we encounter the Aen-phase, where the equation of state is different. Equations (12.7), (29.2), and (29.7) enable one to arrive directly at the expression for the partial energy density and pressure of the neutron gas:

$$\left.\begin{aligned} \rho_n &= 4K_n \left[x_n (1 + 2x_n^2) \sqrt{1 + x_n^2} - \ln \left(x_n + \sqrt{1 + x_n^2}\right)\right], \\ P_n &= \frac{4}{3} K_n \left[x_n (2x_n^2 - 3)\sqrt{1 + x_n^2} + 3 \ln \left(x_n + \sqrt{1 + x_n^2}\right)\right], \end{aligned}\right\} \tag{29.12}$$

where $x_n = p_n/m_n c = (3\pi^2)^{1/3} h n_n^{1/3} m_n c$ is the Fermi momentum of neutrons, measured in units of $m_n c$. If we introduce the parameter

$$t_n = 4 \operatorname{arsh} x_n,$$

then ρ_n and P_n assume the following form:

$$\left.\begin{aligned} \rho_n &= K_n (\operatorname{sh} t_n - t_n), \\ P_n &= \frac{1}{3} K_n \left(\operatorname{sh} t_n - 8 \operatorname{sh} \frac{t_n}{2} + 3t_n\right). \end{aligned}\right\} \tag{29.13}$$

Including also the energy of the nuclei and the partial pressure and energy density of electrons, we obtain for the equation of state of the *Aen*-phase

$$\left.\begin{aligned}\rho &= K_n(\operatorname{sh} t_n - t_n) + m_n c^2 \frac{A}{Z} n_e + \rho_e,\\ P &= \frac{1}{3} K_n\left(\operatorname{sh} t_n - 8\operatorname{sh}\frac{t_n}{2} + 3t_n\right) + P_e.\end{aligned}\right\} \tag{29.14}$$

Here ρ_e and P_e are the energy density and pressure of the electron gas. It should be noted that, soon after the *Aen*-phase threshold, the partial energy density and pressure of the electrons (and even the energy density of the nuclei) are quite small compared to the corresponding values for the neutron gas. Almost throughout this phase, the energy and pressure of the system are mainly determined by the neutron gas.

Generally speaking, during the *Aen*-phase one should also take into account nuclear interactions between neutrons. Their contribution to the energy is negligible, but it is important for the pressure: for a given number of neutrons, inclusion of nuclear forces leads to a decrease in pressure. To our knowledge, for the density range at hand nuclear theory is not yet adequate, therefore we will have to be satisfied with the ideal-gas approximation. The equation of state (29.14) is valid in the density region $2.4\cdot 10^{32} \leqslant \rho \leqslant 5.45\cdot 10^{34}$ $\mathrm{erg\cdot cm^{-3}}$.

30. THE ASYMPTOTIC FORM OF THE EQUATION OF STATE

We shall refrain for a while from discussing the equation of state for nuclear and supernuclear densities, and return to this question in Section 32. It is preferable to study first the asymptotic behavior of the equation of state at extremely high densities. This can be done successfully if one starts with totally general considerations. Experiments on scattering of fast protons by nucleons have shown the existence of highly intensive repulsive forces, which are effective at distances $r \leqslant 2\cdot 10^{-14}$ cm. This experimental fact leads one to believe that, with increasing mass density in the supernuclear region, the state of the baryon plasma (we are talking of a baryon plasma, since the concentration of leptons in it is very low) deviates more and more from a gas, gradually approaching the state of an ideal liquid.

It can be shown that for any type of interaction, whenever the interaction energy between particles is higher than their kinetic energy, the well-known $P \leqslant \rho/3$ law breaks down, which means that at high densities the pressure may be higher than $\rho/3$. The relationship $3P \leqslant \rho$ is valid for an ideal gas, and also when the field is so weak that at all densities the particles' kinetic energy is higher than their interaction energy. Such fields are the electromagnetic, gravitational, and certain types of meson fields.

We shall try to find the form of the equation of state in the limit of infinitely high densities under the following two assumptions: first, that the interaction energy between the particles is much higher than their kinetic energy, and second, that the velocity of light is the limiting value of the velocity of sound ($v \leqslant c$).

Let n be the density of the number of particles, then the average distance between them is $r \sim n^{-1/3}$. We a priori exclude the possibility that the particles have finite dimensions, since in that case at some density the medium would become incompressible, resulting in $\frac{\partial P}{\partial \rho} = \infty$, which is impossible (since then the velocity of sound becomes infinite). Starting from the uncertainty principle, we shall estimate the kinetic energy of the particles. The average momentum of the particle is

$$\bar{p} \approx h/r \approx hn^{1/3}.$$

Because of the high densities involved, the particles are highly relativistic, $hn^{1/3} \gg m_h c^2$. In accordance with this we have for the average kinetic energy

$$\varepsilon \approx \mathrm{ch}\, n^{1/3}. \tag{30.1}$$

The interaction energy of baryons, generally speaking, should have a complicated form. It should depend, beside the distance between particles, also on their velocity, spin, and orbital moment, it apparently has an exchange character, etc. However, keeping in mind one important property of the interaction forces, namely their small action radius, we can replace the real interaction energy by some average which depends only on n. Let $U(n)$ be the average potential energy of one particle in the plasma; then $VnU(n)$ is the interaction energy for the volume V. There is obviously no point in differentiating between the energies $U(n)$ of the particles in an ultrarelativistic plasma. Thus the density of the total energy is approximately

$$\rho(n) \approx \mathrm{ch}\, n^{4/3} + nU(n). \tag{30.2}$$

Let us now calculate the pressure:

$$P = -\left(\frac{\partial(V\rho)}{\partial V}\right)_N \approx \frac{1}{3}\,\mathrm{ch}\,n^{4/3} + n^2U'(n). \tag{30.3}$$

The ratio between pressure and energy density is then

$$\frac{P}{\rho} \approx \frac{3nU'(n) + \mathrm{ch}\,n^{1/3}}{3U(n) + 3\mathrm{ch}\,n^{1/3}}. \tag{30.4}$$

If we now assume that the potential energy is small compared to the kinetic energy, we immediately obtain the well-known result

$$\lim_{n\to\infty} P = \rho/3$$

for an ultrarelativistic ideal gas.

Let us now consider the case when the interaction energy exceeds the kinetic energy:

$$\rho(n) \approx nU(n), \qquad P(n) \approx n^2U'(n). \tag{30.5}$$

From this we find for the velocity of sound

$$v = c\left(\frac{\partial P}{\partial \rho}\right)^{1/2} = c\left(\frac{2nU' + n^2U''}{U + nU'}\right)^{1/2}. \tag{30.6}$$

With increasing baryon density the elasticity of the medium should increase, consequently the velocity of acoustic waves should also increase. With an infinite increase in density, $v(n)$ should tend toward the velocity of light. From (30.6) we thus obtain for high densities n, assuming $v \approx c$,

$$n^2U'' + nU' - U = 0. \tag{30.7}$$

The general solution to this equation is

$$U(n) = an + \frac{b}{n}, \tag{30.8}$$

where a and b are integration constants. From this we obtain for the energy density and the pressure

$$\rho(n) \approx b + an^2, \qquad P \approx an^2 - b,$$

and for $n \to \infty$

$$P \approx \rho \approx \text{const} \cdot n^2. \tag{30.9}$$

This result was obtained in /45, 88/. There are thus strong reasons to believe that, in the supernuclear density range, matter is similar in its elastic properties to a liquid and not to a real gas. With increasing density the elasticity also increases, and in the limit $\rho \to \infty$ it becomes so large that the velocity of sound reaches the velocity of light.

31. PROPAGATION OF ELASTIC WAVES IN A SUPERDENSE MEDIUM

In deriving the asymptotic form (30.9) for the equation of state, we used for the sound velocity the equation

$$v = c\left(\frac{\partial P}{\partial \rho}\right). \tag{31.1}$$

But the above is based on the assumption that the metric of space–time is Euclidean /68/. In Chapter VII we shall see that in superdense media the metric of space–time is no longer "flat." Therefore, strictly speaking, in deriving the limiting equation of state in the last section, we had no right to use equation (31.1) without first checking its validity. We shall now show that this equation applies also to the case of non-Euclidean metrics /92/. To do this, we start with the relativistic equation of hydrodynamics (6.7)

$$\frac{1}{\sqrt{-g}}\frac{\partial}{\partial x^k}\left(\sqrt{-g}\,T_i^k\right) - \frac{1}{2}T^{kl}\frac{\partial g_{kl}}{\partial x^i} = 0, \tag{31.2}$$

where T^{ik} is the energy-momentum tensor. According to (6.6), for a continuous medium

$$\left.\begin{aligned} T^{ik} &= (P+\rho)\,u^i u^k - Pg^{ik}, \\ T_i^k &= (P+\rho)\,u^k u_i - P\delta_i^k, \end{aligned}\right\} \tag{31.3}$$

where u^i is the four-dimensional macroscopic velocity of an element of fluid. We now assume that weak longitudinal waves (acoustic perturbation) propagate in the medium

$$\left.\begin{aligned} \rho^* &= \rho + \varepsilon, & P^* &= P + p, \\ g_{ik}^* &= g_{ik} + \alpha_{ik}, & T_{ik}^* &= T_{ik} + t_{ik}, \end{aligned}\right\} \tag{31.4}$$

where ρ^*, P^*; ρ, P, are the energy density and pressure in the presence and absence of the acoustic perturbation, respectively, ε and p are small perturbations to these quantities, and t^{ik} and α_{ik} are perturbations to the components of the energy-momentum and metric tensors, respectively.

The components of the tensor t^{ik} are obviously of the same order as ε. As to the components of $\rho\alpha_{ik}$ (we should remember that ρ is the intrinsic energy density and therefore a scalar), these have a higher order of smallness. This is more or less evident, since an infinitely small perturbation ε of the energy density cannot cause a relative change of the order of ε/ρ in the metric. The validity of this statement can be verified directly with the example of the metric of a spherically symmetrical gravitational field:

$$ds^2 = g_{ik}dx^i\,dx^k = e^{\nu}\,dx^{0^2} - e^{\lambda}\,dr^2 - r^2\,(d\theta^2 + \sin^2\theta\,d\varphi^2), \qquad (31.5)$$

where ν and λ are unknown functions of the coordinates x^0 and r. If we now substitute (31.4) in (31.3) and calculate t^{ik} and t_i^k in the ε approximation, we see that they satisfy the relationship $t^{ik} = g^{il}t_l^k$ only if $\rho\alpha_{ik} \ll \varepsilon$.

In the first approximation we thus obtain from the equations of hydrodynamics

$$\frac{1}{\sqrt{-g}}\frac{\partial}{\partial x^k}\left(\sqrt{-g}\,t_i^k\right) - \frac{1}{2}\,t^{kl}\frac{\partial g_{kl}}{\partial x^i} = 0. \qquad (31.6)$$

Below, in deriving the velocity equation, it will suffice to limit ourselves to the case of a spherically symmetrical gravitational field, which is described by the quadratic equation (31.5).

We shall assume that before the appearance of the perturbation the field is static, i. e. ν and λ depend only on r. Furthermore, let us also assume that the perturbation propagates from the center of the superdense spherical celestial body in the form of a spherical wave, so that the initial symmetry of the field is not disturbed. However, now ν and λ no longer depend on r alone, but also on the time coordinate x^0. Before the beginning of the perturbation, u^0 is the only component of the four-dimensional velocity to differ from zero, but afterward both u^0 and u^1 are different from zero (we use the notations ct, r, θ, $\varphi \to x^0$, x^1, x^2, x^3). Taking this into account, we have for the components of the tensors t_i^k and t^{ik}:

$$\left.\begin{array}{ll} t_0^0 = -\varepsilon, & t_1^1 = t_2^2 = t_3^3 = p, \\ t_0^1 = -\beta(P+\rho), & t_1^0 = \beta(P+\rho)\,e^{\lambda-\nu}, \\ t^{00} = \varepsilon e^{-\nu}, & t^{11} = pe^{-\lambda}, \\ t^{22} = pr^{-2}, & t^{33} = pr^{-2}\sin^{-2}\theta. \\ t^{01} = \rho(P+\rho)\,e^{-\nu}, & \end{array}\right\} \quad (31.7)$$

The remaining components are equal to zero. Here $\beta = dr/dx^0$ is the macroscopic velocity in the medium related to the propagation of the perturbation.

Substituting (31.7) in (31.6), we obtain the following two equations:

$$\left.\begin{array}{l} \dfrac{\partial \varepsilon}{\partial x^0} + \dfrac{1}{\sqrt{-g}}\dfrac{\partial}{\partial r}\left[\sqrt{-g}\,\beta(P+\rho)\right] = 0, \\ (P+\rho)\,e^{\lambda-\nu}\dfrac{\partial \beta}{\partial x^0} + \dfrac{dp}{dr} + \dfrac{\varepsilon+p}{2}\dfrac{\partial \nu}{\partial r} = 0. \end{array}\right\} \quad (31.8)$$

By using the equation of state $\rho = \rho(P)$, one can express ε through p:

$$\varepsilon = \frac{\partial \rho}{\partial P}\,p. \quad (31.9)$$

Eliminating the quantities ε and β from (31.8), we obtain

$$\frac{1}{\sqrt{-g}}\frac{\partial}{\partial r}\left(\sqrt{-g}\,g^{11}\frac{\partial w}{\partial r}\right) + \frac{1}{v^2}\frac{\partial}{\partial t}\left(\sqrt{-g}\,g^{00}\frac{\partial w}{\partial t}\right) + A(r,t) = 0, \quad (31.10)$$

where the following notations were used:

$$w = pg^{00}, \quad v = c\sqrt{\frac{\partial P}{\partial \rho}},$$

$$A(r,t) = \frac{1}{2\sqrt{-g}}\frac{\partial}{\partial r}\left[\sqrt{-g}\,g^{11}p\frac{dg_{00}}{dr}\left(\frac{v}{c}-1\right)\right].$$

Equation (31.10) describes the propagation of a spherical acoustic wave with a variable amplitude and velocity v. Since the energy density and pressure depend on r, the velocity of sound is also a function of r. At very high densities, when the equation of state (30.9) becomes valid, the sound velocity is close to the velocity of light. In this case the term $A(r, t)$ in the wave equation vanishes.

When the time component g_{00} of the metric tensor varies with distance from the center so slowly that the term containing $\frac{\partial g_{00}}{\partial r}$ in

$A(r, t)$ can be neglected compared to other terms in (31.10), one obtains the regular wave equation written in covariant form:

$$\frac{1}{\sqrt{-g}}\frac{\partial}{\partial x^k}\left(\sqrt{-g}\,g^{ik}\frac{\partial w}{\partial x^i}\right)=0. \tag{31.11}$$

In deriving (31.11) it was assumed that the amplitude of the acoustic wave depends only on the coordinate x^0 and x^1, and also that the metric tensor g_{ik} is diagonal. We have proven that equation (31.1) for the velocity of sound remains in force even in the general case of a Riemann space.

32. CONVENIENT FORMS FOR THE EQUILIBRIUM EQUATIONS

In finding particle concentrations and in calculating stellar configurations, it is convenient to replace the Fermi momenta p_k by the parameters

$$t_k = 4\,\mathrm{arsh}\,(p_k/m_k c). \tag{32.1}$$

From this one obtains for the particle densities

$$n_k = \frac{1}{2} a_k c_k \,\mathrm{sh}^3 \frac{t_k}{4}, \tag{32.2}$$

where $a_k = 2s_k + 1$ and $c_k = m_k^3 c^3/(3\pi^2 h^3)$. The total baryon density is then

$$n = \frac{1}{2}\sum_k \alpha_k c_k \cdot \mathrm{sh}^3 \frac{t_k}{4}. \tag{32.3}$$

For the particles' Fermi energy we have

$$\varepsilon_k = (m_k^2 c^4 + c^2 p_k^2)^{1/2} = m_k c^2 \cdot \mathrm{ch}\,\frac{t_k}{4}. \tag{32.4}$$

The conditions for thermodynamic equilibrium can also be written in a convenient form. Thus, from (23.7)—(23.12) we obtain the following for neutral, positive, and negative baryons and for leptons, respectively:

$$\left.\begin{aligned} m_k \cdot \operatorname{ch}\frac{t_k}{4} &= m_n \cdot \operatorname{ch}\frac{t_n}{4}, \\ m_k \cdot \operatorname{ch}\frac{t_k}{4} &= m_n \cdot \operatorname{ch}\frac{t_n}{4} - m_e \cdot \operatorname{ch}\frac{t_e}{4}, \\ m_k \cdot \operatorname{ch}\frac{t_k}{4} &= m_n \cdot \operatorname{ch}\frac{t_n}{4} + m_n \cdot \operatorname{ch}\frac{t_e}{4}, \\ m_\mu \cdot \operatorname{ch}\frac{t_\mu}{4} &= m_e \cdot \operatorname{ch}\frac{t_e}{4} \end{aligned}\right\} \qquad (32.5)$$

etc. To obtain a closed system of equations, (32.3) and (32.5) have to be supplemented by the condition of neutrality (24.12), which is not difficult to express in terms of the parameters t_k. It is convenient to choose t_n as the independent variable; for a given number of baryons n, equations (32.3) and (32.5) then enable one to express the parameters t_k of the remaining particles through t_n. In a spherically symmetrical mass distribution, t_n itself depends on the distance r from the center of the configuration.

In the case of an ideal baryon gas, the hyperons are in the central part of the star, where $t_n(r) \geqslant 2.1$.

Finally, let us write the equilibrium condition for matter along the radius of the configuration:

$$\left(m_n c^2 \cdot \operatorname{ch}\frac{t_n}{4} + V_n(n)\right)\sqrt{g_{00}(r)} = \text{const.} \qquad (32.6)$$

33. THE EQUATION OF STATE AT NUCLEAR AND SUPERNUCLEAR DENSITIES

The equations of state given in Section 29 pertain to the below-nuclear density region. Now we have to examine the range of nuclear densities and above.

In the ideal-gas approximation we have

$$\left.\begin{aligned} \rho &= K_n \sum_k \delta_k (\operatorname{sh} t_k - t_k), \\ P &= \frac{1}{3} K_n \sum_k \delta_k \left(\operatorname{sh} t_k - 8\operatorname{sh}\frac{t_k}{2} + 3t_k\right), \end{aligned}\right\} \qquad (33.1)$$

where $\delta_k = \left(\frac{m_k}{m_n}\right)^4$. The summation is assumed to be over all particles present in the medium at the given density of matter. The contribution of leptons to (33.1) is totally insignificant and can be left out.

As pointed out before, in the density range at hand the ideal-gas approximation is not quite correct. Equation (33.1) has to be revised, so as to take into account the interaction energy of baryons. This has been done by a number of investigators /96, 250, 35/. Here we shall present the results obtained in /96/, and base our discussion on /137, 138, 139/, which deal with the properties of nuclear material. In /139/ the energy of a neutron gas was calculated for the region $7.10^{35} < n < 7 \cdot 10^{37}$ cm^{-3}. We shall make use of these results. In the nuclear region the binding energy of nucleons is calculated under the assumption that $n_p = n_n$. In using the results of these papers, we shall have to carry out an interpolation for supernuclear densities. It will also be necessary to include corrections related to the symmetry term in the binding energy, since in our case the material consists mainly of neutrons (we are speaking of densities $n \approx 10^{38}$ cm^{-3}). In /138/ the symmetry energy per particle was calculated from

$$\varepsilon = 0.897 T_n (1 - 2n_p/n)^2,$$

where $T_n = p_n^2/2m_n$, and for atomic nuclei the result obtained was too high. Better agreement with experiment is obtained if the coefficient 0.897 in the equation is replaced by 0.592. However, we did not consider it necessary to improve the coefficient, since the above equation was obtained under the assumption that $(1-2n_p/n) \ll 1$, whereas in our case the expression between parentheses is close to unity, and we did not want to underestimate the result. After introducing the proper corrections to the binding energy, the energy per neutron was plotted as a function of $k = (1.5\pi^2 n)^{1/3}$. The average potential energy $U(n)$ per particle was then calculated from this curve, and was later approximated by two polynomials of the type $a_1 k + a_2 k^2 + a_3 k^3$.

In the region $n \lesssim 1.5 \cdot 10^{38}$ cm^{-3}, the part of the energy due to nuclear interactions can be approximated by the formula

$$\rho_v(n) = nV(n) = 1.02 \cdot 10^{-18} n^{4/3} - 8.75 \cdot 10^{-31} n^{5/3} - 3.91 \cdot 10^{-41} n^2. \qquad (33.2)$$

For the range $n \gtrsim 10^{38}$ cm^{-3} the following was obtained:

$$\rho_V(n) = nV(n) = 2.05 \cdot 10^{-17} n^{4/3} - 8.99 \cdot 10^{-30} n^{5/3} + 8.03 \cdot 10^{-43} n^2. \qquad (33.3)$$

We still have to calculate the pressure $P_V = -\rho_V + n \dfrac{\partial \rho_V}{\partial n}$ due to the interaction energy. The pressure of a baryon gas which corresponds to (33.2) and (33.3) is, respectively,

$$P_V = 3.40 \cdot 10^{-19}\, n^{4/3} - 5.83 \cdot 10^{-31} n^{5/3} - 3.91 \cdot 10^{-44}\, n^2 \quad \text{for } n \lesssim 1.5 \cdot 10^{38}\,\text{cm}^{-3}, \tag{33.4}$$

$$P_V = 6.82 \cdot 10^{-18}\, n^{4/3} - 5.99 \cdot 10^{-30}\, n^{5/3} + 8.03 \cdot 10^{-43}\, n^2 \quad \text{for } n \gtrsim 10^{38}\,\text{cm}^{-3}. \tag{33.5}$$

To obtain the equation of state for a real gas at $\rho \gtrsim 10^{13}$ g·cm^{-3}, one has to add ρ_V and P_V, respectively, to the energy density ρ and the pressure P in equations (33.1):

$$\left.\begin{aligned} \rho &= K_n \sum_k \left(\frac{m_k}{m_n}\right)^4 (\operatorname{sh} t_k - t_k) + \rho_V, \\ P &= \frac{1}{3} K_n \sum_k \left(\frac{m_k}{m_n}\right)^4 \left(\operatorname{sh} t_k - 8 \operatorname{sh}\frac{t_k}{2} + 3t_k\right) + P_V. \end{aligned}\right\} \tag{33.6}$$

In the region $10^{37} \lesssim n \lesssim 5 \cdot 10^{38}$ cm^{-3} the energy density ρ_v is small and can be neglected, but the pressure P_V is large from the onset of nuclear forces and has to be taken into account.

The potential energy (33.3) was especially chosen so as to satisfy (30.9), and at sufficiently high densities (33.6) becomes the strict equation of state $P \approx \rho$.

The equation of state given in this section for a degenerate real baryon gas is certainly not a definitive solution to the problem. It should be regarded only as one of the approximate models for a real gas, which are encountered in connection with calculations of degenerate configurations /93, 147/. However, besides providing a rough picture for the complex situation existing in a degenerate baryon gas, the above has the additional advantage of satisfying the law of causality, namely that when $\rho \to \infty$, the sound velocity does not exceed the velocity of light, but approaches it asymptotically from below.

It would certainly be desirable to have an accurate equation of state for all densities. It should, however, be kept in mind that inaccuracies in the equation of state do not strongly affect the mass and radius of superdense configurations (see Chapter VII), and in this sense they do not constitute a serious defect. The proper order of magnitude for the integral parameters of celestial bodies can be obtained even from a model involving only a single type of noninteracting baryon, i.e. by assuming that at all densities we are dealing with an ideal neutron gas. In this simple case we have

$$\left.\begin{aligned}\rho &= K_n (\operatorname{sh} t_n - t_n),\\ P &= \frac{1}{3} K_n (\operatorname{sh} t_n - 8 \operatorname{sh} t_n + 3t_n).\end{aligned}\right\} \tag{33.7}$$

The most characteristic properties of baryon configurations can be obtained by basing the calculations on precisely this equation of state. We should also add that, due to the large volume of calculations involved in using the more accurate equation of state (33.6) (when finding the particle concentrations n_k and their corresponding parameters t_k for a given baryon density n), one cannot be sure of obtaining the sought physical quantity with the required accuracy (the mass defect, for example). The simplified equation of state, on the other hand, does not present this problem.

In the nonrelativistic case $t_n \ll 1$. Expanding the hyperbolic sine in powers of t_n and later eliminating it from (33.7), we obtain the following polytropic equation of state for a nonrelativistic neutron gas:

$$P = \frac{1}{80} \left(\frac{6}{K_n}\right)^{2/3} \cdot \rho^{5/3} = 6.46 \cdot 10^{-26} \rho^{-26} \tag{33.8}$$

In the highly relativistic case $t_n \gg 1$, and

$$P \approx \frac{1}{3} \rho. \tag{33.9}$$

34. THE CHOICE OF A SYSTEM OF UNITS

In numerical calculations it is convenient to use a system of units in which /243/

$$k = 1, \quad c = 1, \quad K_n = \frac{m_n^4 c^5}{32\pi^2 h^3} = \frac{1}{4\pi}, \tag{34.1}$$

where k is the gravitational constant. It can easily be seen that in the new system all physical quantities are dimensionless. We shall now find the relationship between the new and old units.

Let us measure some distance, mass, and time interval and denote the results obtained in the system (34.1) by l', m', t', and in the CGS system by l, m, t. The ratios l/l', m/m', t/t' are obviously independent of the measured quantities, i.e. they are constants:

$$l = \alpha l', \quad m = \beta m', \quad t = \gamma t'. \tag{34.2}$$

TABLE 6. Numerical values of the units of some physical quantities

Name of quantity	Numerical value of (34.1) unit in CGS system	Name of quantity	Numerical value of (34.1) unit in CGS system
Distance	$1.368\cdot10^{6}$ cm	Mass density	$7.189\cdot10^{15}$ g$\cdot$cm^{-3}
Mass	$1.843\cdot10^{34}$ g	Number density of particles	$3.889\cdot10^{-19}$ cm^{-3}
Time	$4.57\cdot10^{-5}$ sec	Electric-field intensity	$2.544\cdot10^{18}$ CGSE
Energy	$1.657\cdot10^{55}$ erg		
Charge	$4.776\cdot10^{30}$ CGSE		

If we now set $l'=1, m'=1$, and $t'=1$, we arrive at the conclusion that α, β, and γ are the values of the new units of length, mass, and time, expressed, respectively, in centimeters, grams, and seconds. The numerical values of α, β, and γ can be obtained from the following obvious relationships between K_n, k, c and K_n', k', c':

$$\left.\begin{aligned} K_n' &= \frac{m_n'^4 c'^5}{32\pi^2 h'^2} = \frac{\alpha\gamma^2}{\beta}\cdot\frac{m_n^4 c^5}{32\pi^2 h^3} = \frac{1}{4\pi}, \\ k' &= \frac{\beta\gamma^2}{\alpha^3} k = 1, \quad c' = \frac{\gamma}{\alpha} c = 1. \end{aligned}\right\} \tag{34.3}$$

From this one obtains three equations for α, β, and γ:

$$\frac{\beta}{\alpha\gamma^2} = \frac{m_n^4 c^5}{8\pi h^3}, \quad \frac{\alpha^3}{\beta\gamma^2} = k, \quad \frac{\alpha}{\gamma} = c, \tag{34.4}$$

from whose solution we find for the unit of length in the new system

$$\alpha = 2c\left(\frac{2\pi}{km_n}\right)^{1/2}\cdot\left(\frac{h}{m_n c}\right)^{3/2} = 1.368\cdot10^3\,\text{cm}, \tag{34.5}$$

for the unit of mass

$$\beta = \frac{c^2}{k}\alpha = 1.843\cdot10^{31}\,г = 9.285\,M_\odot \tag{34.6}$$

and for the unit of time

$$\gamma = \frac{\alpha}{c} = 4.57\cdot10^{-5}\,\text{sec}. \tag{34.7}$$

The elementary charge, neutron mass, and Planck constant are in the units of (34.1), respectively, $e = 1.005 \cdot 10^{-40}$, $m_n = 9.054 \cdot 10^{-59}$, $h = 1.392 \cdot 10^{-75}$.

SUMMARY

In the density region $\rho \lesssim 3 \cdot 10^{11}$ g·cm^{-3}, the state of a plasma consisting of atomic nuclei and a degenerate electron gas (Ae-phase) is described through equation (29.8). In this case the ideal-gas approximation is appropriate.

There are decisive arguments indicating that the state of a plasma of ultrarelativistic baryons is described by an equation of the type $P = \rho \sim n^2$, where n is the baryon density, and not by $P = \rho/3$. This is a very stringent equation. It follows from the assumption that at relativistic densities the interaction energy between particles exceeds their kinetic energy, and from the natural requirement that the sound velocity not exceed the velocity of light. Any equation of the type $P \sim \rho \sim n^k$, where $k > 2$, necessarily leads to violation of the law of causality and is therefore unacceptable.

The situation is worse for densities of the order of nuclear densities, $10^{13} \lesssim \rho \lesssim 10^{15}$ g·cm^{-3}, where the ideal-gas approximation cannot be considered satisfactory. In this region an important role is played by nuclear attraction forces, and they contribute appreciably to the pressure. When $\rho \gtrsim 10^{15}$ g·cm^{-3}, powerful nuclear repulsion forces enter the picture, and these have to be taken into account not only for the pressure, but for the energy density as well. In this region we may be dealing not with a baryon gas, but with a baryon liquid. For the intermediate range close to the nuclear densities ($\rho \gtrsim 10^{13}$ g·cm^{-3}), we derived in this chapter the approximate equations of state, in which the contribution of nuclear interactions was included.

It is necessary to conduct further investigations to find a more accurate equation of state for the range $10^{13} \lesssim \rho \lesssim 10^{16}$ g·cm^{-3}. The solution of this problem is hindered primarily by lack of knowledge regarding the accurate laws for nuclear interactions between particles, or the dependence of the forces on the energy of the particles, their spin, or the form of the baryons. However, even if we knew all these details, still the transition from pairs to many particles, i.e. to the theory of a nuclear plasma, would not be an easy problem. So far, such a theory has been developed only for ordinary nuclear material ($n \approx 2 \cdot 10^{38}$ cm^{-3} for $n_p = n_e$).

Chapter VI

THE NONRELATIVISTIC THEORY OF SUPERDENSE STELLAR CONFIGURATIONS

35. THE CONDITION FOR HYDRODYNAMIC EQUILIBRIUM

So far we have studied the changes occurring in the properties of matter with increasing density. In our discussion we reached infinitely high densities, but we did not stop to wonder whether such high densities do in fact exist in nature. If these do not exist, then the theory of phase changes given in the preceding sections is of academic interest at best. Examples of very high densities are known to us from the microcosm, namely the atomic nuclei. Here, however, we are interested in superdense continuous media on the macroscopic scale. In this and subsequent chapters we shall give an answer to this question. It will be shown that it is not impossible that celestial bodies exist which are in hydrostatic and thermodynamic equilibrium, and in the central region of which the mass density can attain arbitrarily high values.

We shall start our discussion with the Newtonian theory of superdense celestial bodies. Let us consider a spherical distribution of degenerate mass, which is in a state of thermodynamic and hydrodynamic equilibrium, and let $\varphi(r)$ be the potential of the gravitational field at a distance r from the star's center. The function φ then satisfies the Poisson equation

$$\Delta\varphi = \frac{4\pi k}{c^2}\rho(r). \tag{35.1}$$

A body will be in mechanical equilibrium if at each of its points the gravitational attraction toward the center is compensated by the gradient of the pressure

$$\nabla P = -\frac{1}{c^2}\rho\nabla\varphi. \tag{35.2}$$

Assuming that the mass distribution is spherically symmetrical, we find from these two equations that

$$\frac{1}{r^2}\frac{d}{dr}\left(\frac{r^2}{\rho}\frac{dP}{dr}\right)=-\frac{4\pi}{c^4}k\rho(r). \tag{35.3}$$

If we know the equation of state $\rho(P)$, and also the pressure and its derivative at the center, then by integrating the above equation from the center to the surface, where $P(R)=0$, we find the mass distribution and the radius R of the configuration, and later or simultaneously one can also calculate the mass M.

The integration procedure can be made easier by transforming the second-order differential equation (35.3) into a system of ordinary differential equations of the first order. To this end we introduce the following notation:

$$\frac{r^2}{\rho}\frac{dP}{dr}=-\frac{k}{c^2}u(r).$$

In place of (35.3) we then have

$$\left.\begin{aligned}\frac{du}{dr}&=\frac{4\pi}{c^2}r^2\rho(r),\\ \frac{dP}{dr}&=-\frac{k}{c^2}\frac{1}{r^2}\rho(r)u(r).\end{aligned}\right\} \tag{35.4}$$

As we can see, the function $u(r)$ expresses the mass contained in a central sphere of radius r. In the second equation, the left-hand side is the pressure gradient, and the right-hand side is the gravitational attraction; thus the equation is the condition of hydrodynamic equilibrium. To the system (35.4) we also add the equation of state and the boundary conditions at the center and surface of the star:

$$\rho=\rho(P),\ u(0)=0,\ P(0)\equiv P_0,\ P(R)=0. \tag{35.5}$$

The condition $u(0)=0$ means that there is no point mass at the center. For physically meaningful equations of state, the integration of (35.4) with the conditions (35.5) can be carried out only numerically. Before proceeding to discuss details of the integration results, we shall present some estimates, so as to form a clear picture of the celestial bodies concerned.

36. ESTIMATES OF MASS AND RADIUS

Let us estimate the mass and radius of superdense configurations for equations of state of the type

$$P = A\rho^{1+1/n}, \tag{36.1}$$

where A and n are constants. The number n is called the polytropy index. We now substitute (36.1) in the second equation (35.4):

$$\rho' = -\frac{n}{A(n+1)}\frac{ku}{c^2r^2}\cdot\rho^{1-1/n} \tag{36.2}$$

and replace all quantities by their averages:

$$\overline{\rho'} \approx \frac{\overline{\rho}}{R}, \quad \overline{\rho} \approx \frac{Mc^2}{4R^3}, \quad \overline{u} \approx \frac{M}{2}, \quad \overline{r} \approx \frac{R}{2}.$$

From this the following approximate relationship is obtained:

$$R^{3-n} \approx \left[\frac{A(n+1)}{2\cdot 4^{1/n}\cdot n\cdot k}\right]^n \cdot c^{2(n+1)}\cdot M^{1-n}. \tag{36.3}$$

Let us apply (36.3) to two cases. The first case is when matter in the star consists of atomic nuclei and degenerate electrons, i. e. it is in the Ae-phase. In the second case the star consists of a degenerate neutron gas with a small admixture of protons and electrons, which ensure its stability.

1. The Ae-phase; nonrelativistic electron gas. According to (29.10), in this case $n=1.5$ and $A=3.72\cdot 10^{-23}$. Inserting these values into (36.3), we obtain

$$R \approx 1.2\cdot 10^8\left(\frac{M_\odot}{M}\right)^{1/3}, \tag{36.4}$$

where $M_\odot = 1.985\cdot 10^{33}$ g is the sun's mass. For configurations with $M \approx M_\odot$ the radius is of the order of 10^3 km. The numbers thus obtained agree in order of magnitude with the mass and radius of typical white dwarfs.

If the electron gas is relativistic, the equation of state is defined by equation (29.11), for which $n=3$, $A=5.63\cdot 10^{-14}$. The corresponding mass value obtained from (36.3) is

$$M \approx 0.1 M_\odot, \tag{36.5}$$

but the radius cannot be obtained from (36.3) when $n=3$. It can, however, be obtained from the mass if the value of $\bar{\rho}$ is given. The above mass is lower by one order of magnitude than that obtained through accurate calculations for the ultrarelativistic case.

2. The *npe*-phase; nonrelativistic neutrons. From the equation of state (33.8) we know that for this case $n=1.5$ and $A=6.44\cdot 10^{-26}$ and thus

$$R\approx 2\cdot 10^5\left(\frac{M_\odot}{M}\right)^{1/3}. \tag{36.6}$$

If the neutron gas is relativistic then $P=\rho/3$, and consequently $n=\infty$, $A=1/3$. From this one obtains for the mass and radius

$$R\approx 14\cdot 10^6\,\frac{M}{M_\odot}. \tag{36.7}$$

Thus for a mass $M\approx M_\odot$ the radius of neutron configurations should be of the order of ten kilometers.

The estimates given here for the masses and radii give a qualitatively correct picture of the situation.

37. WHITE DWARFS

White dwarfs are stars with a radius of between a few hundred and tens of thousands of kilometers, and a mass of the order of the sun's, so that they belong to the category of superdense celestial bodies. The temperature in their interiors cannot be higher than that of ordinary stars. These temperatures are not sufficient to remove the degeneracy of electrons in their typical representatives (see Section 14).

The theory of white dwarfs was developed over the years by many investigators. In the first studies, calculations of white-dwarf configurations were made without including the effect of neutronization (i.e. the dependence of the ratio A/Z on the electron Fermi energy), for fixed chemical compositions (homogeneous, consisting of layers of different elements, having various mixtures of the elements, etc.). The purpose of these was to study the dependence of the observable parameters (radius, mass, and luminosity) on chemical composition. Sometimes the choice of chemical composition was based on the previous history of the star. A certain cosmogonic theory was used, according to which white dwarfs are the last step in a chain of evolutionary development which started with the gaseous state.

TABLE 7. The parameters of white dwarfs with equilibrium chemical composition /101/

x_e (0)	ρ (0), g·cm^{-3}	Radius R, km	Mass, $M/M_\odot$	Hydrogen envelope	
				thickness, km	mass, $\Delta M/M_\odot$
45	$2.48\cdot10^{11}$	$5.944\cdot10^{2}$	0.9928	1.38	$3.60\cdot10^{-9}$
40	$1.63\cdot10^{11}$	$6.506\cdot10^{2}$	1.034	1.59	$4.99\cdot10^{-9}$
35	$1.05\cdot10^{11}$	$7.240\cdot10^{2}$	1.077	1.89	$7.33\cdot10^{-9}$
30	$6.35\cdot10^{10}$	$8.222\cdot10^{2}$	1.121	2.35	$1.17\cdot10^{-9}$
28	$5.10\cdot10^{10}$	$8.710\cdot10^{2}$	1.138	2.59	$1.45\cdot10^{-8}$
25	$3.55\cdot10^{10}$	$9.581\cdot10^{2}$	1.165	3.06	$2.08\cdot10^{-8}$
20	$1.76\cdot10^{10}$	$1.157\cdot10^{3}$	1.208	4.33	$4.31\cdot10^{-8}$
15	$7.27\cdot10^{9}$	$1.471\cdot10^{3}$	1.247	6.74	$1.08\cdot10^{-7}$
10	$2.06\cdot10^{9}$	$2.046\cdot10^{3}$	1.269	12.9	$3.99\cdot10^{-7}$
7	$6.96\cdot10^{8}$	$2.701\cdot10^{3}$	1.255	22.6	$1.22\cdot10^{-6}$
4	$1.29\cdot10^{8}$	$4.058\cdot10^{3}$	1.153	55.6	$6.76\cdot10^{-6}$
3	$5.42\cdot10^{7}$	$4.935\cdot10^{3}$	1.056	89.9	$1.62\cdot10^{-5}$
1	$1.98\cdot10^{6}$	$9.615\cdot10^{3}$	0.5005	692	$4.56\cdot10^{-4}$
0.8	$1.00\cdot10^{6}$	$1.087\cdot10^{4}$	0.3935	1110	$9.22\cdot10^{-4}$
0.6	$4.23\cdot10^{5}$	$1.268\cdot10^{4}$	0.2784	2070	$2.27\cdot10^{-3}$
0.4	$1.25\cdot10^{5}$	$1.565\cdot10^{4}$	0.1626	4900	$7.53\cdot10^{-3}$
0.2	$1.57\cdot10^{4}$	$2.225\cdot10^{4}$	0.06026	21300	0.120
0.1	$1.96\cdot10^{3}$	$3.152\cdot10^{4}$	0.02158	63000	$8.63\cdot10^{-2}$
0.09	$1.43\cdot10^{3}$	$3.323\cdot10^{4}$	0.01845	66500	$7.38\cdot10^{-2}$
0.07	$6.72\cdot10^{2}$	$3.769\cdot10^{4}$	0.01267	75400	$5.07\cdot10^{-2}$

Note. The figures in the last three rows of the last two columns represent the radius R and mass M of stars consisting wholly of hydrogen.

In Chapter II it was shown that the nuclear parameters Z and A, and thereby the chemical composition, are dependent on density. The number of protons Z in the nuclei diminishes as a result of neutronization. In contrast, the mass numbers A become larger due to special pycnonuclear reactions brought about by the tunnel effect. Light nuclei cannot exist for long in the Ae-phase, but fuse with each other or with heavy nuclei to form still heavier nuclei. The smallest mass number pertaining to a nucleus which can exist in a certain plasma thus depends on the density (see Figure 5). Therefore, if we assume that a star has existed long enough ($\tau \gtrsim 10^7$ years) in the white-dwarf state, then the chemical composition of its nuclei has reached some state of equilibrium in which the average ratio A/Z is a function of density (see Figure 3). As we can see, in calculating the configurations of the densest white dwarfs, it is more appropriate to assume that there is chemical equilibrium between the nuclei, with the average A/Z depending on the Fermi energy as described in Section 18, than it is to start out from some contrived choice of the chemical composition.

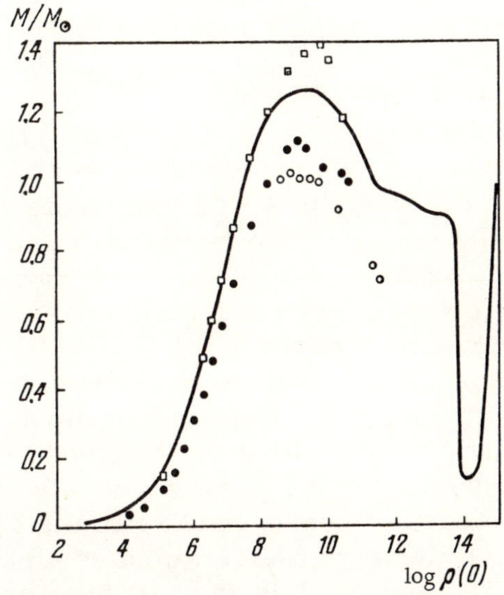

FIGURE 12. The mass of white dwarfs as a function of the common logarithm of their central density for several chemical compositions. The solid line /101/, squares, and black and white circles /255/, respectively, represent configurations at equilibrium, consisting of carbon or iron and with the most probable composition.

We have before us the problem of integrating the differential equations (35.4) together with the equation of state (29.8) and the boundary conditions (35.5). Using (29.8), equation (35.4) can be rewritten in the following form:

$$\left.\begin{aligned} \frac{du}{dr} &= \frac{32}{3}\frac{m_e^3}{m_n^3} r^2 x_e^3 (2 + a_1 x_e + a_2 x_e^2 + a_3 x_e^3), \\ \frac{dx_e}{dr} &= -\frac{m_n}{m_e}\frac{u}{r^2}\frac{1}{x_e}(1 + x_e)^{1/2}\cdot (2 + a_1 x_e + a_2 x_e^2 + a_3 x_e^3). \end{aligned}\right\} \quad (37.1)$$

Here the quantities are measured in the system of units where $c=k=1$ and $K_n = 1/4\pi$. The values of the constants a_1, a_2, a_3

were given after equation (29.8). The boundary conditions (35.5) are now written in the form

$$u(0)=0, \quad x_e(0)\equiv x_0, \quad x_e(R)=0. \tag{37.2}$$

The integration of equations (37.1) was carried out numerically for a series of values $x_e(0)$ at the center, and the results are presented in Table 7 and Figures 12 and 13. It should be remembered that the above is based on relationship (18.5), according to which the average of the ratio A/Z has its largest (i. e. the largest possible for a given density) value at the center of the configuration, and it decreases monotonically with distance from the center, tending toward the value 2 at the surface. However, the ratio A/Z does not necessarily have to be equal to 2 at the surface when the composition of matter is at equilibrium. Even in a sufficiently old superdense star the surface layer, where $x_e(r)<0.145$ ($\rho \lesssim 3000$ g·cm^{-3}), may also include hydrogen, It is therefore of interest to find the parameters of this outer envelope in which hydrogen can be present. The last two columns of Table 7 give the mass and thickness of this external envelope, assuming that it consists solely of hydrogen. Due to the fact that in the case of hydrogen the ratio A/Z is half as large as it is for other light elements (those which could be present in the envelope), the thickness and mass of the hydrogen envelope should be larger by a factor of two and four, respectively, than they are when $A/Z=2$ (see below). From the figures presented in Table 7 it can be seen that for those configurations where $\rho(0)\gg 3\cdot 10^3$ g·cm^{-3} (this being the critical density, above which hydrogen cannot exist long in the medium), the thickness l_H of the hydrogen envelope is very small compared to the radius. Especially small are the values ΔM_H of the envelope's mass. With decreasing central density both l_H and ΔM_H increase.

By performing an approximate integration of equations (35.4) one can obtain an analytical expression for the thickness and mass of the hydrogen envelope, if one assumes that the ratios l_H/R and $\Delta M_H/M$ are small when $x(0)>0.5$. Thus,

$$\left.\begin{aligned} \Delta M_H &= \frac{4}{15}\,\frac{m_e^4 c^5}{\pi k h^3}\cdot\frac{R^4}{M}\,x_H^5 = 5.7\cdot 10^{-7}\frac{M_\odot}{M}\,R^4,\\ l_H &= \frac{m_e c^2 x_H^2}{2m_p k}\cdot\frac{R^2}{M} = 7.72\cdot 10^{22}\,\frac{M_\odot}{M}\cdot R^2. \end{aligned}\right\} \tag{37.3}$$

Here mass is measured in grams and thickness in centimeters, and $x_H=0.145$ (to which corresponds a density of $3\cdot 10^3$ g·cm^{-3}) is the critical value of the parameter x_e, above which hydrogen disappears from the medium as a result of the tunnel effect.

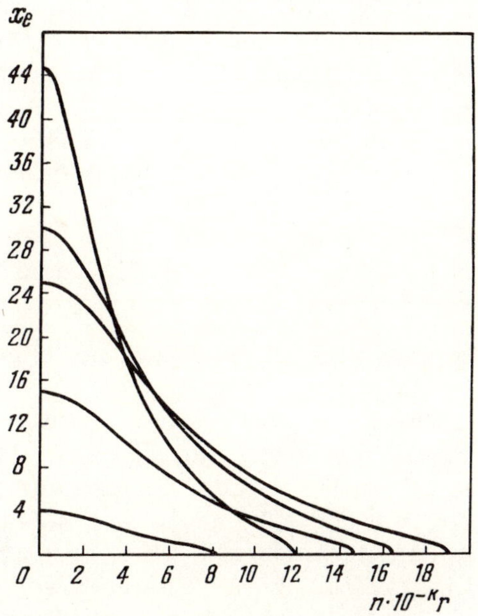

FIGURE 13. Variation of the parameter x_e with distance from the center for several configurations. The variable along the abscissa is $n 10^{-k} r$, where $n = 2, 1, 2, 2, 2$ and $k = 3, 2, 2, 2, 2$ for $x_e(0) = 4, 15, 25, 30, 45$, respectively.

The dependence of the configuration's mass on the central density is shown in Figure 12. The solid line represents the results of numerical integration of (37.1) (equilibrium chemical composition, in which the dependence of A/Z on the electron Fermi energy was taken into account /101/). The squares and the black and white circles represent, respectively, configurations consisting of carbon or iron, or having an equilibrium chemical composition /255/. In /255/ an equilibrium chemical composition was considered to be the following: for each density value the most probable A and Z were calculated (see /125/ and /255/), and in the configuration calculations one nucleus was chosen for each layer (the nucleus which at the given x_e is the most probable in the sense that the energy density ρ is a minimum), and its parameters were regarded as functions of ρ. For light elements (He, C, Mg) the results of /255/ agree with those of /101/ (the latter are shown as the solid line in the figure). Differences are noticed only for $8 < \log \rho < 10.5$, where the points are above the solid line. As to the points which correspond to the heavier elements (iron and the "equilibrium" composition), these are all below the solid line.

In order to understand the reason for the spread of points related to the chemical composition, let us examine the behavior of equations (37.1) with respect to changes in A/Z. For this it is sufficient to consider the simple case when A/Z has the same value from the star's center to its surface. Thus, if we perform the change of variables

$$u(r) = \left(\frac{Z}{A}\right)^2 v(r), \quad r = \frac{Z}{A} R, \tag{37.4}$$

then the resulting equations for v and R will no longer depend on A/Z. The substitution (37.4) can be interpreted as a change in the measuring scales for the mass and radius. With the new scales, at a given $x_e(0)$ these quantities do not depend on the chemical composition. Obviously, to obtain the values of mass and radius in the old units one has to multiply v and R by $(Z/A)^2$ and Z/A, respectively. From this it follows that the mass of a white dwarf is proportional to $(Z/A)^2$ and its radius to Z/A. From Figure 3 we see that the line of the true ratios A/Z for heavy nuclei is above (i. e. Z/A is lower) the continuous curve, which is a good approximation only for nuclei of medium atomic weight), and for light nuclei it is below it. Therefore the mass and radius of configurations consisting of heavy nuclei should be smaller than the values corresponding to an equilibrium chemical composition (for which the average A/Z is expressed through (18.5)), and in the case of light nuclei they should be larger. These arguments enable one to conclude, for example, that stars consisting of an element with $A = 240$ should have a mass which is considerably smaller than that obtained for an equilibrium concentration, since the stepped A/Z curve for it lies above the continuous A/Z curve (see Figure 3).

Now the reason for the appearance of a maximum in the $M = M(\rho_0)$ curve can also be understood. At small $\rho(0) \equiv \rho_0$, when the ratio A/Z has a constant value, the mass goes up monotonically with increasing central density. On the other hand, when densities become high enough for neutronization to set in, Z/A decreases in some part of the star; this is followed by a corresponding decrease in the ordinate of the $M(\rho_0)$ curve and the appearance of a maximum at a finite central density. When the neutronization effect is not taken into account, i. e. when A/Z is considered constant, this maximum does not appear /118/. In this simplified treatment, $M(\rho_0)$ is throughout a monotonically increasing function of the central density, the mass tending toward the value 1.4 $M_\odot$ (called the Chandrasekhar limit) when $\rho_0 \to \infty$.

Figure 13 describes the dependence of the parameter $x_e = p_e/m_e c$ on the distance r from the center for several configurations (i. e. for several values of $x_e(0)$). These curves enable one, with the help of the formula

$$\rho = \frac{32}{3}\left(\frac{m_e}{m_n}\right)^3 K_n x_e^3 (2 + a_1 x_e + a_2 x_e^2 + a_3 x_e^3) \tag{37.5}$$

(see the first of equations (29.8)), to calculate the mass distribution in the corresponding configurations.

38. NEUTRON CONFIGURATIONS IN NEWTON'S GRAVITATION THEORY

The radii of configurations consisting of a degenerate baryon gas (baryon configurations) are of the order of their gravitational radius $r_g = 2kM/c^2$, therefore accurate results can be obtained for them only through the general theory of relativity. However, for purposes of comparison and general consideration of the parameters, the non-relativistic theory of gravitation can also be useful; and since a priori this theory is in our case only an approximation, there is no point in going into details of the chemical composition of the baryon gas. It will here be sufficient to examine the simplest model for the above, namely a degenerate nucleon-electron gas (the npe-phase). Accordingly, we shall start out from the equation of state (33.7) for the hypothetical npe-phase:

$$\left.\begin{aligned} \rho &= \frac{1}{4\pi}(\operatorname{sh} t_n - t_n), \\ P &= \frac{1}{12\pi}\left(\operatorname{sh} t_n - 8\operatorname{sh}\frac{t_n}{2} + 3t_n\right). \end{aligned}\right\} \tag{38.1}$$

We should remember that, since $n_p = n_e \ll n_n$, the partial pressure and energy density of the protons and electrons are quite small, therefore they were not included in the above. The units used are those of (34.1). Substituting the above expressions for ρ and P in (35.4), we obtain

$$\left.\begin{aligned} \frac{du}{dr} &= r^2 (\operatorname{sh} t_n - t_n), \\ \frac{dt_n}{dr} &= -\frac{3u(\operatorname{sh} t_n - t_n)}{r^2\left(\operatorname{ch} t_n - 4 \operatorname{ch} \frac{t_n}{2} + 3\right)} . \end{aligned}\right\} \tag{38.2}$$

The boundary conditions at the center and surface are

$$u(0) = 0, \quad t_n(R) = 0, \quad t_n(0).$$

The neutron configurations will thus be characterized by the values of the parameter t_n at the center.

Equations (38.2) were integrated numerically in /90/ for several values of the parameter $t_n(0)$, which is a single-valued function of $\rho(0)$. The integration was carried out with four-digit precision to the point where saturation of the function $u(r)$ set in. At the same time the pressure decreased by many orders of magnitude but did not vanish. For different configurations the saturation of $u(r)$ set in at different pressures. The result of the integration was approximately $t_n(R) \approx 0.05$ (pressure of about 10^{28} g·cm^{-3}, density of $5 \cdot 10^{10}$ g·cm^{-3}). Further integration changes only the value of the radius by a small amount (2%), while changes in mass are not noticeable.

The integration results are shown in the second half of Table 8 (see page 150) and in Figures 14 and 15. The table also contains data on the extreme configuration, in whose center the energy density is infinite. For this case one can find an asymptotic solution. When the energy density at the center is infinitely high, the solution to the first of equations (38.2) can be sought in the form of the power function

$$\rho_1(r) = a \cdot r^{-n}, \tag{38.3}$$

where a and n are constants and $\rho_1 = 4\pi\rho = \operatorname{sh} t_n - t_n$. Substituting (38.3) in the first equation (38.2) and integrating from $r=0$ to r with the condition $u(0)=0$, we obtain

$$u(r) = \frac{a}{3-n} \cdot r^{3-n} . \tag{38.4}$$

At high densities, when the neutron gas becomes relativistic, the equation of state (38.1) becomes $\rho \approx 3P$. Consequently, for a relativistic degenerate neutron gas

$$P_1 \approx \frac{1}{3} \rho_1 = \frac{a}{3} r^{-n}, \tag{38.5}$$

where $P_1=4\pi P$. Substituting (38.4) and (38.5) in the second equation (38.2), we obtain $n=2$, $a=2/3$. We thus find the asymptotic solution

$$P_1(r)\approx\frac{2}{3}r^{-2},\quad u(r)\approx\frac{2}{3}r. \tag{38.6}$$

This solution is valid for the region starting at $r=0$ where the asymptotic equation of state $3P\approx\rho$ can be used, i. e. for some central region in which the neutron gas is highly relativistic. In the calculations the asymptotic solution was considered valid up to a distance $r_0=0.0584$, where $t_n(r_0)=6$, $\rho_1(r_0)=196$, $P_1=46.5$, and $u(r_0)=0.0388$ (all values here are in units of (34.1)). Beyond $r_0=0.0584$, the calculations for the configuration were done numerically.

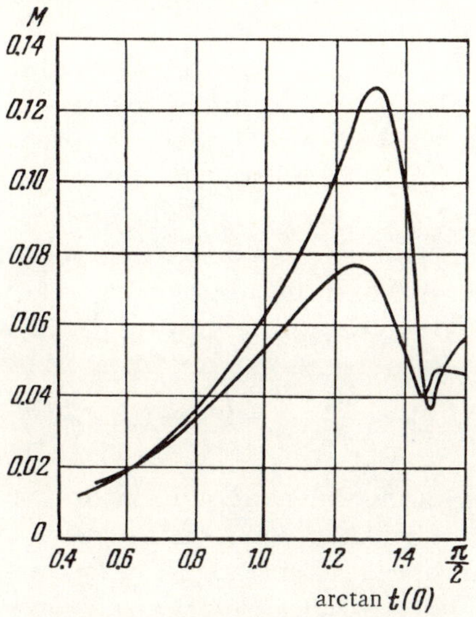

FIGURE 14. The mass of neutron configurations according to the Newton (upper curve) and Einstein gravitation theories. The variable along the abscissa is arctan $t(0)$.

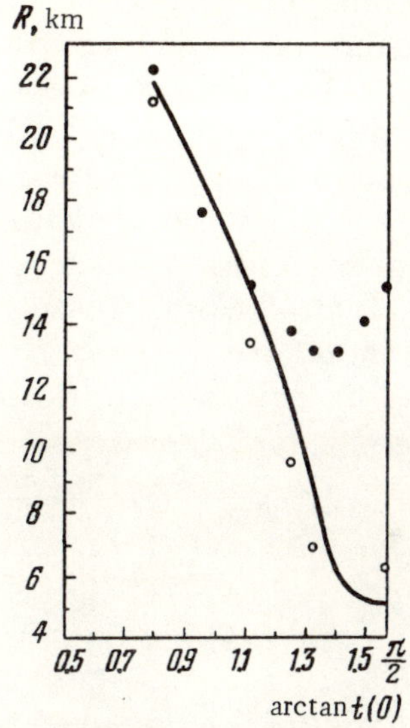

FIGURE 15. The radius of neutron configurations as a function of arctan $t(0)$. The solid line represents the Newton theory, the white circles the Einstein theory, and the black circles the nonrelativistic Landau estimates.

Figure 14 shows the dependence of the configuration's mass on the energy density at the center (upper curve), and for convenience the variable plotted along the abscissa is arctan $t_n(0)$. For comparison, the same figure also includes the curve obtained from the relativistic gravitation theory (lower curve) /243/, which will be discussed in detail later on.

Comparison between the $M(\rho_0)$ curves shows that for relatively low densities, when $t_n(0) \approx 1$, the Newton and Einstein gravitation theories yield almost identical mass values. However, with increasing density the difference between the results becomes larger, and for $2 \lesssim t(0) < \infty$ the mass calculated from the nonrelativistic theory is approximately twice as large as that calculated from the relativistic theory. It is interesting to note that the general behavior of the $M(\rho_0)$ function is similar in the two cases, and the only difference is in the details. Thus, in the Newton theory the maximum of the $M(t_0)$ curve corresponds to $t_0 = t_n(0) \approx 5.2$, while in the Einstein theory it occurs at $t_n(0) \approx 3$. In the second theory there are oscillations when $t_0 \to \infty$ (on the scale chosen these are not seen), whereas in the first theory these are absent.

Figure 15 shows the dependence of the configuration's radius on the parameter $t_n(0)$. The white circles refer to the relativistic theory /243/. As can be seen, the radii obtained from the two theories are not much different from each other. The strongest difference occurs in the region $3.5 \lesssim t(0) < \infty$.

The first to work on the theory of superdense celestial bodies was Landau (see Chapter XI of /67/). For the dependence of radius on mass he obtained the following estimate (see equation (36.6)):

$$R \approx \frac{4.5h^2}{km^{8/3}M^{1/3}} = 15.3 \cdot 10^5 \left(\frac{M_\odot}{M}\right)^{1/3}.$$

The black circles in Figure 15 represent the radii calculated from this equation. The values used for the mass were those given in Table 8. In the range $t_n(0) > 2.3$ the above estimate is very rough.

The material presented in this section indicates that a correct idea about the mass, radius, and internal structure of superdense celestial bodies can be obtained even within the framework of Newton's gravitation theory. However, as we shall see in the next chapter, accurate information on baryon configurations can be obtained only through calculations based on Einstein's equations.

39. THE NONRELATIVISTIC THEORY OF ROTATING SUPERDENSE STELLAR CONFIGURATIONS

The assumption of spherical symmetry is obviously an idealization, since in nature there are no nonrotating celestial bodies.

If one assumes that the angular momenta of stars with similar mass are approximately equal, then among the superdense celestial bodies one should expect to find high rotational angular velocities. It is therefore necessary to find out to what extent one can justify the approximation of a spherically symmetrical mass distribution, and also to find the changes in the configuration parameters when rotation is taken into account.

In the nonrelativistic theory the equations for a rotating configuration are

$$\left.\begin{aligned} \Delta\varphi &= 4\pi k\rho, \\ \Delta P &= -\rho\nabla\varphi + \rho\Omega^2\mathbf{r}, \end{aligned}\right\} \tag{39.1}$$

where $\mathbf{r}$ is the distance from the rotation axis and Ω is the angular velocity. We shall assume that Ω is independent of the coordinates, since in the case of superdense celestial bodies we are dealing only with rigid rotation.

Under rotation the integral characteristics and the configurational structure are determined by two parameters — the central pressure P_0 and the angular velocity Ω. The value of the angular velocity has an upper limit: $0 < \Omega < \Omega_{max}$. The value of Ω_{max} is dictated by the condition that there be no outflow of material at the equator, i. e.

$$\Omega^2_{max} R_e = kM/R_e^2, \tag{39.2}$$

where R_e is the semimajor axis of the pseudospheroid and M is the mass of the star.

We shall limit this discussion to axially symmetrical configurations, since for astrophysics this is the case which interests us most. This kind of configuration is called pseudospheroidal. In this case the functions are independent of the azimuthal angle Φ and the first equation (39.1) can be written in the form

$$\frac{1}{R^2}\frac{\partial}{\partial R}\left(R^2\frac{\partial\varphi}{\partial R}\right) + \frac{1}{R^2}\frac{\partial}{\partial\mu}\left[(1-\mu^2)\frac{\partial\varphi}{\partial\mu}\right] = 4\pi k\rho, \tag{39.3}$$

where R is the distance from the center of the configuration and $\mu = \cos\theta$, where the angle θ is measured from the rotation axis.

We now separate the equation of hydrodynamic equilibrium into its components, multiplying by $\frac{\partial R}{\rho}$ the first equation, which corresponds to the radial component, and by $R\frac{\partial\theta}{\rho}$ the equation corresponding to the meridian component. Summation of the above gives a total differential, whose integral is

$$F = -\varphi + 4\pi k\rho_0\beta R^2(1+\mu^2) + C, \tag{39.4}$$

where C is an integration constant, $\beta = \Omega^2/(8\pi k\rho_0)$, and

$$F = \int \frac{dP}{\rho}. \tag{39.5}$$

In the rotation problem the dimensionless constant β is an important parameter. The potential is determined with accuracy to within a constant term, therefore one could have chosen $C=0$, but in that case $\varphi(\infty)\neq 0$. We prefer to leave the normalization of the potential as $\varphi(\infty)=0$.

In the case of spherical objects it would have sufficed to give the condition at the center. Here, however, we are faced with the following boundary-value problem: in order to integrate the differential equations it is necessary to know the boundary conditions not only at the center, but also at the surface. The conditions at the center cover the values P_0 and Ω, and also the absence of a point mass. Outside the mass distribution the field is determined by the equation

$$\frac{1}{R^2}\frac{\partial}{\partial R}\left(R^2\frac{\partial\varphi}{\partial R}\right) + \frac{1}{R^2}\frac{\partial}{\partial\mu}\left[(1-\mu^2)\frac{\partial\varphi}{\partial\mu}\right] = 0. \tag{39.6}$$

The boundary conditions at the surface are obviously the requirements that both the potential and its first derivative be continuous.

Equation (39.6) has the solution

$$\varphi_e(R,\mu) = \frac{D}{R} + \beta\sum_{l=0}^{\infty}\frac{K_l}{R^{l+1}}P_l(\mu), \tag{39.7}$$

where D and K_l are integration constants. We deliberately isolated the factor β from the constants K_l. These constants are determined by the condition that the exterior solution match the interior solution. At large distances $\varphi \approx -kM/R$, consequently

$$M = -\frac{D+\beta K_0}{k}. \tag{39.8}$$

There are several methods for finding approximate solutions to equations (39.3) and (39.4). In all cases it was established that

$$\beta_{max} = \frac{\Omega^2_{max}}{8\pi k\rho_0} \ll 1.$$

Below we shall present a relatively simple method for solving the problem, giving fairly accurate results /77, 78/. We shall regard the rotation as a perturbation which is characterized by a small β, and in the first approximation in β we shall seek solutions to (39.3) and (39.4) in the form

$$F(R,\mu) = f(R) + \beta\left[f_0(R) + \sum_{l=2}^{\infty} A_l f_l(R) P_l(\mu)\right], \tag{39.9}$$

$$G(R,\mu) = g(R) + \beta\left[g_0(R) + \sum_{l=2}^{\infty} B_l g_l(R) P_l(\mu)\right]. \tag{39.10}$$

Here $G=4\pi k\rho$, and the functions $f(R)$ and $g(R)$ are related to nonrotating configurations. Here, as in the exterior solution (39.7), in the expansion in Legendre polynomials the terms with odd l are absent, as a result of the symmetry of the mass distribution with respect to the equatorial plane.

From the definition of the functions F and G it follows that there is a relation between them. Consequently, there should also be a relation between the constants A_l and B_l. In order to find this relationship, we expand the function $G(F)$ in a Taylor series in the vicinity of $F=f$. With accuracy to terms of the order of β we have

$$\left.\begin{aligned} &G(f)=g, && \gamma f_0 = g_0, \\ &\gamma A_l f_l = B_l g_l, && \gamma = \frac{dg}{df}. \end{aligned}\right\} \tag{39.11}$$

We now substitute the series (39.9) and (39.10) in equation (39.3) and, using (39.4) and (39.11), equate the coefficients of the Legendre polynomials of the same degree. We thus obtain the system

$$\left.\begin{aligned} \Delta_0 f &= -g, \\ \Delta_0 f_0 + \gamma f_0 &= 4G_0, \\ \Delta_l f_l + \gamma f_l &= 0, \quad l > 0, \end{aligned}\right\} \tag{39.12}$$

where $G_0=4\pi k\rho(0)$ and

$$\Delta_l = \frac{1}{R^2}\frac{d}{dR}\left(R^2\frac{d}{dR}\right) - \frac{l(l+1)}{R^2}.$$

Equations (39.12) determine the radial functions $f(R)$, $f_0(R)$, and $f_l(R)$. These are found numerically, and the computations present no special difficulty. Later we shall see that in the approximation which is linear in β the terms with $l>2$ in the series (39.9) and (39.10) disappear. However, the problem can be considered completely solved if we also determine the expansion coefficients A_l, B_l and the shape of the surface.

In the above approximation the shape of the surface is defined by the equation

$$R_\mu = R_0 + \beta \sum_l q_l P_l(\mu), \tag{39.13}$$

where R_0 is the radius of the corresponding configuration, R_μ is the radius of the body of revolution defined by the angle θ, and q_l are constants. For later use, we should separate in (39.4) the contribution of the rotation to the constant C, which determines the normalization of the potential: $C = C_0 + \beta C_1$. At the surface the density, pressure, and consequently also the function F, all vanish. Using the relationships (39.4), (39.7), (39.9), (39.10), and (39.13), we obtain from the conditions $F(R_\mu) = 0$, $\varphi_i(R_\mu) = \varphi_e(R_\mu)$, $\varphi_i(R_\mu) = \varphi_e(R_\mu)$ (the indices i and e denote the interior and exterior solutions) the following equations for the constants A_l, B_l, q_l:

$$\left.\begin{aligned}
&D = R_0^2 f'(R_0), \qquad K_0 = R_0^2 f_0'(R_0) - \frac{4}{3} G_0 R_0^3,\\
&C_0 = R_0 f'(R_0), \qquad C_1 = R_0 f_0'(R) + f_0(R_0) + 2G_0 R_0^2,\\
&A_2 = -\frac{10}{3}\frac{G_0 R_0^2}{R_0 f_2'(R_0) + 3 f_2(R_0)},\\
&K_2 = -\frac{2}{3}\frac{G_0 R_0^5 [R_0 f_2'(R_0) - 2 f_2(R_0)]}{R_0 f_2'(R_0) + 3 f_2(R_0)}\\
&q_0 = -f_0(R)/f'(R_0), \quad q_2 = -A_2 f_2(R_0)/f'(R_0),\\
&f(R_0) = 0, \quad A_l = q_l = K_l = 0 \text{ for } l > 2.
\end{aligned}\right\} \tag{39.14}$$

The expansion coefficients B_l and the functions $q_l(R)$ are found through A_l and $f_l(R)$ with the help of (39.11). The functions $f(R)$, $f_0(R)$, and $f_2(R)$ are obtained from (39.12) through simple numerical integration. In this manner a linear approximation in the small parameter β reduces the rotation problem to the solution of a spherical problem.

By setting $\mu = 0$ and $\mu = 1$ in (39.13), we obtain for the semimajor and semiminor axes of the spheroid:

$$\left.\begin{aligned}
R_e &= R_0 + \beta(q_0 - 0.5 q_2),\\
R_p &= R_0 + \beta(q_0 + q_2).
\end{aligned}\right\} \tag{39.15}$$

Let us also present the equation for the maximum value of β. It is obtained from (39.2) by using (39.7) and (39.15):

$$\beta_{\max} = \frac{D \cdot R_0^2}{1.5\,K_2 + 2DR_0\,(q_0 - 0.5q_2) - K_0R_0^2 - 2G_0R_0^5}. \qquad (39.16)$$

The method described above was employed to calculate the parameters of superdense configurations (white dwarfs and neutron stars). For $P_n > P_e$ (P_n and P_e are the partial pressures of the neutron and electron gas) the equation of state (38.1) was used, and for $P_e > P_n$ equation (29.8) for the Ae-phase. The neutron configuration was assumed to have an electron-nuclear envelope, and the interface between the neutron core and the Ae-envelope was taken to be the distance at which $P_n \approx P_e \equiv P_1$. When $P(0) < P_1$ we obviously have the case of rotating white dwarfs, and when $P(0) > P_1$ we have neutron configurations with an electron-nuclear envelope.

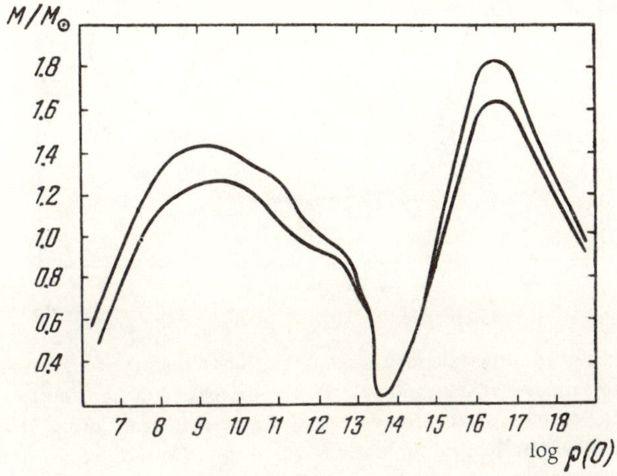

FIGURE 16. Mass as a function of the common logarithm of the central density. The upper curve pertains to configurations which rotate with $\Omega = \Omega_{\max}$, and the lower curve to nonrotating configurations. The parts of the curves where $\rho \gtrsim 10^{12}$ g·cm^{-3} represent neutron configurations, and those where $\rho \lesssim 10^{12}$ g·cm^{-3} represent white dwarfs /78/.

Figure 16 shows the mass as a function of the common logarithm of the central density for configurations which rotate with angular velocity $\Omega_{\max}$ (upper curve). For comparison we also included the curve pertaining to nonrotating stars (lower curve). As a result of the rotation with the highest possible angular velocity, the mass in the vicinity of the two maxima is increased by a factor of about 1.2. The maxima themselves are slightly shifted in the direction of small central densities.

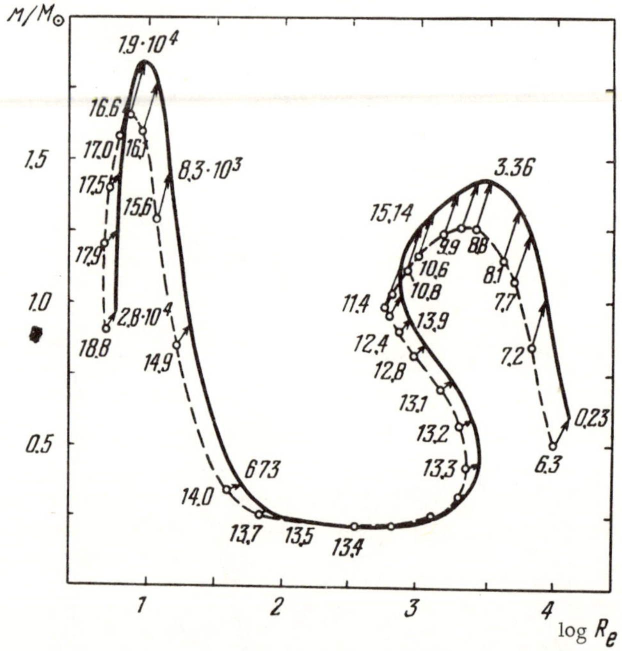

FIGURE 17. The mass of rotating superdense configurations (in units of solar mass) as a function of the major semiaxis R_e (in kilometers). The dashed line pertains to nonrotating configurations, and the numbers next to points along it give the logarithms of the central densities. The solid line describes configurations which rotate with the highest angular velocity. The numbers at some points along it indicate the highest angular velocity which is possible without outflow of matter. In the direction of the arrows the rotational angular velocity increases and the central density remains constant /82a/.

Figure 17 describes the dependence of mass, in units of the solar mass, on the semimajor axis in kilometers. The dashed line pertains to nonrotating configurations (in this case $R_e = R_0$), and the solid line to rotations with angular velocity $\Omega = \Omega_{max}$ (let us remember that Ω_{max} is the highest value of the angular velocity; at this velocity there is the onset of outflow of matter from the equator, resulting from the prevailing of centrifugal forces over the strong gravitational attraction). In the direction of the arrows the rotational angular velocity increases while the value of the central density remains fixed. Thus an arrow originates at a point which defines the mass and radius of a nonrotating body, and traces all values of these quantities for configurations with the same central density but with angular velocities from zero to Ω_{max}. From the distance between the origin of the arrow and some intermediate point on it one can find the angular velocity Ω at that point, since it is simply proportional to the length of the segment. The full length of the arrow is equal to Ω_{max}.

As we see, rotation does not have a strong effect; the highest increase in the mass and dimensions amounts to about 10%.

SUMMARY

The theory of configurations consisting of an electron-nuclear plasma (the Ae-phase of matter) can be based on the Newtonian version of gravitation theory. Such stellar configurations are identified with white dwarfs. Their mass is of the order of the solar mass, and their radius ranges from hundreds (for the densest stars) to tens of thousands (for the less dense stars) of kilometers. An important factor in the theory of white dwarfs is neutronization, due to which the ratio A/Z (where A and Z are the number of all nucleons and of all protons in the nucleus) is dependent on density (or on the electron Fermi energy ε_e). In the case of high densities ($\varepsilon_e \gtrsim 10\, m_e c^2$), this effect leads to an appreciable decrease in mass for an equal central density.

At central densities $\rho_0 \gtrsim 10^{13}$ g·cm^{-3} we have configurations which consist mainly of degenerate baryon gas. The radius of baryon configurations does not differ strongly from their gravitational radius $r_g = 2M$, and in the densest cases it is even of the same order of magnitude. This indicates that accurate calculations of baryon configurations have to be based on Einstein's gravitation theory. However, if one is interested only in approximate values of the parameters, the nonrelativistic gravitation theory is also adequate.

In the last section of this chapter we presented the fundamentals of the nonrelativistic theory of rotating superdense stellar configurations. Inclusion of rotation does not introduce any appreciable changes in the values of the configuration (white dwarfs and baryon configurations) parameters. Even at the highest possible angular velocities, the shapes of the rotating configurations are not very different from spherical. The largest increase in mass due to rotation amounts to about 10%. The semiminor and semimajor axes differ by about the same amount from the radius of the corresponding spherical configuration.

Chapter VII

THE RELATIVISTIC THEORY OF BARYON CONFIGURATIONS

40. STATIC SPHERICALLY SYMMETRICAL MASS DISTRIBUTION

According to (9.1), the metric properties of space in a spherically symmetrical mass distribution are determined by the four-dimensional interval

$$ds^2 = g_{ik}dx^i dx^k = e^{\nu} dx^{0 2} - e^{\lambda} dr^2 - \\ - r^2 (d\theta^2 + \sin^2\theta\, d\varphi^2). \qquad (40.1)$$

In a static field the functions ν and λ depend only on r. For the components of the four-dimensional velocity we have $u^1 = u^2 = u^3 = 0$, $u^0 = e^{-\nu}$, and accordingly we obtain from (6.6) for the nonvanishing mixed components of the energy-momentum tensor:

$$T_0^0 = \rho, \quad T_1^1 = T_2^2 = T_3^3 = -P. \qquad (40.2)$$

The notations used are $(ct, r, \theta, \varphi) \rightarrow (x^0, x^1, x^2, x^3)$. It should be remembered that ρ and P are the intrinsic energy density and the pressure. The corresponding Einstein equations can be copied directly from (9.3), omitting the derivatives with respect to time:

$$e^{-\lambda}\left(\frac{\nu'}{r} + \frac{1}{r^2}\right) - \frac{1}{r^2} = \frac{8\pi k}{c^4} P, \qquad (40.3)$$

$$e^{-\lambda}\left(\frac{\lambda'}{r} - \frac{1}{r^2}\right) + \frac{1}{r^2} = \frac{8\pi k}{c^4} \rho, \qquad (40.4)$$

$$P' = -\frac{P + \rho}{2} \nu'. \qquad (40.5)$$

The last equation was obtained from (6.7), and it expresses the hydrodynamic equilibrium of mass in curved space. These three equations, together with the equation of state, determine the dependence of the functions g_{ik} on r and thereby the distribution of matter in the configuration.

In empty space $P=\rho=0$ and we have the familiar Schwarzschild solution

$$e^{\nu}=e^{-\lambda}=1-\frac{2kM}{c^2r}, \tag{40.6}$$

where M is the star's mass.

Inside the mass distribution

$$e^{\nu(r)}=e^{\nu(R)}\cdot\exp\left[-\int\limits_0^{P(r)}\frac{2dP}{P+\rho}\right], \tag{40.7}$$

where R is the star's radius. Equation (40.7) is obtained by integrating (40.5) with the help of the equation of state $\rho=\rho(P)$. Thus the function $\nu(r)$ is known if we know the dependence of P on r.

Let us introduce a new variable through the relationship

$$e^{-\lambda}=1-\frac{2ku(r)}{c^2r}, \tag{40.8}$$

whereupon equation (40.4) becomes

$$\frac{du}{dr}=\frac{4\pi}{c^2}r^2\rho(r). \tag{40.9}$$

Now, using (40.5) and (40.8), we can write equation (40.3) in the form

$$\frac{dP}{dr}=-\frac{k}{c^2}\cdot\frac{P+\rho}{r\left(r-\frac{2ku}{c^2}\right)}\cdot\left(u+\frac{4\pi}{c^2}r^3P\right). \tag{40.10}$$

Equations (40.9) and (40.10), together with the equation of state $\rho(P)$ and the boundary conditions $u(0)$ and $P(0)$ at $r=0$, form a closed system. The integration is carried out from the center to a distance $r=R$ where $P(R)=0$. The value of the function $u(r)$ at $r=R$ determines $\lambda(R)$ at the surface of the celestial body, and it must match the exterior solution (40.6):

$$u(R)=\frac{c^2R}{2k}\left[1-e^{-\lambda(R)}\right]=M.$$

Thus the mass of a spherical celestial body, as measured by an observer located at an infinitely large distance from it, coincides with the value of the function $u(R)$ at the surface. As to the meaning of

$u(r)$ for $r<R$, equation (40.9) indicates that to a certain degree it characterizes the mass contained inside a sphere of radius r. However, under the conditions at hand, when the gravitation field is very strong and the gravitational mass defect may be large, the concept of the mass contained in a sphere with radius r requires more accurate definition. Here we shall not go into this matter, since we consider the definition (40.8) to be sufficient for our needs.

Here the quantity R is the coordinate radius of the celestial body, and it can be quite different from the true radius. The true radius R_0 of a star is related to its coordinate radius through

$$R_0 = \int_0^R \sqrt{g_{rr}}\, dr = \int_0^R e^{\lambda/2} dr. \tag{40.11}$$

Going over to the dimensionless coordinates $c=k=1$, $K_n = \frac{1}{4\pi}$, we rewrite the basic equations and group them together, obtaining

$$\left.\begin{aligned} u' &= 4\pi r^2 \rho(r), \\ P' &= -\frac{P+\rho}{r(r-2u)}(u + 4\pi r^3 P), \\ \rho &= \rho(P). \end{aligned}\right\} \tag{40.12}$$

Here we also added the equation of state. The boundary conditions are evident:

$$u(0)=0, \qquad P(0), \qquad P(R)=0. \tag{40.13}$$

The condition $u(0)=0$ again indicates the absence of a point mass at the center of the configuration.

Solving the system (40.12) with the boundary conditions (40.13), we find the interior metric of the space, and the mass distribution, mass, and radius of the star. As we can see, the basic parameter characterizing the configuration of degenerate gaseous bodies is the central pressure $P(0)$.

In a weak gravitational field, when

$$2u \ll r, \qquad 4\pi r^2 \ll u, \qquad P \ll \rho, \tag{40.14}$$

equations (40.9) and (40.10) become equations (35.4) of the nonrelativistic theory of gravitation. When the first two conditions are satisfied, the last one is automatically satisfied too. The inverse statement is true only when the mass is not much larger than the solar mass. In principle it is conceivable that there are rarefied celestial bodies in which $P \ll \rho$, whose radius R is nevertheless of the order of the gravitational radius $r_g = 2M$. This, however, is possible in the case of those celestial bodies, whose mass is many orders of magnitude above the sun's. According to relativistic astrophysics, the rarefaction criterion of a configuration is not so much the value of the density itself, as the ratio between its radii r_g and R. Henceforth we shall apply the term superdense to bodies whose r_g is not much different from their R. For the extreme case of the densest celestial body with $R \approx r_g = 2M$, we have as a rough estimate $\bar{\rho} \approx \left(\frac{M}{4R^3}\right) \approx \frac{1}{32M^2}$. From this we obtain for the average density of a celestial body whose radius is not much different from the gravitational radius r_g:

$$\bar{\rho} \approx 2 \cdot 10^{16} \cdot \left(\frac{M_\odot}{M}\right)^2. \tag{40.15}$$

Thus, for example, for a body with $M = 10^{11} M_\odot$ (the Galaxy's mass is of this order) we find that $\bar{\rho} \approx 2 \cdot 10^{-6}$ g·cm^{-3}, and for a star with $M = M_\odot$ the density is extremely high.

41. NEUTRON CONFIGURATIONS IN THE RELATIVISTIC THEORY

Calculations based on Einstein's equations were carried out in 1939 by Oppenheimer and Volkoff /243/ for stellar configurations consisting of a degenerate ideal neutron gas. In the above, manual numerical calculations were performed for only five configurations (for the parameter values $t_n(0) = 1, 2, 3, 4, \infty$). These calculations were later repeated in /18, 90/ with higher accuracy for many values of $t_n(0)$. The higher accuracy is necessary in calculating the anomalous mass defect (see Chapter VIII).

The equation of state of an ideal degenerate neutron gas is given by equation (33.7). Eliminating P and ρ from (40.12) we obtain

$$\left.\begin{aligned}
&\frac{du}{dr} = r^2 (\operatorname{sh} t_n - t_n),\\
&\frac{dt_n}{dr} = -\frac{4\left(\operatorname{sh} t_n - 2\operatorname{sh}\frac{t_n}{2}\right)}{r(r-2u)\left(\operatorname{ch} t_n - 4\operatorname{ch}\frac{t_n}{2} + 3\right)} \times\\
&\times\left[u + \frac{r^3}{3}\left(\operatorname{sh} t_n - 8\operatorname{sh}\frac{t_n}{2} + 3t_n\right)\right].
\end{aligned}\right\} \tag{41.1}$$

TABLE 8. The most important parameters of configurations consisting of a degenerate ideal neutron gas

	According to Einstein's gravitation theory				According to Newton's gravitation theory				
$t_n(0)$	R	M	$n\cdot 10^{-56}$	$\frac{\Delta M}{M_0}$	R	$u(R)$	M	$n\cdot 10^{-56}$	$\frac{\Delta M}{M_0}$
1	1.5133	0.0324595	3.60178	0.008249	1.55668	0.0354367	0.0347413	3.86171	0.009977
2	0.9583	0.0658900	7.44366	0.025880	1.05537	0.0901267	0.0833362	9.53554	0.038240
3	0.6696	0.0766367	8.74018	0.035075	0.805802	0.1391840	0.1171330	14.0320	0.081375
4	0.5074	0.0710653	8.02574	0.025570	0.640732	0.1694480	0.1259460	16.0083	0.134190
5	0.4060	0.0598619	6.56506	—0.003436	0.521618	0.1778630	0.1142420	15.5392	0.190940
6	0.3640	0.0492384	5.18517	—0.045005	0.434748	0.1689490	0.0928140	13.5404	0.245670
7	0.3670	0.0419996	4.26480	—0.083740	0.374142	0.1503140	0.0709330	11.0402	0.292940
8	0.4130	0.0396985	3.97905	—0.097925	0.336988	0.1290290	0.0535240	8.75917	0.327540
10	0.4810	0.0458412	4.72655	—0.067306	0.333320	0.0970354	0.0362572	6.09621	0.345490
11	—	—	—	—	0.365932	0.0918297	0.0371440	5.92353	0.309930
12	0.4530	0.0473537	4.91175	—0.06095	0.404375	0.0948718	0.0424497	6.46310	0.277200
∞	0.4506	0.0458743	4.72432	—0.06858	0.362325	0.1092030	0.0435033	7.43305	0.355920

Note. $t_n(0) = 4\,\mathrm{arsh}\left[(3\pi^2)^{1/3} h n_n^{1/3}(0)/m_n c\right]$, where $n(0)$ is the neutron density at the center, R is the star's radius, M is the observable mass, $u(R)$ is the mass without the Newtonian gravitation, n is the number of neutrons in the star, and $\Delta M/M_0$ is the gravitational packing fraction ($\Delta M = M_0 - M$, $M_0 = n m_n$). The dimensioned quantities are given in the units $c=k=1$, $K_n = 1/4\pi$. To express the mass in units of the solar mass and the radius in kilometers, one has to multiply the values given in the table by 9.29 and 13.7, respectively.

The boundary conditions for the above are

$$u(0) = 0, \quad t_n(0), \quad t_n(R) = 0.$$

The first and last conditions are common to all configurations, therefore they are characterized by one parameter only:

$$t_n(0) = 4\,\text{arsh}\left[(3\pi^2)^{1/3} h n_n^{1/3}(0)/m_n c\right],$$

where $n_n(0)$ is the number density of neutrons at the center.

The results of numerical calculations are given in Table 8. As seen from the tabulated data, the mass of neutron configurations lies in the interval $0.3 \lesssim M \lesssim 0.72\,M_\odot$, and the radius is in the interval $6.2 < R < 22$ km. Thus the masses of neutron configurations are of the order of the solar mass, and the radii are of the order of 10 km; for comparison, let us note that the gravitational radius of the sun is 2.95 km. Higher mass values are not obtained for any values of the parameter $t_n(0)$ (the central pressure $P(0)$).

In Figure 14 (page 137) is shown the dependence of the mass on the parameter arctan $t_n(0)$ according to Einstein's (lower curve) and Newton's gravitation theories. In spite of the substantial quantitative difference between the results obtained from the two theories, there is nevertheless a good qualitative agreement in some of their most important aspects. Thus, the Newtonian theory yields the correct order of magnitude for the mass, radius, and number of baryons in the star.

Moreover, there is a strong similarity in the curves obtained for these parameters. In both theories all configuration parameters are single-valued functions of the central pressure (or its replacing parameter $t_n(0)$). The inverse statement is generally not true. Two and even more values of the central density correspond to a single value of some of the parameters. According to Einstein's gravitation theory, the curves describing the stellar parameters as a function of arctan $t_n(0) = y$ have a series of maxima and minima. On the lower curve of Figure 14 one can clearly discern two maxima and one minimum at $t_n(0) = 3.34$, 12.35, and 8.24 ($n_n(0) = 3.10^{39}$, $4.84 \cdot 10^{42}$, and $2.1 \cdot 10^{31}$ cm^{-3}), respectively. However, after the last maximum at $x = 1.49$ there is a great number of extrema, which with the scale chosen cannot be seen in the figure. At $x > 1.5$ the curve $M(P_0)$, while oscillating (with a strongly attenuating amplitude), approaches a definite limit when the density at the center approaches infinity. These oscillations occurring in the curves for mass and number of particles at relativistic densities were investigated in /41/.

In analogy with the nonrelativistic case (see Section 38), an asymptotic analytical solution was found for the case $t_n(0)=\infty$ (i. e. $\rho(0)=\infty$); this solution is valid for the central region of the extreme configuration, where the neutron gas is highly relativistic /243/. This analytical solution was then extended up to the star's surface through numerical integration. This solution will be further discussed in the following section.

42. CONFIGURATIONS OF A DEGENERATE IDEAL BARYON GAS

By omitting in (33.6) the terms which are related to nuclear interactions, one can obtain the equations of state of a degenerate ideal baryon gas. These equations are

$$\left.\begin{aligned} \rho &= \frac{1}{4\pi}\sum_k \left(\frac{m_k}{m_n}\right)^4 (\operatorname{sh} t_k - t_k), \\ P &= \frac{1}{12\pi}\sum_k \left(\frac{m_k}{m_n}\right)^4 \left(\operatorname{sh} t_k - 8\operatorname{sh}\frac{t_k}{2} + 3t_k\right). \end{aligned}\right\} \tag{42.1}$$

The parameters t_k were defined in (32.1). If the plasma also contains π^- mesons (and we already know from Section 27 that the presence or absence of these particles is determined by the choice of baryon composition), then one has to add the term $n_\pi m_\pi c^2/4\pi K_n$ to the expression for the energy density (here the units of the quantities in the numerator and in the denominator are of no consequence). However, even when π^- mesons can be present, their contribution to the energy is not significant. From the condition of neutrality it follows that at best the concentration of these particles has to be of the order of the concentration of each type of baryon. On the other hand, the π^- mesons are in a state of rest, while the baryons are relativistic. Considering also the large number of baryon varieties, we arrive at the conclusion that the partial energy of the pions is in actual fact small compared to that of the baryons. As to the leptons, their contribution to the energy of the system is even smaller. We already know that all t_k parameters are functions of the baryon density n or the energy ρ. For numerical calculations, though, it is more convenient to choose t_n as the independent variable and regard all others as dependent on it: $t_k(t_n)$. Thus, in analogy with the neutron configuration, we should use the following boundary conditions:

$$u(0)=0, \quad t_n(0), \quad t_n(R)=0.$$

Therefore, in this case too, the configurations will be characterized by a single parameter, namely $t_n(0)$, which is equivalent to specifying the pressure at the center.

Of special interest is the mass of the interior sphere in which there are hyperons *(hyperon core)*. In the case of an ideal gas the hyperons are in the central region where $t_n(r)>2.1$. Some measure of the mass of this region is provided by the quantity $u(r_y)$ (where r_y is the radius of the hyperon nucleus: $t_n(r_y)=2.1$), which we shall refer to below as the mass of the hyperon core.

The main results of the numerical calculations are assembled in Table 9. It contains the baryon and mass densities at the center, and the masses and radii of the stars and their hyperon cores. The first five models are neutron configurations and have no hyperon core.

We shall examine in somewhat more detail the integration procedure for the extreme configuration with an infinitely high central density. At high densities, when the particles of the baryon gas become relativistic, the equation of state $\rho\approx 3P$ becomes valid. We shall write this as

$$P=Z\rho \tag{42.2}$$

so as to be able to apply the result obtained below to a different case, where $Z\neq 1/3$.

Inside the central region, where the relativistic equation of state (42.2) applies, we seek a solution to equations (40.12) in the form

$$4\pi\rho(r)=ar^{-n}. \tag{42.3}$$

Substituting (42.2) and (42.3) in equations (40.12), we immediately obtain

$$\left.\begin{aligned} a&=\frac{2Z}{4Z+(Z+1)^2}, & n&=2,\\ 4\pi\rho(r)&=a\cdot r^{-2}, & u(r)&=a\cdot r.\end{aligned}\right\} \tag{42.4}$$

In particular, for an ideal gas, where $Z=1/3$, we have $a=3/14$, and thus

$$4\pi\rho(r)=\frac{3}{14}\cdot\frac{1}{r^2}, \qquad u(r)=\frac{3}{14}r. \tag{42.5}$$

The solution (42.5) was considered to be valid up to a distance $r=6.9\cdot 10^{-4}$, where $\rho=3.58\cdot 10^4$, $P=1.14\cdot 10^4$, and $u=1.48\cdot 10^{-4}$ (all

TABLE 9. Some of the more important parameters of superdense stellar configurations consisting of an ideal baryon gas

$t_n(0)$	Density at the center		Mass				Coordinate radius			
	of baryons $n(0)$, cm^{-3}	of mass $\rho(0)$, $g \cdot cm^{-3}$	of star		of hyperon core		of star		of hyperon core	
			in units of (34.1)	in units of $M_\odot$	in units of (34.1)	in units of $M_\odot$	in units of (34.1)	in km	in units of (34.1)	in km
0.556	$1.00 \cdot 10^{37}$	$1.62 \cdot 10^{13}$	0.0147	0.139			2.090	28.70		
1.0	$6.50 \cdot 10^{37}$	$1.00 \cdot 10^{14}$	0.0329	0.306			1.537	21.10		
1.3	$1.36 \cdot 10^{38}$	$2.24 \cdot 10^{14}$	0.0443	0.411			1.345	18.40		
1.5	$2.20 \cdot 10^{38}$	$3.62 \cdot 10^{14}$	0.0495	0.460	—	—	1.170	16.00		
1.8	$3.85 \cdot 10^{38}$	$6.60 \cdot 10^{14}$	0.0600	0.557	—	—	1.065	14.60		
2.4	$1.25 \cdot 10^{39}$	$2.34 \cdot 10^{15}$	0.0683	0.634	0.0132	0.123	0.805	11.00	0.245	3.36
2.75	$3.10 \cdot 10^{39}$	$5.92 \cdot 10^{15}$	0.0622	0.578	0.0234	0.217	0.753	10.30	0.270	3.20
3.0	$5.45 \cdot 10^{39}$	$1.09 \cdot 10^{16}$	0.0559	0.519	0.0255	0.237	0.703	9.63	0.254	3.48
3.3	$1.08 \cdot 10^{40}$	$3.62 \cdot 10^{16}$	0.0484	0.450	0.0258	0.240	0.670	9.18	0.232	3.18
4.0	$3.50 \cdot 10^{40}$	$8.29 \cdot 10^{16}$	0.0354	0.329	0.0225	0.209	0.594	8.14	0.185	2.54
5.0	$1.26 \cdot 10^{41}$	$3.45 \cdot 10^{17}$	0.0245	0.228	0.0173	0.161	0.506	6.93	0.137	1.88
6.0	$3.53 \cdot 10^{41}$	$1.15 \cdot 10^{18}$	0.0191	0.177	0.0135	0.125	0.557	7.63	0.115	1.57
7.0	$9.36 \cdot 10^{41}$	$1.39 \cdot 10^{18}$	0.0237	0.220	0.0115	0.107	0.767	10.50	0.112	1.53
∞	∞	∞	0.0349	0.324	0.0130	0.121	0.808	11.10	0.145	1.99

values are in the units of (34.1)). These values were taken as the initial conditions, and the system (40.12) was then integrated numerically up to the surface, i. e. to the distance where $P(R)=0$. In the course of the numerical integration it was found that the asymptotic solution (42.5) remains adequate even at values of r much above the initial value. Thus, for example, at a distance $r=5.5\cdot10^{-3}$ the value of $u(r)$ calculated from (42.5) differs by only 1% from its exact value. The final results obtained for the mass and radius of the extreme configuration are $M=0.324M_{\odot}$ and $R=11.1$ km. For comparison let us remember that the extreme configuration which consists of an ideal neutron gas has a mass $M=0.43M_{\odot}$ and a radius $R=6.2$ km.

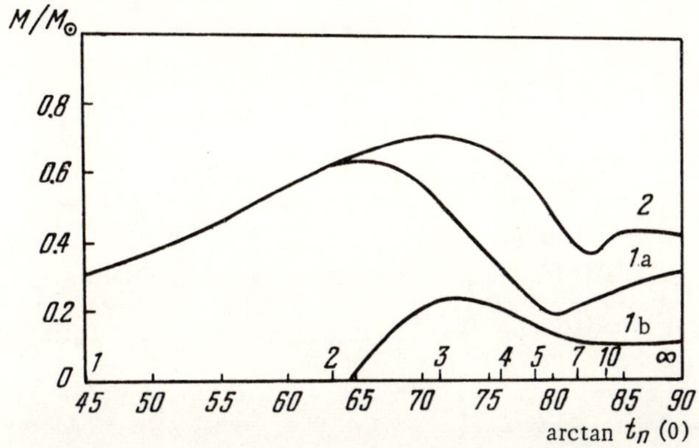

FIGURE 18. The mass of baryon stars (curve 1a) and of their hyperon cores (curve 1b) as a function of the parameter arctan $t_n(0)$ for the case of an ideal baryon gas. Curve 2 describes the mass of neutron configurations (ideal gas).

Figure 18 shows the dependence of the masses of the configurations and their hyperon cores on the parameter arctan $t_n(0)$. Curve 2 pertains to neutron configurations. We note that for configurations with $3 < t_n(0) < 7$ a considerable part of the mass is concentrated in the hyperon core. At $t_n(0)\approx6$ ($n\approx3.5\cdot10^{41}$ cm^{-3}) the mass has a pronounced minimum. In the region $t_n(0)>6$ the mass has a large number of weak, gradually fading maxima and minima (with the scale of Figure 18 these cannot be seen); i. e. with increasing central density the ordinate of the $M(\rho_0)$ curve oscillates with an attenuating amplitude and approaches a finite limit, which is equal to $0.324M_{\odot}$. As we can see, the function $M(\rho_0)$ is single-valued, i. e. to each value of ρ_0 corresponds a single value of the mass, but the inverse function $\rho_0=\rho_0(M)$ is multiple-valued at $t_n(0)\gtrsim3$ (we shall return to this question in Section 43). The dependence of the radius on arctan $t_n(0)$ is shown in Figure 19.

From Figure 18 it is seen that in the density range where there are hyperons, the mass of baryon configurations is appreciably smaller than that of pure neutron configurations. This is due to the fact that the mass of a configuration is approximately inversely proportional to the square of the mass of its constituent elementary particles. To prove this, let us consider some hypothetical configuration consisting of an ideal gas of fermions with rest mass m_k, and let us compare it with a neutron configuration. Choosing the units $c=k=1$, $m_k^4c^5/(32\pi^2h^3)=\frac{1}{4\pi}$ for the one case and $c=k=1$, $K_n=\frac{m_n^4c^5}{32\pi^2h^3}=\frac{1}{4\pi}$ for the second case, we obtain for both cases the same equation of state

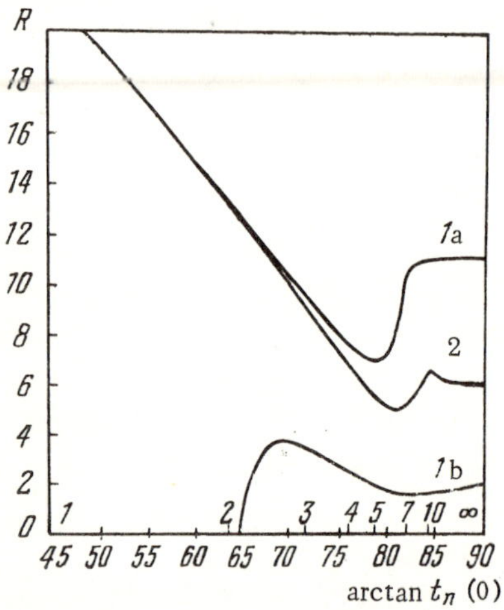

FIGURE 19. The radius of baryon stars (curve 1a) and of their hyperon cores (curve 1b) as a function of the parameter arctan $t_n(0)$ for an ideal gas. Curve 2 describes the radius of neutron configurations (ideal gas). The radius is given in kilometers.

$$\rho=\frac{1}{4\pi}\left(\operatorname{sh}t-t\right),$$

$$P=\frac{1}{12\pi}\left(\operatorname{sh}t-8\operatorname{sh}\frac{t}{2}+3t\right).$$

For the same value of the parameter $t(0)$, we obviously obtain the same mass in both cases. According to (34.6), however, the mass unit is inversely proportional to m_k^2. Therefore, if we now return to the common system of units

$$c=k=1,\ K_n=\frac{1}{4\pi},$$

then it is easily seen that the mass of the hypothetical configuration is multiplied by $(m_n/m_k)^2$. When degenerate matter consists of different particles, the configuration parameters are naturally determined by the average rest mass $\overline{m}_k$, which in the case of baryon configurations is larger than the neutron mass m_n. It is precisely for this reason that, given identical values of the parameter $t_n(0)$, the mass of a baryon star is smaller than that of a neutron star.

From this it also follows that, had the neutrino rest mass been different from zero, then in principle there could have existed configurations consisting of a degenerate neutrino gas and with a mass larger by a factor of $(m_n/m_\nu)^2$ than that of corresponding neutron configurations /236a/.

43. CONFIGURATIONS OF A REAL BARYON GAS

In / 96, 14, 92 / several models of a real baryon gas were studied, and calculations were made for numerous models of "cold" stars. The aim of these studies was to examine the behavior of baryon-configuration parameters through approximations to the nuclear attraction forces in the nuclear density range and the repulsion forces in the supernuclear range. It was shown that when the contribution of nuclear interactions between particles is included in the equation of state, the configuration parameters undergo considerable changes. Thus, for example, it emerged that at densities where an important role is played by nuclear interactions, the mass is larger by a factor of two to three than that of corresponding neutron configurations (the term corresponding here refers to configurations with the same central pressure). On the other hand, it was also established that no reasonable change in the equation of state of a baryon plasma can possibly result in mass values which are appreciably higher than the solar mass. The order of magnitude of the mass does not change, but remains the same as that of an ideal gas, i.e. of the order of the solar mass.

Here we shall present only the results of the calculations, which are based on the equation of state (33.4)—(33.6). In our opinion, these equations incorporate the nuclear forces better than all other equations existing in the literature; moreover, they have the additional advantage of agreeing with the law of causality (i.e. when $\rho \to \infty$ the sound velocity does not exceed the velocity of light).

Table 10 shows the results of numerical integration of the system (40.12) with the equation of state (33.4)—(33.6) and the boundary conditions $u(0)=0$, $P(0), P(R)=0$, and $u(R)=M$.

The first three columns give the mass density, pressure, and number density of baryons at the center of the configurations. These quantities are interrelated through the equation of state. The mass is expressed in solar units, and the radius in kilometers.

The table includes, besides the total mass and the radius of the whole star, also the mass and radius of the hyperon core, nucleon-nuclear phase, and envelope. The term radius of hyperon core here

TABLE 10. Basic parameters of baryon configurations /18/

Parameters at the configuration center				Mass $M/M_\odot$				Radius in kilometers				Total number of baryons in star $\cdot 10^{-57}$
η	mass density $\rho(0)$, $g \cdot cm^{-3}$	pressure $P(0)$, $erg \cdot cm^{-3}$	baryon density $n(0)$, $g \cdot cm^{-3}$	hyperon core	nucleon-nuclear phase	envelope	whole star	hyperon core	nucleon-nuclear phase	coordinate radius of star	true radius of star	
−1.30	$9.06 \cdot 10^{10}$	$1.48 \cdot 10^{29}$	$5.45 \cdot 10^{34}$	—	—	1.07	1.07	—	—	765	766	1.27
−1.15	$2.11 \cdot 10^{12}$	$1.66 \cdot 10^{30}$	$1.24 \cdot 10^{36}$	—	0.0232	0.928	0.951	—	26.4	660	661	1.13
−1.00	$1.01 \cdot 10^{13}$	$1.59 \cdot 10^{31}$	$6.00 \cdot 10^{36}$	—	0.0531	0.859	0.912	—	26.0	886	883	1.08
−0.93	$1.65 \cdot 10^{13}$	$3.48 \cdot 10^{31}$	$9.90 \cdot 10^{36}$	—	0.0647	0.837	0.902	—	25.2	1090	1094	1.07
−0.85	$2.63 \cdot 10^{13}$	$7.16 \cdot 10^{31}$	$1.58 \cdot 10^{37}$	—	0.0759	0.823	0.899	—	24.1	1410	1415	1.07
−0.75	$4.23 \cdot 10^{13}$	$1.49 \cdot 10^{32}$	$2.52 \cdot 10^{37}$	—	0.0899	0.803	0.893	—	22.9	2250	2260	1.06
−0.70	$5.21 \cdot 10^{13}$	$2.04 \cdot 10^{32}$	$3.75 \cdot 10^{37}$	—	0.0974	0.746	0.843	—	22.7	2800	2810	1.00
−0.67	$5.97 \cdot 10^{13}$	$2.56 \cdot 10^{32}$	$3.56 \cdot 10^{37}$	—	0.101	0.632	0.733	—	21.5	3840	3850	0.870
−0.62	$7.10 \cdot 10^{13}$	$3.96 \cdot 10^{32}$	$4.30 \cdot 10^{37}$	—	0.108	0.128	0.236	—	20.6	4500	4510	0.260
−0.60	$7.50 \cdot 10^{13}$	$4.28 \cdot 10^{32}$	$4.55 \cdot 10^{37}$	—	0.110	0.0257	0.136	—	20.6	1350	1360	0.181
−0.59	$7.90 \cdot 10^{13}$	$4.65 \cdot 10^{32}$	$4.78 \cdot 10^{37}$	—	0.112	0.0146	0.127	—	20.6	576	578	0.151
−0.57	$8.27 \cdot 10^{13}$	$4.98 \cdot 10^{32}$	$5.00 \cdot 10^{37}$	—	0.114	0.0110	0.125	—	20.6	364	365	0.148
−0.56	$8.69 \cdot 10^{13}$	10^{32}	$5.25 \cdot 10^{37}$	—	0.116	$8.99 \cdot 10^{-3}$	0.125	—	20.6	263	264	0.149
−0.40	$1.37 \cdot 10^{14}$	$8.85 \cdot 10^{32}$	$8.17 \cdot 10^{37}$	—	0.132	$3.30 \cdot 10^{-3}$	0.136	—	17.0	66	66.3	0.161
−0.30	$1.78 \cdot 10^{14}$	$1.40 \cdot 10^{33}$	$1.04 \cdot 10^{38}$	—	0.145	$1.32 \cdot 10^{-3}$	0.147	—	14.1	42.9	43.1	0.174
−0.11	$2.84 \cdot 10^{14}$	$5.11 \cdot 10^{33}$	$1.74 \cdot 10^{38}$	—	0.258	$2.69 \cdot 10^{-4}$	0.259	—	12.1	17.7	18.1	0.317
0.18	$5.51 \cdot 10^{14}$	$2.56 \cdot 10^{34}$	$3.23 \cdot 10^{38}$	0.314	0.325	$8.92 \cdot 10^{-5}$	0.639	7.12	11.6	13.0	13.8	0.813
0.37	$8.78 \cdot 10^{14}$	$7.16 \cdot 10^{34}$	$4.90 \cdot 10^{38}$	0.737	0.247	$3.44 \cdot 10^{-5}$	0.984	8.68	11.4	12.1	13.2	1.28
0.46	$1.14 \cdot 10^{15}$	$1.28 \cdot 10^{35}$	$6.27 \cdot 10^{38}$	0.998	0.186	$2.51 \cdot 10^{-5}$	1.18	9.11	11.1	11.8	13.2	1.55
0.58	$1.65 \cdot 10^{15}$	$2.56 \cdot 10^{35}$	$8.64 \cdot 10^{38}$	1.26	0.131	$1.40 \cdot 10^{-5}$	1.39	9.18	10.6	10.9	11.9	1.85
0.69	$2.44 \cdot 10^{15}$	$5.11 \cdot 10^{35}$	$1.17 \cdot 10^{39}$	1.42	0.0892	$8.81 \cdot 10^{-6}$	1.51	8.88	9.92	10.2	12.1	2.03
0.78	$3.58 \cdot 10^{15}$	$1.06 \cdot 10^{36}$	$1.64 \cdot 10^{39}$	1.49	0.0576	$5.46 \cdot 10^{-6}$	1.55	8.31	9.11	9.28	11.5	2.08

0.79	$3.71\cdot10^{15}$	$1.13\cdot10^{36}$	$1.68\cdot10^{39}$	1.49	0.0557	$5.20\cdot10^{-6}$	1.55	8.25	9.03	9.19	11.5	2.08
0.82	$4.26\cdot10^{15}$	$1.39\cdot10^{36}$	$1.83\cdot10^{39}$	1.50	0.0483	$4.46\cdot10^{-6}$	1.55	8.07	8.76	8.92	11.7	2.09
0.87	$5.48\cdot10^{15}$	$2.05\cdot10^{36}$	$2.20\cdot10^{39}$	1.50	0.0381	$3.56\cdot10^{-6}$	1.54	7.72	8.36	8.49	10.9	2.07
0.98	$1.16\cdot10^{16}$	$6.14\cdot10^{36}$	$3.60\cdot10^{39}$	1.37	0.0242	$2.13\cdot10^{-6}$	1.39	6.78	7.27	7.37	10.5	1.81
1.03	$1.69\cdot10^{16}$	$1.02\cdot10^{37}$	$4.50\cdot10^{39}$	1.29	0.0204	$1.83\cdot10^{-6}$	1.31	6.40	6.87	6.97	9.70	1.66
1.11	$3.53\cdot10^{16}$	$2.56\cdot10^{37}$	$6.95\cdot10^{39}$	1.15	0.0177	$1.63\cdot10^{-6}$	1.17	5.92	6.40	6.51	9.33	1.41
1.13	$4.70\cdot10^{16}$	$3.58\cdot10^{37}$	$8.10\cdot10^{39}$	1.11	0.0177	$1.63\cdot10^{-6}$	1.13	5.81	6.30	6.41	9.45	1.33
1.16	$6.80\cdot10^{16}$	$5.42\cdot10^{37}$	$9.90\cdot10^{39}$	1.06	0.0186	$1.73\cdot10^{-6}$	1.08	5.74	6.27	6.38	9.28	1.26
1.17	$8.55\cdot10^{16}$	$7.01\cdot10^{37}$	$1.12\cdot10^{40}$	1.04	0.0186	$1.81\cdot10^{-6}$	1.06	5.72	6.27	6.39	9.31	1.23
1.19	$1.21\cdot10^{17}$	$1.02\cdot10^{38}$	$1.36\cdot10^{40}$	1.02	0.0195	$1.93\cdot10^{-6}$	1.04	5.74	6.30	6.43	9.39	1.19
1.21	$1.65\cdot10^{17}$	$1.43\cdot10^{38}$	$1.58\cdot10^{40}$	1.01	0.0204	$2.13\cdot10^{-6}$	1.03	5.78	6.40	6.54	9.55	1.17
1.23	$2.27\cdot10^{17}$	$2.05\cdot10^{38}$	$1.85\cdot10^{40}$	1.00	0.0223	$2.27\cdot10^{-6}$	1.03	5.84	6.48	6.63	9.67	1.17
1.24	$2.80\cdot10^{17}$	$2.56\cdot10^{38}$	$2.06\cdot10^{40}$	1.01	0.0232	$2.36\cdot10^{-6}$	1.03	5.89	6.53	6.69	9.75	1.17
1.27	$5.47\cdot10^{17}$	$5.11\cdot10^{38}$	$2.86\cdot10^{40}$	1.02	0.0260	$2.60\cdot10^{-6}$	1.05	6.04	6.71	6.87	10.0	1.20
1.29	$1.09\cdot10^{18}$	$1.02\cdot10^{38}$	$3.98\cdot10^{40}$	1.04	0.0269	$2.65\cdot10^{-6}$	1.07	6.15	6.82	6.99	10.2	1.24
1.30	$1.61\cdot10^{18}$	$1.53\cdot10^{39}$	$4.85\cdot10^{40}$	1.06	0.0269	$2.69\cdot10^{-6}$	1.08	6.18	6.85	7.00	10.2	1.26
1.34	$5.35\cdot10^{18}$	$5.11\cdot10^{39}$	$8.70\cdot10^{40}$	1.09	0.0269	$2.59\cdot10^{-6}$	1.10	6.18	6.84	6.98	10.3	1.29
1.37	$2.70\cdot10^{19}$	$2.56\cdot10^{39}$	$1.91\cdot10^{41}$	1.07	0.0251	$2.49\cdot10^{-6}$	1.10	6.13	6.77	6.92	10.2	1.28
1.38	$5.41\cdot10^{19}$	$5.11\cdot10^{40}$	$2.67\cdot10^{41}$	1.07	0.0232	$2.58\cdot10^{-6}$	1.09	6.12	6.81	6.96	10.3	1.28
1.39	$1.09\cdot10^{20}$	$1.02\cdot10^{41}$	$3.76\cdot10^{41}$	1.06	0.0251	$2.52\cdot10^{-6}$	1.09	6.12	6.77	6.92	10.2	1.27
1.40	$2.70\cdot10^{20}$	$2.56\cdot10^{41}$	$5.90\cdot10^{41}$	1.07	0.0251	$2.52\cdot10^{-6}$	1.09	6.13	6.78	6.93	10.2	1.27
1.41	$5.48\cdot10^{20}$	$5.11\cdot10^{41}$	$8.30\cdot10^{41}$	1.06	0.0251	$2.54\cdot10^{-6}$	1.09	6.12	6.79	6.93	10.2	1.27
1.43	$5.68\cdot10^{21}$	$5.11\cdot10^{42}$	$2.52\cdot10^{42}$	1.06	0.0260	$2.54\cdot10^{-6}$	1.09	6.13	6.79	6.94	10.2	1.27
1.44	$2.84\cdot10^{22}$	$2.56\cdot10^{43}$	$5.64\cdot10^{42}$	1.07	0.0251	$2.53\cdot10^{-6}$	1.09	6.13	6.79	6.93	10.4	1.27
1.45	$5.68\cdot10^{22}$	$5.11\cdot10^{43}$	$7.98\cdot10^{42}$	1.06	0.0260	$2.53\cdot10^{-6}$	1.09	6.12	6.79	6.93	10.2	1.27
1.57	∞	∞	∞	1.06	0.0242	$2.46\cdot10^{-6}$	1.09	6.11	6.74	6.89	9.48	1.26

refers to the distance r_Σ from the star's center at which the last hyperon, namely Σ^-, disappears; the mass of the hyperon core is the quantity $u(r_\Sigma)$. Correspondingly, the radius of the nucleon-nuclear phase is the distance r_n at which the neutrons disappear and the Ae-phase sets in; the mass of the above phase is

$$\Delta u = u(r_n) - u \cdot (r_\Sigma).$$

The term nucleon-nuclear phase here refers to the Aen-phase. Finally, the mass of the envelope is understood to be the mass of that layer of the star where matter consists of atomic nuclei and free electrons (Ae-phase). The calculation of the envelope mass will be discussed in Section 45.

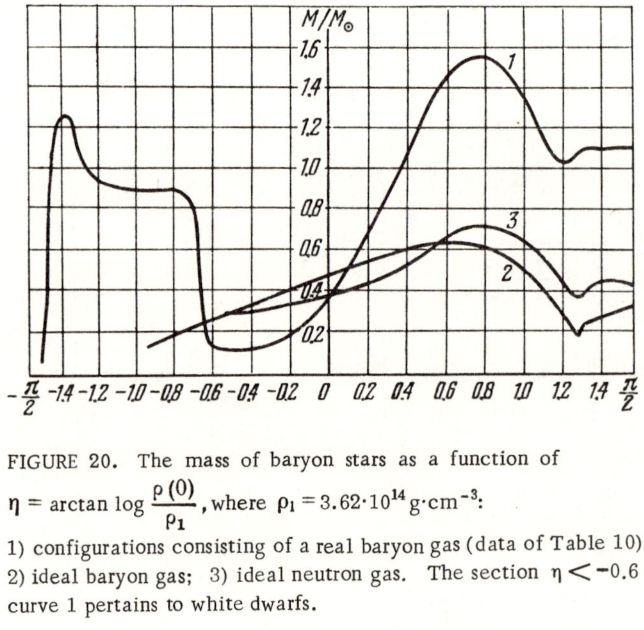

FIGURE 20. The mass of baryon stars as a function of $\eta = \arctan \log \frac{\rho(0)}{\rho_1}$, where $\rho_1 = 3.62 \cdot 10^{14}\,\mathrm{g \cdot cm^{-3}}$: 1) configurations consisting of a real baryon gas (data of Table 10); 2) ideal baryon gas; 3) ideal neutron gas. The section $\eta < -0.6$ of curve 1 pertains to white dwarfs.

From the table it is seen that when the central density is above the hyperon-generation threshold, the main part of the mass is concentrated in the hyperon core, and only an insignificant fraction is in the nucleon-nuclear phase; an even smaller fraction is in the envelope.

The configuration with $\rho(0) = \infty$ was treated like the corresponding ideal-gas configuration. In this case the equation of state has the asymptotic form $P = \rho$. From (42.4) we then have the solution

$$\left.\begin{aligned} 4\pi\rho(r) &= \frac{1}{4}\frac{1}{r^2}, \\ u(r) &= \frac{1}{4}r. \end{aligned}\right\} \qquad (43.1)$$

This solution was considered adequate to a distance $r = 0.47$ km, where $\rho = 1.08 \cdot 10^{38}$ erg·cm^{-3}, $P = 1.02 \cdot 10^{38}$ erg·cm^{-3}, and $u = 0.08\, M_\odot$. These values were then taken as the initial conditions, and equations (40.12) were integrated numerically to the edge of the star by the method used for other configurations.

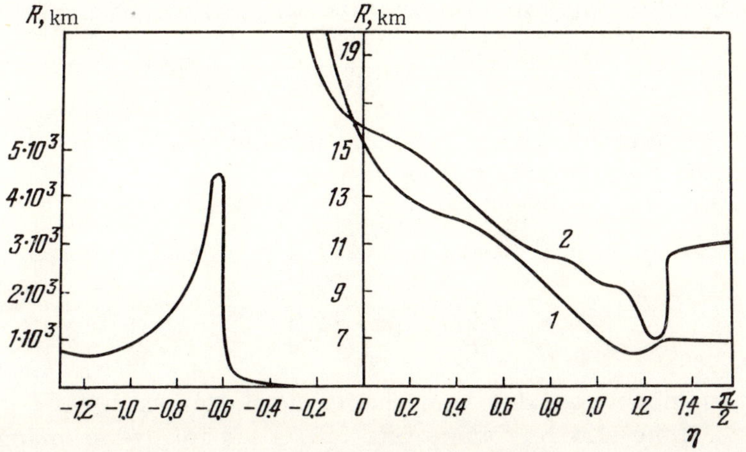

FIGURE 21. The coordinate radius of baryon stars as a function of η. Curve 1 is for a real gas, and curve 2 for an ideal baryon gas.

Figures 20 and 21 show the dependence of mass and radius on the mass density at the center. To obtain convenient curves and cover the whole density range, the variable plotted along the abscissa was

$$\eta = \arctan\left[\log\frac{\rho(0)}{\rho_1}\right],$$

where $\rho_1 = 3.62 \cdot 10^{14}$ g·cm^{-3} is the nuclear density. For comparison, the figures also include the curves $M(\eta)$ and $R(\eta)$ for an ideal baryon gas (line 2) and an ideal neutron gas (line 3). The section $\eta < -0.2$ of the $R(\eta)$ curve was drawn on a smaller scale to fit into the figure. The $M(\eta)$ curve in the section $\eta \lesssim -0.6$ represents white dwarfs.

The curves of the mass and radius as a function of $\rho(0)$ display an irregular behavior.. This is caused by certain singularities of the equation of state in different density ranges. When the density increases from its regular values to infinitely large values, the equation of state undergoes a series of substantial changes. The configuration parameters are, on their part, sensitive to any changes in the equation of state. The validity of this statement can be proven even by comparing the mass and radius curves calculated for an ideal gas with those calculated for a real gas. However, to obtain a better understanding, we think it useful to present the following simplified treatment.

Let us consider an equation of state which is written in the form

$$P = bf(\rho),$$

where b is a constant. We now substitute the above in (40.12) and proceed to the Newtonian limit, obtaining

$$\left.\begin{aligned} \frac{du}{d\rho} &= -4\pi b\frac{r^4 f'(\rho)}{u}, \\ \frac{dr}{d\rho} &= -b\frac{r^2 f'(\rho)}{\rho u}, \end{aligned}\right\} \qquad (43.2)$$

where the unknown variable was chosen as the mass density ρ instead of r. If we now introduce into (43.2) the following changes:

$$\left.\begin{aligned} r(\rho) &= \sqrt{b}\,\chi(\rho), \\ u(\rho) &= \sqrt{b^3}\,v(\rho), \end{aligned}\right\} \qquad (43.3)$$

then the resulting equations will no longer contain the constant b. From (43.3) one can therefore conclude that if the pressure is changed by a factor of b, the radius and mass of the configuration change by a factor of $b^{1/2}$ and $b^{3/2}$, respectively. This schematic examination shows that the configuration parameters are indeed sensitive to any changes in the form of the equation of state.

We shall now try to understand the behavior of the mass curve $M(\eta)$, starting our discussion with the left-hand maximum in the region of white-dwarf configurations (its abscissa is $\eta = -1.38$). Matter at that point consists of atomic nuclei and a degenerate electron gas (Ae-phase). Had the ratio A/Z (A is the mass number and Z the atomic number) remained constant in the region $\eta > -1.38$, then the $M(\eta)$ curve would have continued to rise slowly up to the Chandrasekhar limit $M = 1.4\,M_\odot$. We know, however, that with rising density the ratio A/Z increases, causing the pressure to drop below the value it would have had in the case of a constant A/Z value (≈ 2). A decrease in pressure, according to (43.3), then leads to a decrease in mass. This decline in the $M(\eta)$ curve continues up to the point where the Aen-phase sets in. With the appearance of neutrons, the partial energy density of the nuclei and the partial pressure of the electrons become negligible compared to those of the neutron gas. As a result, the effect of A/Z on the decrease in mass is partially compensated, and the drop in $M(\eta)$ slowed down, until the nuclear forces come into play. In this region (i.e. $\eta < -0.7$) curves 1 and 2 should have coincided, since in it the contribution of nuclear forces is still small and therefore the real- and ideal-gas models should not differ from each other. Figure 20 shows, though, that the mass of an ideal gas (curve 2) is smaller than that of a real gas. The reason for this discrepancy lies in the fact that the calculations for the case of an ideal baryon gas were carried out without taking into consideration the existence of the Aen- and Ae-phases. For the real-gas model, on the other hand, the numerical integration was done in the proper manner up to the star's surface and including all possible phases.

The effect of nuclear attraction forces leads to some decrease in pressure compared to the pressure of an ideal gas. This slowing down of the pressure rise results in a very fast, almost abrupt, change in mass within a narrow central-density interval ($\eta = -0.63$). In the vicinity of this point, the radius also undergoes a significant change, namely a narrow maximum in the $R(\eta)$ curve. As seen from the table, this fast change in the mass and radius is related to the mass and dimensions of the envelope, i.e. the Ae-phase.

With further increase in density toward infinitely high values, the mass curve $M(\eta)$ oscillates with an attenuating amplitude, and when $\rho(0) \to \infty$ it converges to the limit of $1.09 M_{\odot}$. As was shown in /41/, these oscillations in the mass (which also occur in the radius and the total number of baryons) are a result of the specific form $P \approx \text{const} \cdot \rho$ of the equation of state in the central region of the configuration at relativistic densities.

When densities are supernuclear, the forces are repulsive. Therefore for one and the same energy density, the pressure of a real gas is higher than that of an ideal gas (in the limit $\rho \to \infty$ the ratio between the pressures of a real and ideal gas is equal to three). This is the reason why in this region the mass obtained for a real-gas configuration is larger than that obtained for the model of an ideal gas.

44. MODELS FOR CONFIGURATIONS CONSISTING OF AN INCOMPRESSIBLE LIQUID

In our earlier discussions we underlined the important role of the equation of state, namely its effect on the configuration parameters. It was noted at the same time that no reasonable change in these equations can possibly lead to a change in the orders of magnitude of the mass and radius obtained for celestial bodies consisting of degenerate matter. Later we shall see (Chapter X) that the reason for the absence of superdense configurations with mass much above the sun's lies not in the equation of state, but rather in Einstein's equations or, more precisely, in the Newton-Einstein gravitation theory itself. The Einstein equations do not allow for masses which are much larger than the sun's and which at the same time have a radius $R \leqslant 2M$. Nevertheless, in order to fully understand the possibilities permitted by the equation of state, we shall consider the extreme state of matter, as represented by an incompressible liquid. As is a priori clear, this case should be the one to yield large values for the configuration mass, since an incompressible liquid provides the upper limit for the possibilities inherent in the equation of state.

Thus, let us examine the extreme case of a real gas in which the particles are ascribed finite dimensions. When the particles are adjacent to each other, the interaction energy is zero. On the other hand, the particles are completely solid and impenetrable. Let us also assume that baryons of all types have the same proper radius (we should also take into account the Lorentz contraction, if it is sizable). In this kind of gas there should exist some density ρ_m at which the space has the densest packing of particle balls (the balls are then contiguous). It may happen that in a given configuration this maximum density is already reached at some distance R_0 from the center. We then have the following expressions for the density:

$$\left.\begin{aligned}\rho(r) &= \frac{1}{4\pi}\sum_k \left(\frac{m_k}{m_n}\right)^4 (\operatorname{sh} t_k - t_k) \quad &&\text{at} \quad r > R_0,\\ \rho(r) &= \rho_m = \text{const} &&\text{at} \quad r < R_0.\end{aligned}\right\} \tag{44.1}$$

As to pressure, at $r > R_0$ it is determined by the equation

$$P(r) = \frac{1}{12\pi}\sum_k \left(\frac{m_k}{m_n}\right)^4 \left(\operatorname{sh} t_k - 8\operatorname{sh}\frac{t_k}{2} + 3t_k\right), \tag{44.2}$$

and for $r < R_0$ it increases to some maximum value at the center.

Let us now find the values of the parameter t_n at which incompressibility sets in. Clearly, for this t_n we should have

$$\tau(t_n) = 1,$$

where τ is the combined volume of the particles enclosed in unit volume of the star, the Lorentz contraction being included in the calculation. The quantity τ should obviously also include the spaces left between the particles during maximum packing. The empty space left between the balls in dense packing comprises 26% of the total volume. For the volume τ we then have

$$\tau(t_n) = b \sum_k \frac{1}{2} a_k \cdot \int_0^{p_k} \frac{m_k c^2}{\varepsilon_k(p)} \cdot \frac{p^2 dp}{\pi^2 h^3} = 1,$$

where b is the proper volume pertaining to one baryon (including the empty space for one particle), $a_k = 2s_k + 1$ is the number of spin states, p_k is the Fermi momentum of the k-th particle, and the factor $m_k c^2/\varepsilon_k(p)$ is related to the Lorentz flattening of the particles in the direction of motion. After the integration we obtain

$$b \sum_k \alpha_k \left(2 \operatorname{sh} \frac{t_k}{2} - t_k\right) = 1, \tag{44.3}$$

where $\alpha_k = \frac{(2s_k + 1) m_k^3 c^3}{16\pi^2 h^3}$. Equation (44.3) thus defines the value of t_n at which incompressibility sets in (we should remember that the parameters t_k are functions of t_n).

In order to solve equation (44.3) it is necessary to know the value of the parameter b. As an example, we take $b = 4.5 \cdot 10^{-40}$ cm^3 (this corresponds to a radius $r = 4.75 \cdot 10^{-14}$ cm for the balls, which is approximately twice as large as the distance between two baryons at which nuclear repulsion forces take effect). Substituting this value of b in equation (44.3), we find for $t_n(R_0)$:

$$t_n(R_0) \approx 2.65. \tag{44.4}$$

This result was obtained in /15, 92/ for a special chemical composition, in which baryon resonances were not included. At this value of t_n, relativistic contraction of the particles' volume is negligible.

We shall now find the mass and radius of the central core consisting of incompressible matter. According to (40.12),

$$\frac{du}{dr} = 4\pi r^2 \rho_m, \tag{44.5}$$

$$\frac{dP}{dr} = -\frac{P + \rho_m}{r(r - 2u)}(4\pi r^3 P + u), \tag{44.6}$$

where $\rho_m \approx 0.63$ is the maximum density, at which incompressibility sets in. This value is found from the energy-density equation:

$$\rho_m = \frac{1}{4\pi} \sum_k \left(\frac{m_k}{m_n}\right)^4 [\operatorname{sh} t_k(R_0) - t_k(R_0)],$$

where $t_k(R_0)$ is the k-th particle parameter corresponding to $t_n\ (R_0) = 2.65$.

Remembering that $u(0) = 0$, and integrating (44.5) from zero to r, we obtain

$$u(r) = \frac{4\pi}{3} r^3 \rho_m. \tag{44.7}$$

We now substitute the expression for $u(r)$ in (44.6) and integrate from the center to the surface of the incompressible sphere; from the result we then obtain the radius of this sphere:

$$R_0 = \left\{\frac{3}{8\pi\rho_m}\left[1 - \left(\frac{3P_m + \rho_m}{P_m + \rho_m} \cdot \frac{P_0 + \rho_m}{3P_0 + \rho_m}\right)^2\right]\right\}^{1/2}, \tag{44.8}$$

where P_0 and P_m are the pressures at the star's center and at the surface of the incompressible sphere, respectively. Substitution of the value $t_n = 2.65$ in equation (44.2) yields $P_m = 0.0342$.

From (44.7) and (44.8) we obtain for the mass of the incompressible sphere

$$u(R_0) = \frac{0.613}{\sqrt{4\pi\rho_m}}\left[1 - \left(\frac{3P_m + \rho_m}{P_m + \rho_m} \cdot \frac{P_0 + \rho_m}{3P_0 + \rho_m}\right)^2\right]^{3/2}. \tag{44.9}$$

Since in the case of an incompressible fluid there is no relation between P and ρ, the mass, radius, and other configuration characteristics now depend on two parameters, namely P_0 and ρ_m. For a fixed mass density, however, we again have a one-parameter solution. Assuming that $\rho_m = 0.63$ and $P_m = 0.342$, we arrive at the following final result:

$$\left.\begin{aligned} R_0 &= 0.436\left[1 - 1.22\left(\frac{0.63 + P_0}{0.63 + 3P_0}\right)^2\right]^{1/2}, \\ u(R_0) &= 0.22\left[1 - 1.22\left(\frac{0.63 + P_0}{0.63 + 3P_0}\right)^2\right]^{3/2}. \end{aligned}\right\} \tag{44.10}$$

The maximum mass value is obtained when $P_0 \gg \rho_m$, and in this case

$$R_0 = 0.406, \quad u(R_0) = 0.175. \tag{44.11}$$

Subsequently, taking into account that at $r > R_0$ the equation of state is given by equation (42.1), we can integrate equations (40.12) with boundary conditions (44.11) up to the star's surface, i. e. to the distance R where $P(R) = 0$. This integration can obviously be carried out only numerically, and it yields the result

$$R = 6 \text{ km}, \quad M = 1.7\, M_\odot. \tag{44.12}$$

We could have obtained a larger mass by choosing a higher value for the volume of the baryon balls. However, even the value already taken for the volume b is unjustifiably high, and we can therefore consider the possibilities for obtaining a larger mass as exhausted. It should not be forgotten, though, that the strictest limitation on the mass value is imposed not by the equation of state, but by the current gravitation theory, according to which the mass cannot reach the limit $R/2$.

45. THE ENVELOPE OF BARYON STARS

When discussing baryon configurations, we mentioned that the integration has to be continued up to those $t_n(r)$ where the coordinate r and the function $u(r)$ become saturated. These saturation values of r and $u(r)$ were taken as the radius and mass of the configuration. In practice, the integration is continued up to the distance where the parameter $t_n(r)$ vanishes, and together with it also the baryon pressure $P(r)$. Strictly speaking, an error may be incurred if the integration is based on this. In actual fact, the star does not really end where the free neutrons disappear. In the above procedure, the integration covers the hyperon core and the neutron layer, i. e. it is carried out up to the interface between the Aen- and Ae-phases. Whereas up to the outside boundary of the Aen-phase the gravitational attraction is almost balanced by the pressure gradient of the baryon gas, from the beginning of the Ae-phase this role is taken over by the electron gas. Thus the external layer, where hydrodynamic equilibrium is secured through the pressure of the electron gas, has so far remained outside our field of interest. We shall now examine this region of the star, and verify that it does not contribute appreciably to the resultant mass and radius. Of course, by suitable changes in the equation of state the numerical integration can be continued all the way to the edge, where the pressure of the electron gas also vanishes. This, however, is not necessary, since an approximate integration can be used to obtain an analytical expression for the envelope.

In the exterior envelope

$$P(r) \ll \rho(r), \quad 4\pi r^3 P(r) \ll u(r), \tag{45.1}$$

which permits an appreciable simplification of the second equation in the system (40.12). We thus have

$$\left.\begin{aligned} \frac{du}{dr} &= 4\pi r^2 \rho, \\ \frac{dP}{dr} &= -\frac{\rho u}{r(r-2u)}. \end{aligned}\right\} \tag{45.2}$$

We keep the term $2u$ in the denominator, since it is not sufficiently small and its omission may in some cases lead to an error of a few percent in the star's radius.

Using the equation of state (29.8) and equation (18.5) for the average of the ratio A/Z, the system (45.2) can be converted into the following form:

$$\left.\begin{aligned} \frac{du}{dr} &= \frac{32}{3}\left(\frac{m_e}{m_n}\right)^3 \frac{A}{Z}\, r^2 x^3, \\ \frac{dx}{dr} &= -\frac{m_n}{m_e} \cdot \frac{A}{Z} \cdot \frac{u\sqrt{1+x^2}}{x \cdot r(r-2u)}. \end{aligned}\right\} \tag{45.3}$$

Here A/Z is the average value of the ratio between the numbers A and Z. This equation differs from equation (37.1) for white-dwarf configurations only in the term $2u$ in the denominator. The initial conditions here are

$$r = R_0, \qquad x_0 \equiv x(R_0), \qquad u(R_0) = u_0, \tag{45.4}$$

where R_0 is the coordinate radius of the interface between the Aen- and Ae-phases, $x_0 = 45$ is the electron Fermi energy (in units of $m_e c^2$)

at this surface, and u_0 is the "mass" of the baryon core. It should be remembered that the average of A/Z is determined by (18.5). With the term x^2 included, analytical integration is impossible. Nevertheless, this equation can be replaced by the following:

$$\frac{A}{Z} = 2 + 5.44 \cdot 10^{-3} x + 2.47 \cdot 10^{-4} x^2, \qquad (45.5)$$

which is also a fairly accurate approximation to the continuous curve for $\overline{A/Z} \equiv A/Z$ shown in Figure 3.

Equations (45.4) permit further simplifications. Thus, considering the fact that the envelope's mass is very small compared to the total mass of the star, one can neglect its negligible variation when calculating the envelope thickness. The validity of this statement has been verified through numerical integration, but it becomes obvious after obtaining the final result. Considering these remarks, as well as the relationship (45.5), and integrating the second equation (45.3) from $x(R_0)$ to $x(R)=0$, we find for the star's radius R:

$$R = R_0 \left[1 - \alpha \frac{R_0}{r_g} \left(1 - \frac{r_g}{R_0} \right) \right], \qquad (45.6)$$

where $r_g = 2M$ is the gravitational radius of the star, and $\alpha = 38.72 \frac{m_e}{m_n} = 0.02109$. From this we note that in configurations where R_0 is close to r_g (this occurs in the case of configurations with a hyperon core), the thickness $R-R_0$ of the envelope is very small compared to the radius R. For less dense configurations, where $\rho(0) < 10^{14}$ g·cm^{-3}, we have $\frac{R_0}{r_g} > 25$, and the envelope thickness is comparable with the star's radius.

Now let us estimate the envelope's mass. From equations (45.3) it follows that

$$\frac{du}{dx} = -\frac{32}{3} \left(\frac{m_e}{m_n} \right)^4 \frac{r^3 (r - 2u)}{u \sqrt{1 + x^2}} \cdot x^4. \qquad (45.7)$$

Integrating (45.7) between x_0 and $x=0$, we obtain

$$\int_{u_0}^{M} \frac{u\, du}{r^3 (r - 2u)} = \frac{4}{3} \left(\frac{m_e}{m_n} \right)^4 \left[x_0 (2x_0^2 - 3) \sqrt{x_0^2 + 1} + 3 \ln \left(x_0 + \sqrt{x_0^2 + 1} \right) \right],$$

TABLE 11. Parameters of configurations consisting of an ideal baryon gas

Density at center, g·cm^{-3}	Baryon core		Envelope				Star		Parameters of hydrogen envelope	
			Aen-phase		*Ae*-phase					
	mass in units of $M_\odot$	radius, km	mass in units of $M_\odot$	thickness, km	mass in units of $M_\odot$	thickness, km	mass in units of $M_\odot$	radius, km	mass in units of $M_\odot$	thickness, m
$1.00\cdot10^{14}$	0.1812	11.755	0.1050	5.863	$6.76\cdot10^{-4}$	12.575	0.2869	30.193	$1.58\cdot10^{-8}$	493
$2.24\cdot10^{14}$	0.3428	12.522	0.0649	2.739	$2.39\cdot10^{-4}$	5.240	0.4074	21.201	$6.38\cdot10^{-9}$	265
$3.62\cdot10^{14}$	0.4162	11.467	0.0356	2.326	$1.12\cdot10^{-4}$	3.362	0.4519	17.155	$3.15\cdot10^{-9}$	174
$2.34\cdot10^{15}$	0.6299	9.974	0.0114	1.000	$3.24\cdot10^{-5}$	1.227	0.6413	12.201	$4.61\cdot10^{-9}$	54
$1.09\cdot10^{16}$	0.5110	8.343	0.0069	0.898	$1.66\cdot10^{-5}$	1.093	0.5179	10.334	$3.35\cdot10^{-10}$	51
$8.29\cdot10^{16}$	0.3220	6.672	0.0049	0.979	$9.01\cdot10^{-6}$	1.301	0.3269	8.952	$2.83\cdot10^{-10}$	61
$1.15\cdot10^{16}$	0.1721	5.452	0.0047	1.390	$8.24\cdot10^{-6}$	2.328	0.1768	9.169	$4.33\cdot10^{-10}$	105
∞	0.3066	8.439	0.0161	2.110	$4.04\cdot10^{-5}$	2.826	0.3227	13.375	$1.05\cdot10^{-9}$	118

TABLE 12. Parameters of configurations consisting of a real baryon gas /24/

Density at center, $g \cdot cm^{-3}$	Baryon core		Envelope				Star	
			Aen-phase		*Ae*-phase			
	mass in units of $M_\odot$	radius, km	mass in units of $M_\odot$	thickness, km	mass in units of $M_\odot$	thickness, km	mass in units of $M_\odot$	radius, km
$1.38 \cdot 10^{15}$	0.9388	8.399	$7.3 \cdot 10^{-4}$	0.1924	$8.76 \cdot 10^{-6}$	0.3951	0.9395	8.9865
$1.78 \cdot 10^{15}$	1.2328	8.579	$5.0 \cdot 10^{-4}$	0.1230	$7.68 \cdot 10^{-6}$	0.2611	1.2333	8.9631
$2.30 \cdot 10^{15}$	1.5702	8.550	$3.0 \cdot 10^{-4}$	0.0769	$5.77 \cdot 10^{-6}$	0.1584	1.5705	8.7853
$3.22 \cdot 10^{15}$	1.7304	8.266	$2.0 \cdot 10^{-4}$	0.0546	$4.12 \cdot 10^{-6}$	0.1110	1.7306	8.4315
$5.14 \cdot 10^{15}$	1.7367	7.697	$1.0 \cdot 10^{-4}$	0.0396	$2.68 \cdot 10^{-6}$	0.0833	1.7369	7.1898
$8.28 \cdot 10^{15}$	1.6953	7.284	$1.0 \cdot 10^{-4}$	0.0352	$2.01 \cdot 10^{-6}$	0.0715	1.6924	7.3907
$1.37 \cdot 10^{16}$	1.5567	6.727	$1.0 \cdot 10^{-4}$	0.0317	$1.48 \cdot 10^{-6}$	0.0671	1.5568	6.8259
$7.42 \cdot 10^{16}$	1.3124	6.088	$1.0 \cdot 10^{-4}$	0.0362	$1.15 \cdot 10^{-6}$	0.0754	1.3124	6.1995
$1.08 \cdot 10^{17}$	1.2911	6.076	$1.0 \cdot 10^{-4}$	0.0401	$1.22 \cdot 10^{-6}$	0.0784	1.2912	6.1945
∞	1.3147	6.277	$1.0 \cdot 10^{-4}$	0.0412	$1.37 \cdot 10^{-6}$	0.0841	1.3147	6.4023

where $M \equiv u(R)$, $u_0 = u(x_0)$, $x_0 = 45$, and $x(R) = 0$. Let us denote the left-hand side by I. If we replace the denominator of the expression under the integral sign by its smallest value $R_0^3(R_0 - 2u_0)$ and by its largest value $R^3(R-2M)$, we obtain

$$\frac{I_{max}}{I_{min}} = \frac{R_0^3(R_0 - 2u_0)}{R^3(R - 2M)} \approx \left(\frac{R_0}{R}\right)^4.$$

For configurations containing a hyperon core ($\rho(0) > 3 \cdot 10^{14}$ g·cm^{-3}), the radius R is not much different from R_0, consequently $(I_{max} - I_{min})/I_{max} \ll 1$. As a result, we obtain for such configurations

$$M = u_0 + P(R_0)\frac{\bar{r}^3(\bar{r} - 2u_0)}{u_0}. \qquad (45.8)$$

Here $P(R_0) = 9.55 \cdot 10^{-7}$ is the pressure at the interface between the Ae- and Aen-phases, where $x_0 = 45$ and $\bar{r} = \frac{R_0 + R}{2}$. In baryon configurations u_0 is about 0.05—0.2, and the second term in (45.8) is of the order of 10^{-6}, therefore the envelope's mass is smaller by more than three to four orders of magnitude than the mass of the whole star. For less dense configurations, where the above approximation does not hold, the mass is found through numerical integration.

Tables 11 and 12 show the parameters of a series of baryon configurations for the ideal- and real-gas versions of an elementary-particle gas. The mass and dimensions of the various star regions are given in the middle columns of Table 11. In the last two columns of Table 11 are also presented the envelope parameters obtained under the assumption that they consist solely of hydrogen /89/. For the case of a real gas, the parameters of the baryon core differ from those given in Table 10, since they were calculated for a different version of a real gas /14, 92/. The parameters for the Ae- and Aen-phases were taken from /101/. The envelope parameters were calculated under the assumption that in advancing toward the star's surface, A/Z varies continuously. In reality, at certain distances, where the charge Z of certain nuclei increases owing to β-decay, the ratio A/Z, and thereby also the energy density, should undergo an abrupt change. However, calculations indicate that these small jumps in the value of ρ do not affect the results obtained for the star's mass and radius. The pressure, of course, changes continuously.

46. ROTATING SUPERDENSE STARS

Consider an observer be in a fixed reference system, and let the star rotate relative to him in clockwise direction. By definition, the angular velocity of the rotation is

$$\Omega = c\frac{d\varphi}{dx^0}.$$

We shall use spherical coordinates $x^0=t, x^1=R, x^2=\theta, x^3=\varphi$.

In the case of rotating superdense stellar configurations, the mass distribution, and consequently also the gravitational field, are axially symmetrical:

$$g_{ik} = g_{ik}(R, \theta, \Omega).$$

It is thus assumed that Ω is independent of the coordinates, and the field and mass distribution are stationary.

Let us perform the coordinate transformation $x'^0=-x^0, x'^1=x^1, x'^2=x^2, x'^3=-x^3$. This does not change the sign of the angular velocity. As a result, the g_{ik} will undergo no changes. Then, from the invariance of ds^2, it follows directly that $g_{01}=g_{13}=g_{23}=g_{02}=0$. Thus for constant Ω the four-dimensional interval can be written in the form

$$ds^2 = g_{00}(dx^0)^2 + g_{11}(dx^1)^2 + g_{22}(dx^2)^2 + g_{33}(dx^3)^2 + \\ + 2g_{12}dx^1dx^2 + 2g_{03}dx^0dx^3.$$

Through suitable transformation of the coordinates R and θ one can reduce the number of components of the metric tensor by another two. We thus require that $g'_{21}=0, g'_{33} = g'_{22}\sin^2\theta$; then after introducing the following notations:

$$g'_{11} = e^{\lambda}, \quad g'_{22} = e^{\mu}, \quad g'_{03} = \omega e^{\mu}\sin^2\theta, \quad g'_{00} = \omega^2 e^{\mu}\sin^2\theta - e^{\nu}$$

the interval is written in the form

$$ds^2 = (e^{\nu} - \omega^2 e^{\mu}\sin^2\theta)(dx^0)^2 - e^{\lambda}dR^2 - \\ - e^{\mu}(d\theta^2 + \sin^2\theta\, d\varphi^2) - 2\omega e^{\mu}\sin^2\theta\, d\varphi dx^0, \qquad (46.1)$$

where ν, μ, λ, and ω are functions of R, θ, and Ω. It is important to note that in the case of rotation, the component g_{03} cannot be eliminated through any coordinate transformation (stationary field). This can be done only in the case of a static (i. e. nonrotating) axially symmetrical mass distribution, but this situation does not occur in nature.

Since time reversal changes the sign of the angular velocity, but ds^2 remains unchanged, it follows that all components of the metric tensor, to the exclusion of g_{03}, are even functions of Ω. On the other hand, ω is an odd function of Ω.

The exterior solution for the case of rotation was found in /221/ This is an accurate, but partial, solution, depending on the mass and the total relativistic moment of inertia. The interior problem here is far more difficult than in the case of nonrotating bodies, since we are then dealing with partial differential equations. There is one fact, though, which enables one to find approximate numerical solutions, namely that even in relativistic theory the rotational angular velocity cannot be arbitrarily high. When the angular velocity exceeds a certain maximum value Ω_{max}, there is outflow of matter from the equator. In analogy with the nonrelativistic case, here also the integral characteristics and the functions determining the internal structure of the configuration (and naturally also the metric of the space) are dependent not on Ω, but on the dimensionless parameter $\beta = \Omega^2/8\pi k\rho_0$. The maximum value of β for a given central density is fairly small: $\beta_{max} = \frac{\Omega^2_{max}}{8\pi k\rho_0} \ll 1$. This fact permits the development of approximate methods for solving the problem, in which the rotation is regarded as a small perturbation. The sought functions are expanded in powers of the small parameter β. The series for the diagonal elements of the metric tensor will contain integral powers of β (since they are even functions of Ω), and the series for the g_{03} functions will contain half-integral powers of β.

In the zeroth approximation the mass distribution is spherically symmetrical, and in free space the metric is determined by the Schwarzschild solution. In the following approximation, we should first of all consider those corrections to the unperturbed spherical problem which are proportional to $\beta^{1/2}$. In this approximation the diagonal elements of the metric tensor naturally remain unperturbed, since they depend on integral powers of β (even functions of Ω), and therefore the first correction to them should be of the order of β. In the $\beta^{1/2}$ approximation the mass distribution continues to be spherically symmetrical, since it is determined by the diagonal elements of g_{ik}; or, expressed more clearly, in this approximation the configuration maintains its central symmetry because the pressure is also an even function of Ω. This approximation was studied in /246, 42a, 79—81/. The result found for g_{03} outside the mass distribution is

$$g_{03} = -\frac{2I}{r}\sin^2\theta, \tag{46.2}$$

where I is the total relativistic moment of inertia. Thus in the $\beta^{1/2}$ approximation, the only addition to the expression for ds^2 is g_{03}.

In the second approximation we should consider corrections of the order of β. This approximation was investigated in /193—196, 108, 109/. Here we should write

$$\left.\begin{aligned} &e^{-\lambda} = e^{-\lambda_0}(1+\beta f), \quad e^{\nu} = e^{\nu_0}(1+\beta\Phi),\\ &e^{\mu} = R^2(1+\beta U), \quad \omega = \sqrt{\beta}\, q,\\ &P = P^0 + \beta P_1, \end{aligned}\right\} \tag{46.3}$$

where f, Φ, U, and q are unknown functions of R and θ, and e^{λ_0}, e^{ν_0} are the corresponding functions in the zeroth approximation; they are dependent only on R. Outside the mass distribution $e^{\nu_0} = e^{-\lambda_0} = 1 - r_g/R$.

In /108/ the independent equations were chosen as the following combinations:

$$\left.\begin{aligned} &G_1^1 - G_0^0 = 8\pi(T_1^1 - T_0^0), \qquad G_2^1 = 0,\\ &G_2^2 + G_3^3 = 8\pi(T_2^2 + T_3^3), \qquad G_0^3 = 8\pi T_0^3 \end{aligned}\right\} \tag{46.4}$$

together with the two equations of hydrodynamics. Here $G_i^k = R_i^k - \frac{1}{2}R\delta_i^k$.

The equations of hydrodynamics $T_{i;k}^k = 0$ reduce to the following /196/:

$$\left.\begin{aligned} &u^3 = \Omega u^0 = [e^{\nu} - e^{\mu}(\omega+\Omega)^2\sin^2\theta]^{-1/2},\\ &\int\frac{dP}{P+\rho} = -\frac{1}{2}\nu -\\ &-\frac{1}{2}\ln[1 - e^{\mu-\nu}(\omega+\Omega)^2\sin^2\theta] + C = \Psi(R,\theta). \end{aligned}\right\} \tag{46.5}$$

The first of equations (46.5) implies that the configuration should rotate with a constant angular velocity Ω. The second equation is the integral of the relativistic equations of hydrodynamic equilibrium, and C is an integration constant. The function Ψ also has to be expanded in powers of β. In the present approximation,

$$\Psi(R,\theta) = \Psi^0(R) + \beta\Psi_1(R,\theta).$$

Expression (46.2) for g_{03} obviously does not change, since the power series for this function includes no term which is proportional to β (g_{03} is an odd function of Ω). From this it also follows that in the β approximation the function q depends only on R. In /108, 193/ an analytical expression was found for this function outside the mass distribution.

The functions f, Φ, U, P were expanded in Legendre polynomials. It was found that in the β approximation the variables R and θ in the Einstein equations become separated. The rest of the solution procedure is identical to that of the nonrelativistic case (see Section 39). In the exterior region we have the following solution:

$$\left.\begin{aligned} e^{-\lambda} &= 1-\frac{r_g}{R}+\beta\sum_l\left[A_l\,\Phi_l(R)-\frac{24C_1^2}{R^4}\delta_{0l}-\frac{3C_1^2}{R^4}\delta_{2l}\right]P_l(\gamma),\\ e^{\nu} &= 1-\frac{r_g}{R}+\beta\sum_l A_l\,\Phi_l(R)\,P_l(\gamma),\\ e^{\mu} &= R^2\left[1+\beta\sum_l U_l(R)\,P_l(\gamma)\right],\\ \omega &= \sqrt{\beta}\cdot\frac{C_1}{R^3}, \end{aligned}\right\} \tag{46.6}$$

where

$$\Phi_l(R)=\frac{1}{R^{l+1}}F\left(l+1,\ l-1,\ 2l+2,\ \frac{r_g}{R}\right)+$$
$$+\frac{C_1^2}{R^4A_l}\left(\frac{1}{2}-\frac{r_g}{R}\right)\delta_{0l}-\frac{C_1^2}{2r_gR^3A_l}\left(\frac{r_g^2}{R^2}+\frac{r_g}{R}-2\right)\delta_{2l},$$
$$U_l(R)=\int\left\{-\frac{A_l}{1-r_g/R}\frac{d\Phi_l}{dR}-\right.$$
$$\left.-\frac{6C_1^2}{l(l+1)-2}\cdot\frac{1}{R^5}\left(\frac{r_g/R}{1-r_g/R}+2\right)(\delta_{0l}-\delta_{2l})\right\}dR,$$

A_l and C_l are constants, $\gamma=\cos\theta$, and F is the hypergeometric function. Subsequently, the conditions at the center of the mass distribution were defined and the equations (which are ordinary differential equations in R) were integrated numerically. At the surface of the configuration the interior solution was matched to the exterior solution (46.6). This operation determines all constants which entered into the interior and exterior solutions. It was found that in the β approximation all expansion terms with $l>2$ drop out. Figure 22 shows the dependence of the mass on the logarithm of the central density for configurations which rotate with maximum angular velocity. The value of Ω_{max} was determined by equating the centrifugal and centripetal forces on the equator's surface. The equation of state for a real gas, given by (33.6), was employed. As can be seen from this figure, the maximum increase in mass is approximately 30%. Figure 23 shows the dependence of the relativistic moment of inertia I on the central density ρ_0. As was shown in /81/, I is determined by the formula

$$I = \frac{1}{2} \frac{R_0^4 q'(R_0)}{3q(R_0) + R_0 q'(R_0)} . \tag{46.7}$$

Up to the second approximation, i.e. to terms of the order of β, I is independent of the angular velocity and is the same as for a rigid sphere rotating around the z-axis. The $I(\rho_0)$ curve is similar in shape to the mass curve $M(\rho_0)$.

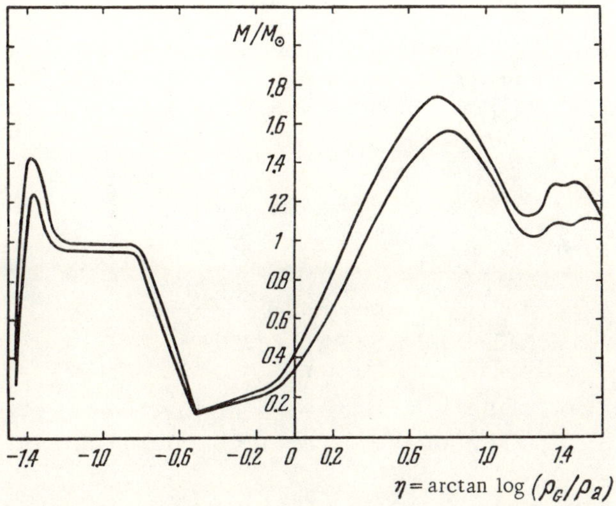

FIGURE 22. The mass as a function of central density for baryon configurations which rotate with the maximum angular velocity Ω_{max} (upper curve). The lower curve describes nonrotating configurations.

As we have seen, the inclusion of rotation does not introduce any substantial changes in the values of the parameters of baryon configurations. In the case of rotation, all configuration characteristics depend on two parameters: the density ρ_0 at the center and the angular velocity Ω.

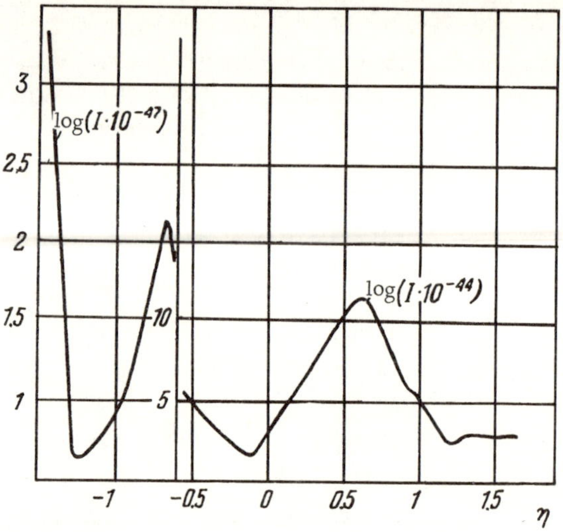

FIGURE 23. The total relativistic moment of inertia I as a function of the central density. I is measured in CGS units. The variable plotted along the abscissa is $\eta = \arctan \log \frac{\rho_0}{\rho_1}$, where $\rho_1 = 3.62 \cdot 10^{14}$ g·cm^{-3} is the nuclear density.

47. HOT SUPERDENSE CELESTIAL BODIES

The observation of baryon stars is a highly topical problem in astrophysics. In view of the extremely small radii of these stars, their luminosity

$$L_\gamma = 4\pi R^2 \sigma T_R^4 \tag{47.1}$$

is very low even for the surface temperatures T_R which are characteristic of white dwarfs. At higher temperatures, the radiation peak is shifted into the X-ray region. Therefore the observation of such objects may be expected only with methods of extra-atmospheric astronomy. Nowadays, with the help of satellites, a whole series of X-ray sources, unidentified in the optical region, have been discovered in the Galaxy. It is quite possible that some of these are baryon stars.

The temperature inside dense stars cannot be sufficiently high to remove the degeneracy. High temperatures are excluded because of the catastrophic nature of neutrino losses and the high heat conductivity. The possible values of T are far from sufficient for any appreciable change in the internal structure, mass, or radius of the configurations. In accordance with this, in the hot-configuration problem one first integrates equations (40.12) to determine the mass M, radius R, and the functions $\nu(r)$, $\lambda(r)$, $u(r)$, $P(r)$, for completely degenerate matter ($T=0$). Subsequently, as a second step, one can proceed to finding the temperature-dependent parameters of the star.

In degenerate matter there is no convection /123/. The transport of energy from the interior to the surface is accomplished through heat conduction. The temperature gradient is determined by the equation /230a, 267/

$$\frac{d}{dr}(Te^{\nu/2}) = -\frac{3\chi\rho}{16\sigma T^3}\cdot\frac{L_\gamma e^{\nu/2}}{4\pi r^2}\left(1-\frac{2u}{r}\right)^{-1/2}. \qquad (47.2)$$

Here $L_\gamma(r)$ is the photon luminosity (energy flux of the electromagnetic radiation) at a distance r from the center, and $\chi(r)$ is the effective coefficient of opacity, which is expressed through the absorption coefficient $\chi_r(r)$ of the radiation and the electron heat conductivity $\lambda_c(r)$ /267, 258/:

$$\frac{1}{\chi} = \frac{1}{\chi_r} + \frac{3\rho\lambda_c}{16\sigma T^3}. \qquad (47.3)$$

The first term predominates in the nondegenerate surface layer, and the second predominates in the degenerate interior. In the general case both χ_r and λ_c are functions of the density and temperature:

$$\chi_r = \chi_r(\rho, T), \quad \lambda_c = \lambda_c(\rho, T).$$

Detailed information concerning these quantities can be found in /257, 118/.

In white dwarfs and baryon stars the thermonuclear energy sources are assumed to be exhausted, and the radiation is due only to reserves of thermal energy. Let us denote the thermal-energy density by $\varepsilon_T(\rho, T)$. According to (13.6), in the T^2 approximation this quantity is equal to

$$\varepsilon_T = \frac{1}{2}T^2\sum_i \frac{\pi^2k^2}{2m_i c^2}\frac{\sqrt{x_i^2+1}}{x_i^2}n_i + \frac{3}{2}n_A kT, \qquad (47.4)$$

where k is the Boltzmann constant, n_i is the density of the i-th type of fermion in the degenerate particle gas, $x_i = p_i/m_i c$, p_i is the Fermi momentum, and n_A is the density of atomic nuclei. In the case of baryon stars the nuclei are found only in a relatively thin layer at the surface, and in white dwarfs n_A is everywhere different from zero. The first term in (47.4) is obtained from the second term in (13.6). For the total internal energy we then have

$$u_T = 4\pi \int_0^R \varepsilon_T e^{\nu/2}\left(1-\frac{2u}{r}\right)^{-1/2} r^2 dr \qquad (47.5)$$

Here u_T is the energy which reaches an observer at infinity through photon or neutrino radiation, when the star cools down from its original state with temperature T to a state with $T=0$.

From the energy-balance equation we have

$$\frac{d}{dr}\left[(L_\gamma + L_\nu)e^\nu\right] = -4\pi r^2 e^{\nu/2}\left(1-\frac{2u}{r}\right)^{-1/2}\left(\frac{d\varepsilon_T}{dt}\right)_{r=\text{const}}, \qquad (47.6)$$

where L_ν is the energy flux due to neutrinos (neutrino luminosity). This quantity is expressed through the energy q_ν removed by a neutrino from unit volume in unit time, by the relationship

$$\frac{d}{dr}(L_\nu e^\nu) = 4\pi r^2 e^\nu \left(1-\frac{2u}{r}\right)^{-1/2} q_\nu\,. \qquad (47.7)$$

By using the universal Lagrangian for weak interactions, q_ν can be expressed in terms of the cross sections of all processes in which a neutrino is generated. We do not have the possibility of discussing this matter here, but to those who are interested we recommend the papers /132a, 270/. The above quantity, in analogy with λ_c and χ_r, is a function of the density and temperature: $q_\nu = q_\nu(\rho, T)$.

We shall thus assume that the mass M and radius R have already been found by integrating the equation system (40.12) for completely degenerate matter. Then, given the luminosity L_γ, one can calculate from (47.1) the effective surface temperature T_R. We also have to find the values of the pressure P_R and the density ρ_R at the configuration's surface. For determining these quantities, we shall first of all use the definition of the optical depth of the surface layer. In the theory of stellar atmospheres, an atmosphere's optical thickness is the term applied to the quantity

$$\tau_R = \int_R^\infty \chi\rho\, dr \approx \chi(\rho_R, T_R)\int_0^\infty \rho\, dr = \frac{2}{3}. \qquad (47.8)$$

On the other hand, from equation (40.10) we have

$$P_R = \frac{GM}{R^2}\left(1+\frac{P_R}{\rho_R c^2}\right)\left(1+\frac{4\pi R^3 P_R}{Mc^2}\right)\left(1-\frac{2GM}{c^2 R}\right)^{-1}\int_R^\infty \rho dr.$$

Considering (47.8), we find that

$$\chi(\rho_R, T_R)\cdot P_R = $$
$$= \frac{2GM}{3R^2}\left(1+\frac{P_R}{\rho_R c^2}\right)\left(1+\frac{4\pi R^3 P_R}{Mc^2}\right)\left(1-\frac{2GM}{c^2 R}\right)^{-1}. \tag{47.9}$$

We still have one unused equation, namely the equation of state for the nondegenerate exterior layer

$$P = \frac{k\rho}{\mu m_p} T + \frac{1}{3} a T^4, \tag{47.10}$$

where μ is the average molecular weight, measured in units of the proton mass. If the form of the functions is known, then from (47.9) and (47.10) we can determine the pressure P_R and density ρ_R at the surface for given M, R, and T_R (M and R are the parameters of the cold baryon configuration, and L_γ, and consequently also T_R, are independent parameters).

Knowing the boundary values of the basic quantities at the star's surface, one can proceed to the integration of the equations (from the surface toward the center) which determine the temperature distribution. At the outset, (47.10) is used as the equation of state. Then the pressure and density of the degenerate Ae-phase are calculated in parallel from equation (29.8). Deep inside, where the pressure calculated from equation (47.10) is equated to the pressure of the degenerate electron gas, a transition is made to the equation of state (29.8), which is also used subsequently until the onset of the Aen-phase. With further advance toward the center, we have to deal with the same equations of state of degenerate matter as were used in calculations of completely cold configurations.

In integrating equation (47.2), as a result of the high heat conductivity of the degenerate electrons, from some distance r_0 the quantity $Te^{\frac{\nu(r)}{2}}$ remains practically constant:

$$T(r)\, e^{\frac{\nu(r)}{2}} \approx \text{const.} \tag{47.11}$$

From this we obtain for the temperature at the star's center

$$T(0) = T(r_0)\, e^{\frac{\nu(r_0)-\nu(0)}{2}}. \tag{47.12}$$

In a rough approximation, the surface $r = r_0$ can be taken as that where the pressures determined by equations (29.8) and (47.10) are the same, i. e. where the degenerate electrons become noticeable. In /270/ it was shown that this surface is located only slightly deeper, somewhere within the layer of the Ae-phase. However this might be, though, the surface layer where there is an appreciable temperature gradient is very thin, so that the main part of matter is in the central sphere, where (47.11) is valid. Therefore, in integrating equation (47.2) from the star's surface, one can a priori disregard the variation of $L_\gamma e^\nu$, which is in the general case determined from (47.6).

TABLE 13. Thermal parameters of a neutron star with $M = 1$ $M_\odot$ and $R =$ $= 5.184$ km, having an iron envelope

$T(R)$	$\log L_\gamma$	$\log L_\nu$	$\log T(0)$	$\log U_T$	$\log t$	$\lambda_{max} = \frac{0.29}{T_R}$
$5\cdot 10^7$	39.078	45.86	10.019	49.48	—4.26	0.584
$2\cdot 10^7$	37.486	41.25	9.435	48.51	—1.10	1.459
10^7	36.282	38.91	9.035	47.51	0.615	2.918
$5\cdot 10^6$	35.078	36.98	8.704	46.84	2.10	5.836
$2\cdot 10^6$	33.486	34.03	8.215	45.84	4.08	14.59
10^6	32.282	31.03	7.738	44.80	5.16	29.18
$5\cdot 10^5$	31.078	28.38	7.293	44.01	5.78	58.36
$2\cdot 10^5$	29.486	—	6.621	42.71	6.25	145.9
10^5	28.282	—	6.178	41.78	6.54	291.8
$5\cdot 10^4$	27.078	—	5.703	40.86	6.81	583.6
$2\cdot 10^4$	25.486	—	5.070	39.61	7.19	583.6

Note. The energy is measured in erg and the luminosity in erg/sec, t is the total cooling time of the star (in years), and λ_{max} is the wavelength, in angströms, which corresponds to the maximum radiation.

After finding the temperature distribution $T(r)$, the neutrino luminosity $L_\nu(R)$ and the total thermal energy $U_T(R)$ can also be determined by integrating equations (47.5) and (47.7) from the star's center with the conditions $L_\nu(0) = 0$ and $U_T(0) = 0$. By varying the photon luminosity L_γ, one can through integration of equations (47.2), (47.5), and (47.7) construct the curves of U_T, L_ν, and $T(r)$ as a function of L_γ. At the same time one can also determine the cooling time of the star from luminosity $L_\gamma^{(1)}$ to $L_\gamma^{(2)}$. This can be obtained from equation (47.6). We have

$$L_\gamma(R) + L_\nu(R) = -\frac{dU_T}{at} e^{-\nu(R)}.$$

From this

$$\Delta t = -\frac{\Delta U_T e^{-\nu(R)}}{L_\gamma(R) + L_\nu(R)}, \tag{47.13}$$

where Δt is the time in which the luminosity decreases from L_1 to L_2, and $\Delta U_T = U_T^{(2)} - U_T^{(1)}$.

In /270/ were calculated the main characteristics of a hot neutron star with $M = 1\,M_\odot$ and $R = 5.184$ km, with the assumption that its atmosphere (the nondegenerate surface layer) consists of iron. These calculations were made without the requirements of the general theory of relativity, i. e. in the equations for the temperature characteristics it was assumed that $e^\nu = e^\lambda = 1$. The results are presented in Table 13. The thickness of the atmosphere does not exceed one meter.

SUMMARY

In the relativistic theory of gravitation, parameters such as mass, radius, and others, are single-valued functions of the central density ρ_0 (or pressure P_0). The inverse functions $\rho_0(M)$, $\rho_0(R)$, etc., are in general not single-valued. The function $M(\rho_0)$, for example, has a series of maxima and minima. In the intermediate region (between the white-dwarf branch at $\eta < -0.6$ and the baryon-star branch at $\eta < -0.2$), the mass displays a deep minimum, with $M \approx 0.125 M_\odot$. To the left of this minimum, at $\eta = -1.36$ ($\rho_0 \approx 7 \cdot 10^9$ g·cm^{-3}), there is the white-dwarf maximum with $M = 1.25 M_\odot$, and to the right there is the highest baryon-star maximum, with $M = 1.55 M_\odot$ at $\eta \approx 0.8$ ($\rho_0 \approx 4 \cdot 10^{15}$ g·cm^{-3}). Above this maximum, the $M(\rho_0)$ curve has an infinite number of oscillations, whose amplitude attenuates strongly with increasing ρ_0, and when $\rho_0 \to \infty$ the mass converges toward the limit $M = 1.09 M_\odot$.

From Figure 20 it is seen that for relatively dense ($\eta > 0$) baryon configurations, the mass of those consisting of a real gas is considerably larger than the mass of ideal-gas configurations. This is caused by the fact that, at supernuclear densities, we are dealing with powerful nuclear repulsion forces, which at a given density obviously lead to an increase in pressure.

Comparison between curves 2 and 3 of Figure 20, which describe an ideal baryon gas (in which hyperons are included) and an ideal neutron gas, respectively, shows that at $\eta > 0.6$ ($\rho_0 \gtrsim 2 \cdot 10^{15}$ g·cm^{-3})

the latter mass curve $M(\rho_0)$ lies much above the former. This can be understood in view of the fact that in going over to the system of units $c=k=1$, $m_k^4c^5/(32\pi^2h^3)=\frac{1}{4\pi}$ (see (34.1)), where m_k is the mass of the elementary particle which is the main constituent of the star, the star's mass becomes independent of m_k, this dependence having been transferred to the unit of measurement. Thus, according to (34.5) and (34.6), in this system the mass unit is $\beta_k\sim\frac{1}{m_k^2}$. For neutron configurations, $m_k=m_n$, and for baryon configurations where none of the baryons predominate numerically, the determining factor is some mean mass value $\overline{m_k}$. Since $\overline{m_k}>m_n$, the mass of a baryon configuration should be smaller, because the value of its unit of measurement is lower.

In the region $0.5>\eta>0.45$ the situation is reversed (see Figure 20): the $M(\eta)$ curve for neutron configurations lies above that of baryon configurations. Strictly speaking, this is the result of incorrect treatment, since in calculating neutron configurations it was assumed that up to the surface itself the matter consists of a degenerate neutron gas; that is, no account was taken of the change in chemical composition related to the presence of the *Ae*- and *Aen*-phases. In this, the most important role is played by the *Ae*-phase. In the case of baryon configurations, all this was taken into account.

Of special interest are configurations with $-0.6\lesssim\eta\lesssim-0.2$ ($8\cdot10^{13}\lesssim\rho_0\lesssim2\cdot3\cdot10^{14}$ g$\cdot$cm^{-3}). Here the masses are smaller than the sun's by almost one order of magnitude (see Table 10), and the radii are large. For example, at $\rho_0=7.1\cdot10^{13}$ g$\cdot$cm^{-3}, $M=0.236\,M_\odot$ and $R=4500$ km.

A baryon configuration with a supernuclear central density consists of a massive core, in which there are hyperons and baryon resonances. The concentrations of the various baryons are all of the same order of magnitude, and a substantial part of the star's mass is concentrated here. Beyond the hyperon core there is a layer with density of the order of nuclear densities (*Aen*-phase). This is followed by the surface layer, where matter is in a state typical for white dwarfs (the *Ae*-phase, consisting of a degenerate electron gas and bare atomic nuclei). Finally, in hot baryon stars there is also a dense atmosphere of small thickness, where, depending on temperature, there may be ions, free electrons, neutral excited and unexcited atoms, and electromagnetic radiation. In this kind of baryon star the dimensions of the outer layers, including the thickness of the atmosphere, are small compared to the radius of the dense core. The main part of the mass is also concentrated in this core. Neutrons predominate numerically only in the nuclear layer

(i. e. in the *Aen*-phase), whose contribution to the star's mass and radius is quite small. In this sense the term neutron stars, widespread in the literature, is unjustified and completely arbitrary. The term "baryon star" is a more correct reflection of the actual situation.

When the central densities are below-nuclear, we also have a central dense section, consisting of nuclear matter (*Aen*-phase), an envelope (*Ae*-phase), and an atmosphere. In this range, when ρ_0 decreases the role of the envelope and atmosphere becomes significant. Thus, for the above-mentioned configuration with $\rho_0 = 7.1\cdot10^{13}$ g·cm^{-3}, the mass of the central core (the nucleon-nuclear phase) is $0.108\,M_\odot$ and that of the *Ae*-envelope is $0.128\,M_\odot$; the radius of the core is 20.6 km, and the thickness of the envelope is about 4500 km.

The configuration parameters depend on the equation of state. However, no reasonable change in the latter can lead to a change in the order of magnitude of the mass, which is always of the order of the sun's.

Inclusion of rotation introduces no significant changes in the situation described. The possibilities here are limited, because the angular velocity Ω cannot become arbitrarily high. When $\Omega > \Omega_{max}$, where $\Omega_{max} \approx \left(\frac{GM}{R^3}\right)^{1/2}$, the configuration loses stability, and the first thing to happen is outflow of matter from the equator. The maximum angular velocity is not sufficient to cause substantial changes in the configuration's structure and parameters. Thus, even when the star rotates with angular velocity $\Omega = \Omega_{max}$, the highest increase in mass does not exceed 30% (see Figure 22), and the semimajor and semiminor axes of the spheroid differ from the radius of the corresponding spherical configuration by about 10%.

The observation of baryon stars is difficult because of their low luminosity in the visible range (owing to their small radius). Their observation is to be expected in the X-ray region of the spectrum, if one assumes the presence of high temperatures. However, particularly high temperatures are not possible, due to the high energy losses, primarily those related to neutrinos.

Chapter VIII

THE INTERNAL STRUCTURE AND MASS DEFECT OF BARYON CONFIGURATIONS

48. MASS DISTRIBUTION AND GEOMETRY INSIDE THE CONFIGURATIONS

In calculating the external parameters of baryon configurations, during the integration process one can also obtain the values of $t_n(r)$ and $u(r)$ as a function of distance r from the center. The knowledge of these functions is sufficient for solving all problems related to the geometry and internal structure. In the case of configurations which are in thermodynamic and hydrodynamic equilibrium, the internal geometry and structure are equivalent, in the sense that knowledge of one of them is sufficient for determining the other.

According to (26.1), the condition of thermodynamic equilibrium along the radius is given by the relationship

$$[\varepsilon_n(r) + V(r)]\sqrt{g_{00}(r)} = \text{const.} \tag{48.1}$$

When the star's temperature is different from zero, it is necessary to add a second condition:

$$T(r)\sqrt{g_{00}(r)} = \text{const.} \tag{48.2}$$

The values of the constants can be determined from the conditions at the surface. Below we shall examine only bodies with an absolute-zero temperature.

At the star's surface $\varepsilon_n = m_n$, $g_{00} = 1 - \frac{2M}{R}$ (M is the mass, R the radius, and we are using the system of units (34.1)), and there is no interaction energy: $V_n(R) = 0$. Using also formula (32.4), the equilibrium condition along the radius can be written in the following form:

$$g_{00}(r) = \frac{1 - 2M/R}{V_n(r)/m_n + \text{ch}^2\,\frac{t_n(r)}{4}}, \tag{48.3}$$

and in the case of configurations consisting of an ideal baryon gas

$$g_{00}(r) = \left(1 - \frac{2M}{R}\right) \mathrm{ch}^{-2} \frac{t_n(r)}{4}. \tag{48.4}$$

For the radial components of the metric tensor we have

$$g_{rr}(r) = \left[1 - \frac{2u(r)}{r}\right]^{-1} \quad (r < R), \tag{48.5}$$

where $u(r)$ is the cumulative mass (see definition (40.8)). Thus, the problem of determining the metric has been reduced to finding the functions $t_n(r)$ and $u(r)$ which, as was indicated earlier, are obtained during the process of numerical integration.

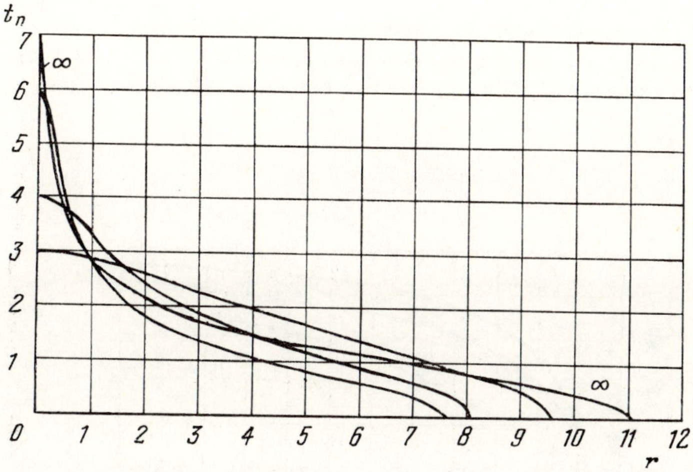

FIGURE 24. The function $t_n(r)$ for four models of baryon configurations (the ideal-gas version). The abcissa is the distance from the center in kilometers.

The curves of the functions $t_n(r)$ and $u(r)$ for several models of an ideal baryon-gas configuration are shown in Figures 24 and 25. Since $t_k = t_k(t_n)$, from these curves one can also calculate the mass density $\rho(r)$. Figures 26 and 27 show the curves $g_{00}(r)$ and $g_{rr}(r)$ for the components of the metric tensor. The values corresponding to the Euclidean case are $g_{00}(r) = -g_{rr}(r) = 1$. At the center the ordinates of $g_{00}(r)$ for all models are less than unity, with distance from the center they rise monotonically, and at infinity they tend toward 1. An interesting case is the $g_{00}(r)$ curve for a configuration with $t_n(0) = \infty$, and consequently with an infinitely high central pressure.

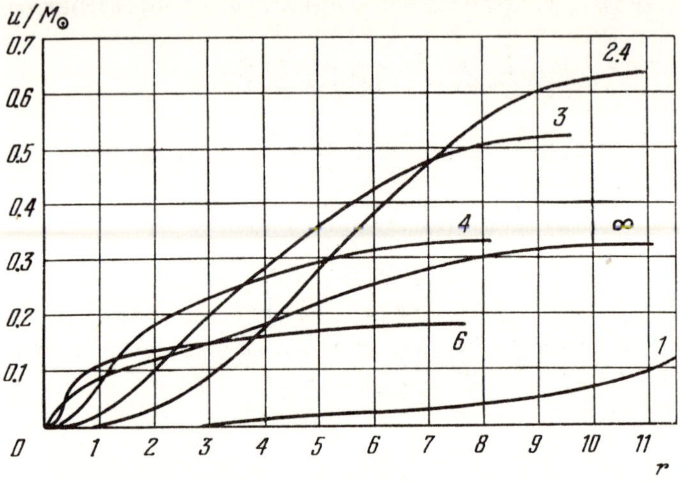

FIGURE 25. The function $u(r)$ for several models of baryon configurations (the ideal-gas version). The numbers on the curves indicate the values of $t_n(0)$. The curve for the model with $t_n(0) = 1$ extends to $r = 21.1$ km. The distance is measured in kilometers.

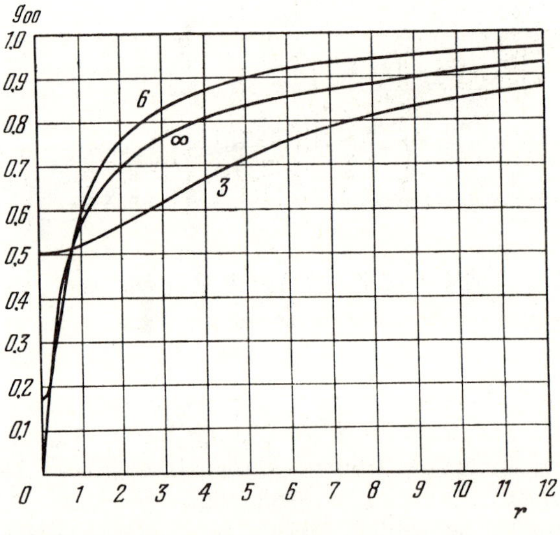

FIGURE 26. The time component g_{00} (r) of the metric tensor as a function of the distance from the center for several configurations of an ideal baryon gas. The distance is measured in kilometers. The numbers on the curves indicate the values of $t_n(0)$ for the given configurations.

Here $g_{00}(0)=0$, which means that the four-dimensional interval becomes purely spacelike. From the point of view of an outside observer, events at the center of such a configuration should proceed extremely slowly.

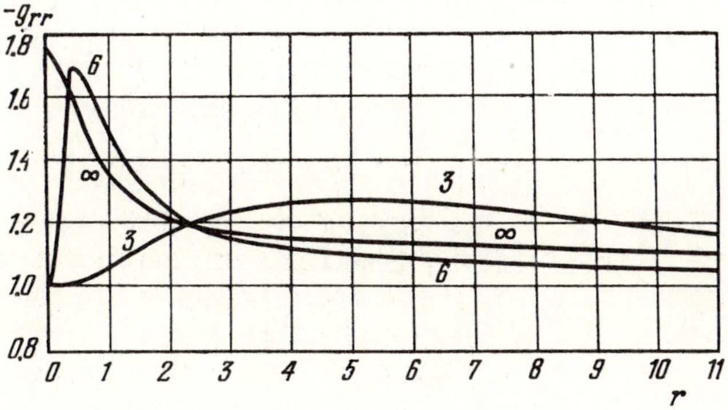

FIGURE 27. The function $g_{rr}(r)$ for several configurations of an ideal baryon gas. The numbers on the curves indicate the values of $t_n(0)$. The abscissa is the distance from the center in kilometers.

The $g_{rr}(r)$ curves start at the center with the Euclidean value, then rise up to a definite (for each configuration) distance, after which they descend slowly and tend toward the limit $g_{rr}=1$ outside the configuration. With increasing $\rho(0)$ the maximum of the function $g_{rr}(r)$ shifts toward the left, and in the limit $\rho(0)\to\infty$ it reaches the ordinate axis. For this limiting configuration, the $g_{00}(r)$ curve decreases monotonically from the center and tends asymptotically toward the limit 1. According to (42.5), in this case the function g_{rr} has the constant value 1.75 up to the distances where the solution $u=3r/14$ is valid (strictly speaking, the above value is reached only at the center).

The behavior of the functions $g_{00}(r)$ and $g_{rr}(r)$ clearly indicates that, inside the baryon configurations, the metric properties of the space are substantially different from those of the "flat" case, the space in the region of the mass distribution being strongly curved.

Figure 28 shows the $g_{rr}(r)$ curves for six configurations consisting of a real baryon gas (using the equation of state given by equations (33.4)—(33.6)). The shape of these $g_{rr}(r)$ curves, as well as their variation between configurations, is somewhat different from the corresponding ideal-gas case (the term corresponding here means configurations with the same central energy density). For $\rho(0) < 5\cdot10^{14}\ \mathrm{g\cdot cm^{-3}}$, the behavior of the curves is the same in

Figures 27 and 28, since in both cases we are then actually dealing with an ideal gas. The $g_{rr}(r)$ curves inside the star here also have a flat maximum. With increasing central density, the maximum shifts in the direction of the center, its height diminishing at the same time. With increasing $\rho(0)$ the $g_{rr}(r)$ curves also tend toward the curve of the limiting configuration with $\rho(0) \to \infty$. Here the maximum of $g_{rr}(r)$ for this extreme configuration does not occur at $r=0$, but rather somewhere around $r=3.5$ km. For this configuration, in the region $r < 0.47$ km, where the asymptotic equation of state $P \approx \rho$ is valid, according to (43.1) $g_{rr}(r) = \frac{1}{1-2u/r} = 2$ for $r \to 0$. The curves of Figure 28 extend beyond the configuration's surface to a distance $r = 11$ km. The vertical lines on the curves indicate the star's boundary. The real-gas configurations are more compact, therefore the non-Euclidean character of space is more pronounced in this case than in the case of ideal-gas configurations.

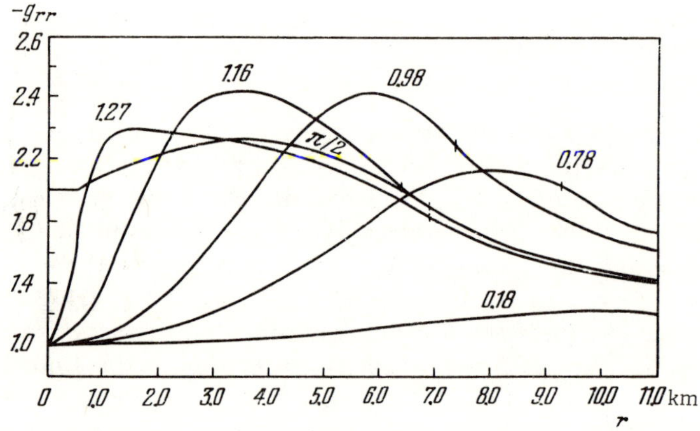

FIGURE 28. The function $-g_{rr}(r)$ for several configurations of a real baryon gas. The numbers above the curves indicate the values of the parameters $\eta = \arctan\log\frac{\rho(0)}{\rho_1}$ ($\rho_1 = 3.62\cdot 10^{14}\,\mathrm{g\cdot cm^{-3}}$ is the nuclear density).

Let us note that, according to (4.8), the properties of three-dimensional space are determined by the tensor

$$\gamma_{\alpha\beta} = -g_{\alpha\beta} + \frac{g_{0\alpha}g_{0\beta}}{g}.$$

In our case (static field), this three-dimensional tensor has only diagonal elements. In accordance with (4.7), the proper or true distance in the direction of the radius is

$$R_0 = \int_0^R \sqrt{g_{rr}}\, dr. \tag{48.6}$$

The values given in Table 10 for the true radius of baryon configurations were calculated from this formula. In this case R is the coordinate radius of the star. When $\rho(0)$ is of the order of the nuclear density or less, the above radius is not much different from R, but for denser configurations it is considerably larger.

49. THE NUMBER OF BARYONS IN A STAR

A very important characteristic of baryon configurations is the total number of baryons in the star. It is determined by the integral

$$N = 4\pi \int_0^R n(r) \sqrt{g_{rr}(r)}\, r^2 dr, \tag{49.1}$$

where $n = \sum n_k$ is the density of baryons; the summation is taken over all forms of these particles. The values of N for configurations of a real baryon gas are presented in the last column of Table 10.

Figure 29 shows the number of baryons as a function of the density at the center. From this figure it is clearly seen that the shape of the $n(\eta)$ curve is similar to that of the mass curve $M(\eta)$. This similarity is characteristic not only for real-gas models, but it also occurs in models of an ideal gas of baryons and neutrons. The locations of the extrema coincide in the two curves. This can be seen from the figures, but it can also be proven directly. To this end, let us examine the change occurring in the mass of a configuration of degenerate baryon gas when we add to it a small number of particles, without disturbing the thermodynamic equilibrium at each point and along the radius. The change in the mass of a baryon star during such an adiabatic change in the number of baryons is

$$dM = \Big(\sum_k \varepsilon_k^0 dN_k^0 + \sum_k \varepsilon_k^+ dN_k^+ + \sum_k dN_k^- \varepsilon_k^- + \sum \varepsilon_e\, dN_e\Big) \sqrt{g_{00}(r)}. \tag{49.2}$$

Here $\varepsilon_k^0, \varepsilon_k^+, \varepsilon_k^-$ are the Fermi energies of the k-th type of neutral, positive, and negative baryons, respectively, dN_k^0, dN_k^+, dN_k^- are the changes in the numbers of the corresponding baryons when the total

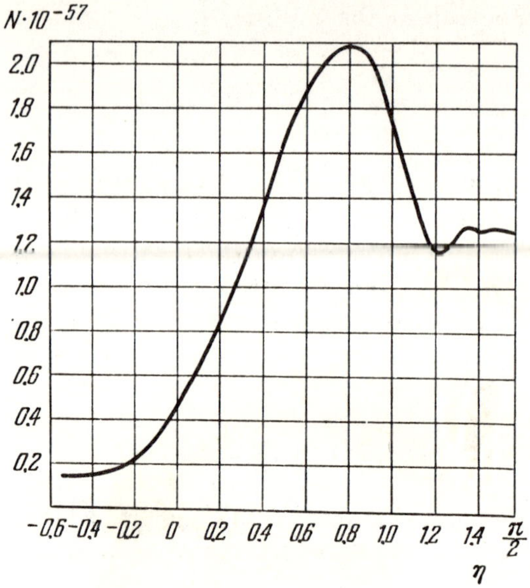

FIGURE 29. The number of baryons N as a function of the parameter $\eta = \arctan \log \frac{\rho(0)}{\rho_1}$, where $\rho_1 = 3.62 \cdot 10^{14}$ g·cm^{-3}.

number of baryons is changed by dN, and ε_e and dN_e are, respectively, the Fermi energy of the leptons in the star and the change in their number. Naturally the addition or removal of particles has to take place at the Fermi surface, so as not to disturb the equilibrium. One also has to bear in mind that in a gravitational field the quantity conserved is not ε, but $\varepsilon\sqrt{g_{00}}$. Figuratively speaking, the factor $\sqrt{g_{00}}$ allows for the gravitational potential energy of the particles. This is seen even from the fact that in the nonrelativistic case $\varepsilon\sqrt{g_{00}}$ becomes $\varepsilon(1+\varphi/c^2)$. Using the conditions of local equilibrium and neutrality

$$\varepsilon_k^0 = \varepsilon_k, \quad \varepsilon_k^+ = \varepsilon_n - \varepsilon_e, \quad \varepsilon_k^- = \varepsilon_n + \varepsilon_e, \quad \varepsilon_\mu = \varepsilon_e,$$

$$\sum_k dN_k^+ = \sum_k dN_k^- + \sum_l dN_l,$$

equation (49.2) can be reduced to the following:

$$dM = \varepsilon_n(r)\sqrt{g_{00}(r)}\,dN, \tag{49.3}$$

where dN is the change in the total number of baryons, $dN = \sum_k dN_k$. On the other hand, from the condition of equilibrium along the radius we have

$$\varepsilon_n(r)\sqrt{g_{00}(r)} = \varepsilon_n(R)\sqrt{g_{00}(R)} = m_n c^2 e^{\nu(R)/2}.$$

Taking account of this last relationship, we obtain from (49.3)

$$\frac{dM}{dN} = m_n c^2 e^{\nu(R)/2} = \text{const}. \qquad (49.4)$$

The change in mass is thus proportional to the change in the number of baryons. Since $M'(\eta) = \text{const}\ N'(\eta)$, the similarity between the $M(\eta)$ and $N(\eta)$ curves becomes obvious.

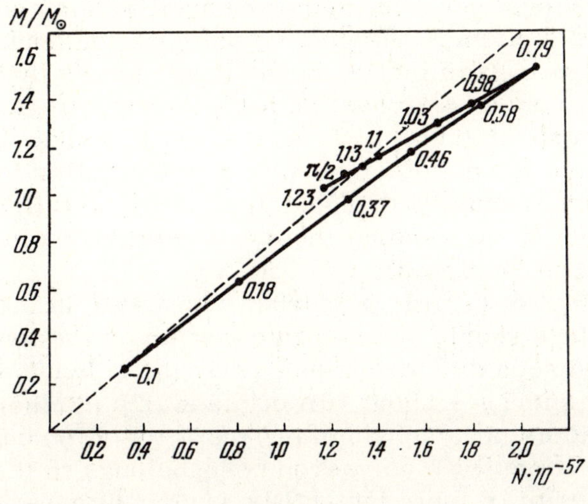

FIGURE 30. The mass of a configuration as a function of the total number of baryons N (for a real gas). The numbers next to the points indicate the values of the parameter η for the corresponding configurations.

Of special interest is the dependence of the mass of a configuration on the total number of baryons in it (Figure 30). Above a certain density, the $M(N)$ curves for all models (real and ideal baryon gas, neutron gas) are multiple-valued: for each value of N there are two, and in some regions even more, values of the mass. At the transition points between different branches of these curves, the derivative $\frac{dM}{dN}$ has a discontinuity (this was shown in /47/). In fact, in a smooth transition between the lower and upper branches, at the turning point we should have had $M'(N) = \infty$, whereas from (49.4) it follows that the derivative is finite, since $\nu(R) < 0$. Thus, at the turning points, the derivative $M'(N)$ must have different values for the upper and lower branches.

Above $\eta=1.33$ the $M(N)$ curve has a great number of turning points, which are so close to each other that they cannot be shown on the figure. The only point indicated in the figure is $\eta=\pi/2$, which corresponds to a configuration with $\rho(0)\to\infty$.

The dashed line represents the function $M=m_n N$. As we can see from the figure, only a small section at the end of the curve lies above the $M=m_n N$ line. The mass of configurations which correspond to points on this section of the $M(N)$ curve is larger than $m_p N$. The gravitational mass defect for these has an anomalous value (see Section 52). Obviously, of two (or more) equilibrium configurations corresponding to a given N, the stable configuration is the one with the smallest mass. Examination of the graph shows that a configuration with larger mass also has the higher concentration at the center. It may appear that a high central density should signify a large gravitational mass defect, resulting in a smaller total mass of the configuration. Actually, though, the situation is different. This is a special effect of the general theory of relativity, which we shall analyze in detail in Section 53.

Under the effect of certain perturbations, a star in the upper branch of the curve should convert into a state on the lower branch. Such a transition should be accompanied by the release of an enormous amount of energy — about 10% of the star's intrinsic energy.

Let us also note that a substantial part of the $M(N)$ curve, representing configurations with a hyperon core, belongs to the unstable branch. As a result of their instability, such configurations can obviously not be formed from a diffuse state of matter. They can arise only out of denser celestial bodies.

50. THE GRAVITATIONAL MASS DEFECT

According to the first equation (40.12), the mass of a spherical celestial body is

$$M=4\pi\int_0^R \rho(r)\,r^2dr. \qquad (50.1)$$

At first glance it may appear that this expression is identical with the mass equation of the nonrelativistic theory of gravitation. However, this correspondence is only formal. Equation (50.1) covers all forms of energy, including the gravitational energy, whereas in the corresponding Newtonian mass formula the energy of the gravitational field is left out. We should remember that in Einstein's theory the gravitational field is replaced by the corresponding curving of space. This clarifies the manner in which the energy is taken into account; namely, that in (50.1) the coordinate distances are used in place of the true distances.

In order to obtain the proper mass of a star, one has to multiply the proper energy density $\rho(r)$ by the element dV_0 of proper volume and integrate over the coordinates ϑ, r, φ. The element of proper volume is $dV_0 = dl_r\, dl_\vartheta\, dl_\varphi$, where, according to (4.6) and (4.7),

$$dl_r = \sqrt{g_{rr}}\, dr, \quad dl_\vartheta = r d\vartheta, \quad dl_\varphi = r \sin\vartheta d\varphi$$

are the true distances along the coordinate axes r, ϑ, φ. Thus the proper mass of a star (all energy forms, excluding the gravitational field) is

$$M_0 = 4\pi \int_0^R \rho(r) \sqrt{g_{rr}}\, r^2 dr. \tag{50.2}$$

In a weak gravitational field (when the curved space approaches flat space), the difference between the coordinate and true distances vanishes, and (50.1) then becomes (50.2).

The difference $M_0 - M = \Delta M_1$ is called the gravitational mass defect:

$$\Delta M_1 = 4\pi \int_0^R r^2 \rho(r) (1 - e^{\lambda/2})\, dr. \tag{50.3}$$

Here we have used the notation $g_{rr} = e^{\lambda/2}$. We also introduce the concept of the gravitational packing fraction

$$\alpha_1 = \frac{\Delta M_1}{M}, \tag{50.4}$$

which represents the mass defect per unit mass of the configuration. Figure 31 shows the curves for the packing fractions of configurations consisting of an ideal and a real (from the data of Table 10) baryon gas. As we can see, in both cases the mass defect is very large. For an ideal gas the packing fraction varies from a few percent to 20% for the densest configurations, and for a real gas it reaches 55%.

Beside the mass defect (50.3), an important property of the configurations is also the quantity

$$\Delta M = Nm - M, \tag{50.5}$$

which we shall call the absolute gravitational mass defect. Here N is the number of baryons in the star, and m is the sum of the rest masses of a proton and an electron (it may perhaps be more accurate to define m as the mass of a hydrogen atom). This quantity represents the difference between the mass of a diffuse hydrogen cloud and that of a baryon star with an identical number of baryons.

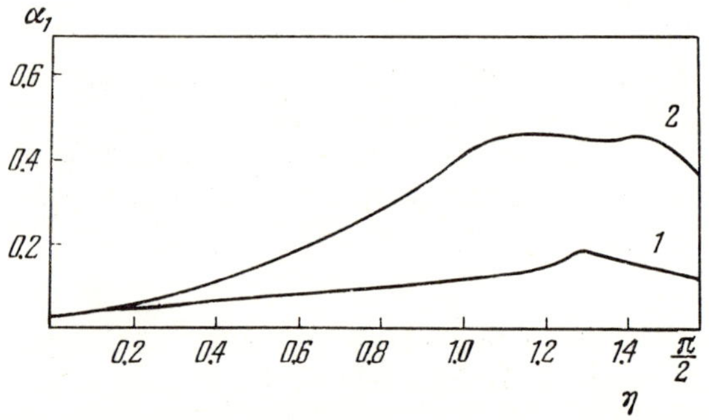

FIGURE 31. The gravitational packing fraction $\alpha_1 = \frac{\Delta M}{M}$ as a function of the parameter η. Curve 1 is for configurations consisting of an ideal baryon gas, and curve 2 is for a real baryon gas.

For ordinary celestial bodies obviously always $\Delta M > 0$. Calculations of baryon configurations have shown /15, 92, 16/ that when the central density exceeds a certain value $\rho_c(0)$, the absolute mass defect becomes negative. The value of $\rho_c(0)$ depends on the form of the equation of state used in the calculations. In models of a real baryon gas $\rho_c(0)$ is lower than in ideal-gas models. The reason for this is that in a real gas, at supernuclear densities an important role is played by nuclear repulsion forces between baryons. The repulsion forces naturally assist the appearance of the above effect (this can be seen from the fact that one has to perform some work in order to bring the particles together to the distance where the attraction forces are replaced by repulsion). From the last remark it follows that the definition (50.5) of the gravitational mass defect is not altogether correct for the real-gas model, since the effect is in this case caused not only by gravitation, but by nuclear forces as well (attraction and repulsion). Therefore, in order to exclude the effect of nuclear forces on the absolute mass defect, and to be able to study the phenomenon in purer form, we shall consider below only the models of superdense stars consisting of an ideal baryon gas. In this case, (50.5) actually represents the absolute gravitational mass defect.

The existence of baryon configurations with an anomalous (negative) mass defect was first observed in /15/. In this paper the importance of such stellar configurations for astrophysics was underlined, but the matter was not studied in detail. This problem was later studied in /47/, and it was shown that $M < fNm_n$, where M is the observed mass of the star, N is the number of neutrons in it, and $f \approx 1.5\sqrt{g_{rr}(R)}$, R being the star's radius and g_{rr} the component of the metric tensor. This estimate of the upper limit of the mass M does not exclude the possibility of the existence of equilibrium baryon configurations with $M > Nm_p$. A thorough investigation of this effect was later conducted in /18/.

51. THE ABSOLUTE MASS DEFECT IN NEWTON'S GRAVITATION THEORY

We shall begin the study of the absolute mass defect with the nonrelativistic theory of gravitation. The return to this theory may seem strange, since we already know that a consistent and correct study of the properties of baryon configurations is possible only within the framework of the general theory of relativity. We shall, though, knowingly admit a certain degree of inconsistency by examining configurations of a relativistic neutron gas by means of Newton's gravitation theory. This will be done with a definite purpose, namely, through comparison between accurate and approximate calculations to determine the role of the relativistic nature of baryon gas and the curvature of space in the problem of the absolute gravitational mass defect.

Therefore, to exclude secondary effects due to hyperons and nuclear forces, we deem it advisable to carry out this study only for configurations consisting of an ideal gas of neutrons. When employing Newton's law of gravitation, the parameters of neutron configurations are determined from the equations

$$\left.\begin{aligned} \frac{du}{dr} &= r^2(\operatorname{sh} t_n - t_n), \\ \frac{dt_n}{dr} &= -\frac{3u(\operatorname{sh} t_n - t_n)}{r^2\left(\operatorname{ch} t_n - 4\operatorname{ch}\frac{t_n}{2} + 3\right)}, \end{aligned}\right\} \tag{51.1}$$

$$\left.\begin{aligned} N &= A\int_0^R \operatorname{sh}^3 \frac{t_n}{4} r^2 dr, \\ M &= u(R)\left[1-\frac{u(R)}{2R}\right]-\frac{1}{2}\int_0^R \frac{u^2(r)\,dr}{r^2}. \end{aligned}\right\} \qquad (51.1)$$

The first equation determines the star's mass (it should be remembered that the mass density of a neutron gas is $\rho = (\operatorname{sh} t_n - t_n)/4\pi$, without taking into account the gravitational interaction between particles). For ordinary stars, $u(R)$ is very close to the observed mass. For neutron configurations, where the mass defect is comparable with the mass, $u(R)$ differs considerably from the latter. Therefore, even within the framework of Newton's gravitation theory we have to include the gravitational interaction energy between particles. The second equation was obtained from the condition of hydrodynamic equilibrium. These two equations are already familiar to us from (38.2). The third equation gives the total number of neutrons in the configuration, with $A = m_n^3 c/(3\pi^2 h^3)$ (see equation (32.3) for the particle density). In the last equation, M is the true mass of the star (i. e. the mass which includes the gravitational interaction). It is determined in the following fashion:

$$M = 4\pi \int_0^R \left[1+\frac{1}{2}\varphi(r)\right]\rho(r)\,r^2 dr, \qquad (51.2)$$

where φ is the gravitational potential at a distance r (it should be remembered that $c=k=1$). It is easy to show that

$$\varphi(r) = -\frac{u(r)}{r} - 4\pi\int_0^R \rho(r)\,r\,dr. \qquad (51.3)$$

After substituting (51.3) in (51.2) and performing a series of simple transformations, we arrive at the last equation in (51.1). The conditions at the center and at the surface are

$$u(0)=0, \quad N(0)=0, \quad t_n(0), \quad t_n(R)=0.$$

The results of numerical calculations /18/ are given in Table 8. The values of the absolute gravitational packing fraction $\Delta M/M$ are presented in the last column of the table. The dependence of the mass on arctan $t_n(0)$ is shown in Figure 14 (upper curve), and the number of neutrons in Figure 32 (upper curve). We shall return to a discussion of the results in the following section.

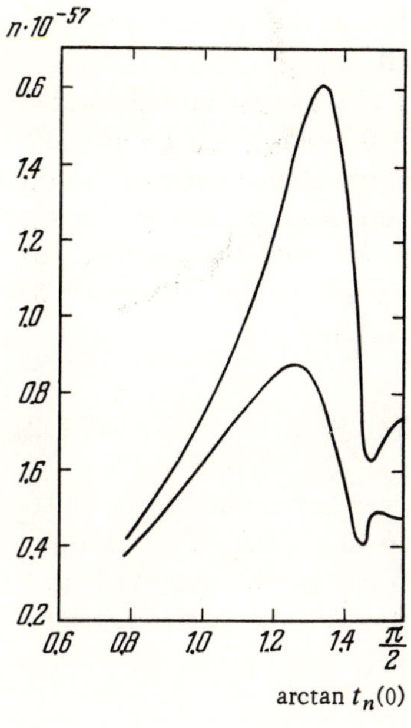

FIGURE 32. The number of particles in neutron stars as a function of arctan $t_n(0)$ according to the gravitation theories of Einstein (lower curve) and Newton.

52. THE ABSOLUTE MASS DEFECT IN EINSTEIN'S THEORY

When starting from Einstein's theory of gravitation, configurations consisting of an ideal neutron gas are determined by the following system of equations:

$$\left.\begin{aligned} &\frac{du}{dr} = r^2(\operatorname{sh} t_n - t_n),\\ &\frac{dt_n}{dr} = -\frac{4\left(\operatorname{sh} t_n - 2\operatorname{sh}\frac{t_n}{2}\right)}{r(r-2u)\left(\operatorname{ch} t_n - 4\operatorname{ch}\frac{t_n}{2}+3\right)}\times\\ &\qquad\times\left[u+\frac{1}{3}r^3\left(\operatorname{sh} t_n - 8\operatorname{sh}\frac{t_n}{2}+3t_n\right)\right],\\ &N = A\int_0^R\left(1-\frac{2u}{r}\right)^{-1/2}\operatorname{sh}^3\frac{t_n}{4}\,r^2dr. \end{aligned}\right\} \tag{52.1}$$

The first two equations we have already encountered in (41.1). The last equation gives the number of neutrons. The expression under the integration sign is the product of the particle density and the element of proper volume $4\pi\sqrt{g_{rr}}\,r^2dr$. The boundary conditions are $u(0)=0$, $N(0)=0$, $u(R)=M$, $t_n(0)$, $t_n(R)=0$.

The numerical integration was carried out by computer, and the results are given in the first half of Table 8. Figures 14 and 32 show the dependence of the mass and number of baryons of a configuration on the parameter arctan $t_n(0)$, which describes the density at the center (see (32.3)). The similarity between the curves for the mass and for the number of neutrons is striking. Figures 33 and 34 show the curves for the packing fraction

$$\frac{\Delta M}{M_0}=\frac{M_0-M}{M_0}, \quad M_0=Nm_n. \tag{52.2}$$

The legends under the figures give sufficient information concerning the main points of the graphs, therefore we shall forgo their detailed discussion. Let us only point out some of the general and most important features of the functions described in the figures.

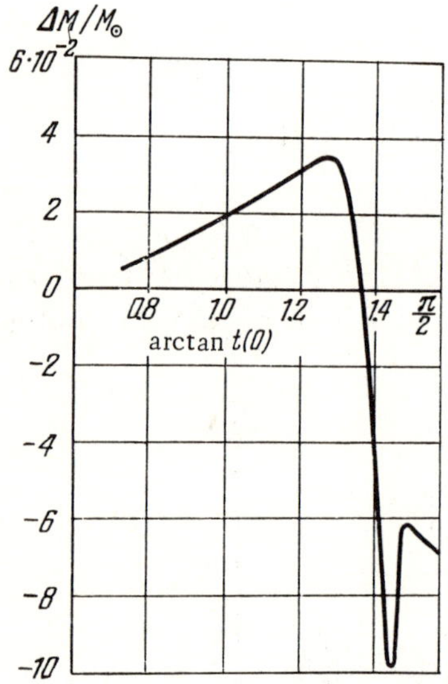

FIGURE 33. The packing fraction of neutron configurations as a function of arctan $t_n(0)$ according to Einstein's gravitation theory. $\Delta M=Nm_n-M$ is the absolute gravitational mass defect.

We shall not dwell on a discussion of the mass and radius curves, which were already considered in Sections 38 and 41. The $N(y)$ curves, where $y=\arctan t_n(0)$, for the number of particles, are similar in the Newton and Einstein gravitation theories (in analogy with the $M(y)$ and $R(y)$ curves), although there are considerable differences in the values and in some other details. In both cases N is a single-valued function of y. The inverse statement, as we see from

Figure 32, is generally not correct, since in certain sections $y(N)$ is a multiple-valued function. The characteristic maxima, minima, and attenuating oscillation at the end of the mass and radius curves are preserved in the curves for the number of particles and the packing fraction (Figure 33). Here we shall not show the curve for the $M(N)$ function — it is similar to Figure 30 for a real baryon gas. Instead, we give the graph for the packing fraction as a function of the number of particles. The characteristic cusps occurring in the $M(N)$ curve have evidently been preserved here, and their cause was explained in Section 49.

In all Newtonian models of neutron configurations, the absolute gravitational mass defect has a positive value. As we can see from the last column of Table 8, the packing fraction increases with the central density, except in the region in which the latter is extremely high (see row before last in Table 8, where $t_n(0) \approx 12$), where a noticeable minimum occurs.

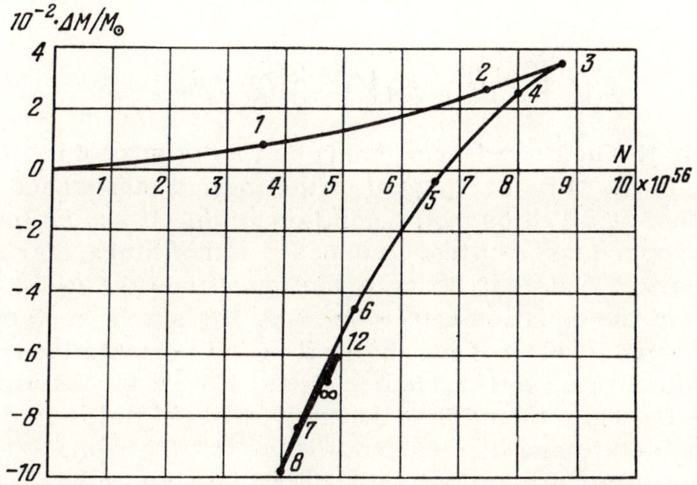

FIGURE 34. The packing fraction of neutron configurations ($\Delta M = Nm_n - M$) as a function of their total number of neutrons. The numbers next to the heavy dots indicate the values of the parameter $t_n(0)$ for the corresponding configurations.

In the case of relativistic models, the packing fraction displays an anomalous behavior (see Figure 33). At the outset it increases with the central density, reaching a maximum at $y = 1.27$ ($t_n(0) = 3.34$; $n(0) = 3 \cdot 10^{39}$ cm^{-3}), after which it starts decreasing. At $y > 1.36$ ($t_n(0) > 4.67$; $n(0) > 1.12 \cdot 10^{40}$ cm^{-3}) the packing fraction becomes negative, with a deep minimum of about -0.1 at the point $y = 1.45$ ($t_n(0) = 8.2$;

$n(0)=2\cdot 10^{41}$ cm^{-3}). After this, oscillating with a small and attenuating amplitude, and remaining negative when $\rho(0)\to\infty$, it converges toward the limit -0.069. In the following section we shall attempt to explain this anomaly in the absolute mass defect.

53. THE NATURE OF THE ANOMALOUS MASS DEFECT

We shall now consider in more detail the anomalous behavior of the mass defect. In the mass-defect problem, the results obtained from the Newton and Einstein gravitation theories differ not only quantitatively, but qualitatively as well. This indicates clearly that the curvature of space is here an important factor, i. e. we are dealing with an effect which is specific to the general theory of relativity. Let us look into this matter in somewhat more detail.

From the condition of equilibrium along the star's radius it follows that

$$\varepsilon_n(r)\sqrt{g_{00}(r)} = m_n\sqrt{1-2M/R} < m_n, \tag{53.1}$$

where $\varepsilon_n(r)$ is the neutron Fermi energy at a distance r in the units of (34.1), g_{00} is the time component of the metric tensor, and M and R are the mass and radius of the configuration. It should be noted that in cold superdense bodies, as opposed to hot stars, there is a sharp boundary. Thus, for neutron configurations the radius is determined from the equation $\varepsilon_n(r) = m_n$, i. e. the star's surface is where the Fermi level crosses the wall of the "gravitational potential well." According to (53.1), the highest Fermi level is inside the potential well. This means that the individual particles in the star are in a bound state and hence cannot leave it. To remove a particle from the star, a certain amount of work would have to be performed. On the other hand, calculations show that for some of the densest configurations

$$M > \sum_k m_k N_k \tag{53.2}$$

(for neutron configurations this is evident from the data in the fifth column of Table 8). How can these two facts be reconciled? Actually, though, there is no contradiction between (53.1) and (53.2), if it is remembered that the total internal energy of a star is not equal to the sum of the energies (kinetic and potential) of its constituent particles.

If the internal energy were additive, then the mass would be

$$M_1 = 4\pi \int_0^R \psi(r)\,\rho(r)\,r^2\,dr < Nm_n, \tag{53.3}$$

where

$$\psi(r) = \sqrt{-g_{00}(r)\,g_{rr}(r)}.$$

On the other hand, according to the first equation in (52.1) the observed mass of the star is

$$M = 4\pi \int_0^R \rho(r)\,r^2 dr. \tag{53.4}$$

Comparison between (53.3) and (53.4) shows that the energy is not additive, the function ψ being a measure of the deviation from additivity. Figure 35 shows this function for configurations with $t_n(0) = 4, 5, 6, 7, \infty$. These curves are similar to those of $g_{00}(r)$ (see Figure 26). At the center ψ has a minimum, then it rises monotonically with r, and at the surface it is close to unity. Had the energy been strictly additive we would have had $\psi = 1$, with the maximum deviation from additivity occurring when $\psi = 0$. The value of $\psi(r)$ in the central region diminishes with increasing $t_n(0)$. Hence the largest contribution to the nonadditivity of the energy is from the central region.

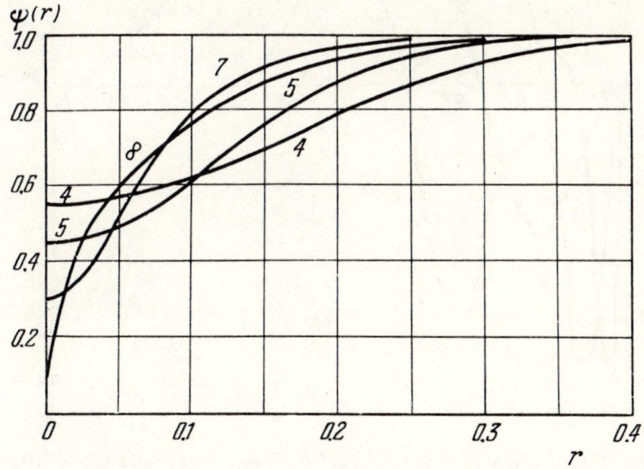

FIGURE 35. The function $\psi(r) = \sqrt{-g_{00}(r)\,g_{rr}(r)}$, where r is the radial distance and g_{00} and g_{rr} are the time and radial components of the metric tensor. Distance is measured in the units $c = k = 1$, $K_n = \frac{1}{4\pi}$. The values of $\psi(r)$ indicate the extent to which the internal energy is nonadditive.

The clearest picture of the region in the star accounting for the negative mass defect can be obtained from Figure 36, which describes the function

$$\Phi(r) = \left[1 - \left(1 - \frac{2u(r)}{r}\right)^{1/2} \frac{\rho(r)}{m_n\, n(r)}\right] \left(1 - \frac{2u(r)}{r}\right)^{-1/2} n(r)\, r^2, \qquad (53.5)$$

where $n(r)$ is the density of the number of neutrons. The integral of this function determines the absolute mass defect:

$$\Delta M = 4\pi m_n \int_0^R \Phi(r)\, dr. \qquad (53.6)$$

The function $\Phi(r)$ is negative for small r, has one minimum and one maximum, and becomes zero at the center and at the surface. As we can see, the area below the abscissa increases with $t_n(0)$, and already at $t_n(0) > 4.7$ the algebraic sum of the areas between $\Phi(r)$ and the abscissa becomes negative.

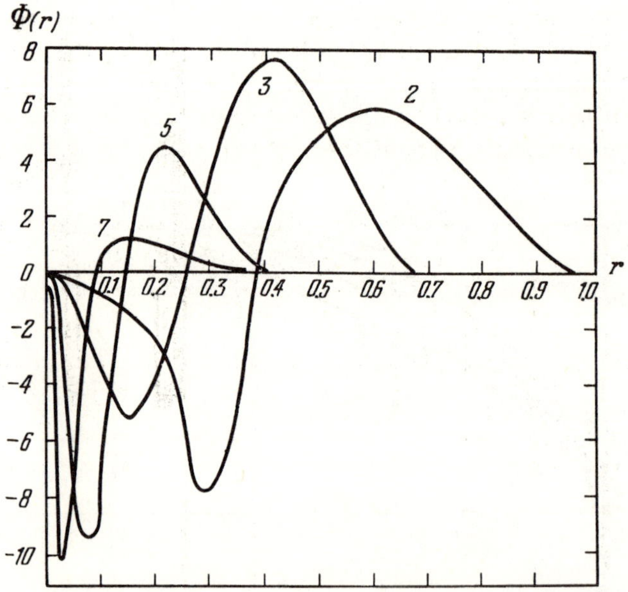

FIGURE 36. The function $\Phi(r)$, where r is the distance from the star's center in the units $c=k=1$, $K_n = \frac{1}{4\pi}$. The numbers on the curves indicate the values of $t_n(0)$ for the corresponding configurations. The ordinate scale differs from curve to curve; for $t_n(0) = 2, 3, 5, 7$, a unit scale is a, $2a$, $5a$, $10a$, respectively, a being a known constant.

We thus arrive at the conclusion that the anomaly in the mass defect is due to the catastrophic deviation from additivity in the internal energy, caused by curving of the space metric in the most compact baryon configurations. In the case of these configurations there is a strong deviation from energy additivity even in the Newtonian approximation (here the kinetic energy is additive and the potential energy is not), but it is not sufficient to change the sign of the mass defect. In fact, from the virial theorem it follows that /66/

$$\Delta M = \sum_k m_k c^2 \left(1 - \frac{m_k c^2}{\varepsilon_k}\right) > 0. \tag{53.7}$$

We shall show that, to prove this relationship, periodicity or quasi-periodicity of the particles' motion does not necessarily have to be assumed.

Let us return to Figure 34. It is obvious that configurations with $t_n(0) < 4.67$ on the lower branch of the $\Delta M/M$ curve are unstable with respect to transitions to the upper branch, where the mass defect displays a normal behavior. On the other hand, configurations with $t_n(0) > 4.67$ having a negative absolute mass defect are unstable not only with respect to transitions to the upper branch, but also with respect to dissociation into a diffuse state (escape of matter into infinity). Since the mass defect constitutes several percent of the star's mass, the amount of energy released during such transitions would be inconceivably high. This energy, per gram of stellar matter, is higher by an order of magnitude than the corresponding energy released by thermonuclear reactions in the burning of hydrogen. It should be noted that the binding energy of each individual particle in the star is negative, therefore the particles cannot escape singly to infinity. The removal of baryons from the star requires an addition of outside energy to the remaining configuration. Hence this process cannot take place spontaneously. This means that the system can convert into the more stable state only under the effect of very strong perturbations. At the same time there is an expansion, accompanied by heating of the star. The corresponding transition will have the nature of a cosmic explosion.

54. MODELS OF SUPERDENSE MASSIVE BODIES

The anomalous mass defect can have an important bearing on the cosmogonic concepts developed by V. A. Ambartsumyan /3—13/, if there exist celestial bodies with this property, but with a mass many orders of magnitude above the sun's. These could serve as models for prestellar bodies. To avoid confusion, let us note that in Ambartsumyan's theory the prestellar bodies do not necessarily have to be superdense in the same sense as used in the present work. He essentially postulates only that the evolution of matter in our epoch is from dense to rarefied, as opposed to the alternative concept (diffuse prestellar bodies), in which the opposite is believed. Now the hypothesis of arbitrarily dense prestellar bodies, far from being in conflict with these concepts, is even in harmony with them. From the material presented in the last three chapters, we have seen that the mass of static baryon configurations is only of the order of the sun's. But prestellar bodies are obviously nonstationary objects, with fairly long lifetimes. It should therefore be checked whether it is possible to develop models of supermassive stellar bodies which are at the same time also superdense and nonstationary, and have the necessary lifetimes. Generally speaking, this problem is very difficult, since it involves solving a system of several differential equations involving partial derivatives with respect to r and t. With the aim of determining the main features, we shall consider here only a highly simplified model for the initial state of such configurations /101/.

In constructing a model for superdense celestial bodies whose masses exceed the sun's, there is a difficulty related to a singularity in the metric, which imposes a limitation on the density. For a body not to be inside the Schwarzschild sphere, the average density should not exceed the value $\bar{\rho} \approx 2 \cdot 10^{16} (M_\odot/M)^2 \, g \cdot cm^{-3}$ (see 40.15). However, this difficulty is due to the inapplicability of the Schwarzschild metric to distances below the gravitational radius. It is a known fact /183/ that a body cannot be in a static state if its radius is smaller than the gravitational radius, hence in this case the metric depends on time (see /66/, Section 100). In /186/ it was shown that, in contrast to the case of a static metric, where the singularity cannot be removed through any coordinate transformation, with a nonstationary metric this is entirely feasible. Examples of nonstatic metrics without the singularity were given in /85/. For the region outside the mass distribution we shall use here the Lemaître metric (9.6):

$$ds^2 = dt^2 - \left(\frac{3}{2}\,\frac{R+t}{2M}\right)^{-2/3} dR^2 - 4M^2 \left(\frac{3}{2}\,\frac{R+t}{2M}\right)^{4/3} (d\vartheta^2 + \sin^2\vartheta\, d\varphi^2), \qquad (54.1)$$

where M is the mass of the star ($c=k=1$). The reference system determined by the metric (54.1) is defined by test bodies which move freely along the geodesics in the direction opposite to that of free fall (see end of Section 9). At infinity, it asymptotically approaches the Galilean frame of reference.

Let us examine nonstationary objects which are inside the singular sphere of the Schwarzschild system. We shall show that, in this case, it is possible that there are bodies with arbitrarily large densities and masses, which are subject to anti-collapse. To write the metric of the space occupied by the mass distribution, we shall use the accompanying system of reference:

$$ds^2 = e^{\sigma} d\tau^2 - e^{\omega} dR^2 - r^2 (d\vartheta^2 + \sin^2\vartheta\, d\varphi^2), \tag{54.2}$$

where σ, ω, r are functions of the coordinates R and τ.

In solving the evolutional problem of cosmogony, one should have taken into account the energy sources and also its processes of transport and loss. In a qualitative discussion of the general behavior of configurations with respect to collapse and anti-collapse, though, it is not necessary to get involved in such complex problems. In the simplified problem before us, we have for the energy-momentum tensor

$$T_1^1 = T_2^2 = T_3^3 = -P(R,\tau), \quad T_0^0 = \rho(R,\tau). \tag{54.3}$$

The Einstein equations which correspond to (54.2) and (54.3) can be reduced to the following convenient form /84/:

$$\dot{u} = -4\pi P r^2 \dot{r}, \tag{54.4}$$

$$u' = 4\pi\rho r^2 r', \tag{54.5}$$

$$\sigma' = -\frac{2P'}{P+\rho}, \tag{54.6}$$

$$\dot{\omega} = -\frac{2\dot{\rho}}{P+\rho} - \frac{4\dot{r}}{r},$$

$$\rho = P(\rho). \tag{54.7}$$

The dots and primes signify partial derivatives with respect to τ and R, respectively. As before, and with certain reservations, the function $u(R,\tau)$ defines the mass contained in a sphere of radius R:

$$u(R,\tau) = \frac{1}{2} r \left(1 + \dot{r}^2 e^{-\sigma} - r'^2 e^{-\omega}\right). \tag{54.8}$$

In the case of stationary Schwarzschild metric, $r=R$, $\omega=\lambda$, $\dot{r}=0$, $r'=1$ and (54.8) becomes (40.8).

The solution of the system (54.4)—(54.7) for the internal region must be matched with the Lemaître metric at the surface. From the requirement that the components of the metric tensor and their first derivatives be continuous at the surface we obtain

$$u(a, t)=M, \tag{54.9}$$

where a is the radius of the star.

The initial conditions can be conveniently written in the form

$$r(R,0)=2M\left(\frac{3R}{4M}\right)^{2/3} f(R),\ e^{\omega(R,0)}=\left(\frac{4M}{3R}\right)^{2/3}\psi(R),\ R\leqslant a. \tag{54.10}$$

There is a great degree of arbitrariness in the choice of the functions $f(R)$ and $\psi(R)$. The only restriction imposed on them is related to the continuity of the metric-tensor components and their derivatives at the configuration's surface. Below we shall assume that $f(R)=\psi(R)=1$. With this choice, a continuous transition between the interior and exterior metrics is automatically ensured.

The function $\sigma(R, 0)$ is determined by the exterior solution (54.1) and equation (54.6). Thus, integrating (54.6) for $\tau=0$ from the star's surface to the distance R, we find

$$\sigma(R,0)=-\int_0^{P(R,0)}\frac{dP}{P+\rho}. \tag{54.11}$$

From (54.10) we have

$$r'^2(R,0)\,e^{-\omega(R,0)}=1. \tag{54.12}$$

It then follows from (54.8) that

$$\dot{r}(R,0)=\left[\frac{u(R,0)}{M}\right]^{1/2}\left(\frac{4M}{3R}\right)^{1/3} e^{\frac{1}{2}\sigma(R,0)}, \tag{54.13}$$

which assumes the value $(4M/3R)^{1/3}$ at the surface, as was expected from (54.1). A solution with $\dot{r}<0$ is also possible, but it cannot be matched with the exterior solution (54.1); it corresponds to the metric of a collapsing celestial body (see equation (9.6), upper sign).

It now remains to assign the initial mass distribution $\rho(R, 0)$. It can, however, be easily verified that, regardless of the actual form of the function ρ $(R, 0)$ chosen for the initial condition of the celestial body, it subsequently has to expand. This was practically predetermined by the fact that the interior metric was matched at the surface to the external metric (54.1). We shall show that the expanding celestial body should have an anomalous mass defect $\Delta M = Nm_p - M$:

$$M = 4\pi \int_0^a \rho(R,0)\, r^2(R,0)\, r'(R,0)\, dR,$$

$$M_0 = 4\pi m_p \int_0^a n(R,0)\, e^{\frac{1}{2}\omega(R,0)}\, r^2(R,0)\, dR.$$

Remembering (54.12), we obtain from this for the absolute mass defect

$$\Delta M = 12\pi M \int_0^a \left(\frac{m_p}{\bar{\varepsilon}} - 1\right) \rho(R,0)\, R\, dR, \qquad (54.14)$$

where $\bar{\varepsilon} = \rho(R, 0)/n(R, 0)$ is the average energy of the baryons and $n(R, 0)$ is their initial density. Since $\varepsilon > m_p$, it is obvious that $\Delta M \leqslant 0$. Thus the model under consideration has an anomalous mass defect, and is therefore unstable with respect to a transition into a diffuse state. We see that there is in fact no restriction on the choice of ρ $(R, 0)$. At densities which are appreciably higher than nuclear we have $m_p/\bar{\varepsilon} \ll 1$, and it then follows from (54.14) that ΔM is of the order of M itself. Such configurations will hold enormous reserves of internal energy. In Section 69 we shall return to the question of internal energy sources in the cosmogony of superdense celestial bodies.

The above model can serve as an illustration for expanding massive celestial bodies with an anomalous mass defect. Needless to say, many such models of nonstatic configurations can be constructed by changing the initial conditions in the internal region in a suitable manner, so as not to disturb the match with the exterior metric.

An important problem is also the time spent by a massive body in the superdense state, from the point of view of an observer at a very large distance. To examine this question, let us consider the motion in the Schwarzschild system of a test body, which is at the surface and moves with it along a four-dimensional geodesic in the direction of increasing r. The dependence of the radial coordinate r on t in the Schwarzschild field is given by

$$t = \frac{\varepsilon}{m} \int \frac{dr}{\left(1 - \frac{r_g}{r}\right) \sqrt{\frac{\varepsilon^2}{m^2} + \frac{r_g}{r} - 1}} \qquad (54.15)$$

(see /66/, equation (100.5)). Here ε and m are the energy and rest mass of the test body, and $r_g = 2M$ is the gravitational radius of the celestial body. This equation can be applied only to $r \geqslant r_g$, where the Schwarzschild metric is not devoid of physical meaning. As we can see, when $r \to r_g$, i. e. when the test body, and with it also the expanding surface of the configuration, are infinitely close to the singular Schwarzschild surface, the time diverges: $t \to \infty$. This means that, from the point of view of an outside observer, the detachment of the star's expanding surface from the singular Schwarzschild surface takes an infinitely long time. It is not difficult to see that the corresponding proper time will be finite. If the mass is of the order of the Galaxy's mass, then the body will be in an extremely rarefied state, in the usual sense of this word, long before it reaches the singular Schwarzschild surface at $r = r_g$. From the point of view of a Lemaître observer, the time spent by the celestial body in a superdense state will be smaller than r_g/c. It is difficult to judge what the value of this time is from the point of view of an outside observer, since the Schwarzschild metric is not valid for $r < r_g$ (a formal application of the transformation formula (9.5) leads to the conclusion that the time interval during which the star's surface moves from $r = 0$ to $r = r_g$ is equal to $0.281\, r_g/c$).

There is still one weak spot in our arguments, which we have tacitly avoided. This is the fact that the rate of change of the spherical component of the metric tensor at the surface of the configuration exceeds the velocity of light. In fact, from (54.13) we find $\dot{r}(a,\ 0) = (4M/3R)^{1/3} > 1$ (let us remember that $c = 1$), since when $M \gg M_\odot$ we have $R < M$ for a superdense body.

In our opinion, the singularity in the Schwarzschild solution is not removed or avoided when using the Lemaître system (or another system which is equivalent to it), but only concealed. Actually, this singularity is transferred from one component of the metric tensor to another. In our opinion, these difficulties are inherent in Einstein's gravitation theory itself.

55. THE STABILITY OF EQUILIBRIUM BARYON CONFIGURATIONS

Not all equilibrium baryon configurations corresponding to the $M(\eta)$ curve of Figure 20 are stable. For certain ranges of the central density ρ_0 the configurations are unstable, and with some perturbations these can pass into other, more stable, states.

This statement is obvious for configurations with $\eta > 0.79$ (let us remember that $\eta = \arctan \log \frac{\rho_0}{\rho_1}$, where $\rho_1 = 3.64 \cdot 10^{14}$ g·cm^{-3}), if we return to Figure 30. For $\eta > 0.79$ the function $M(N)$ becomes double-valued, and even multiple-valued at still higher values of η. For a given number of baryons N, the stable state is of course the one with the smaller mass. However, from merely examining the $M(N)$ curves it is difficult to reach a conclusion regarding the stability of the remaining configurations, for which $\eta < 0.79$ (these belong to the lower branch of the $M(N)$ curve). As mentioned earlier, in this kind of analysis the question of the lifetime of unstable configurations remains completely open.

The stability of baryon configurations has been studied by many investigators /269a, 237a, 270a, 267/. There are two basically different approaches to this problem. The first is the dynamic method, in which the behavior of an equilibrium stellar configuration is analyzed with respect to radial oscillations. The second is the static approach, in which the stability problem is solved through comparison between adjacent configurations on the $M(\rho_0)$, $R(\rho_0)$, or $M(R)$ curves. One particular example of the static method is the analysis carried out with the help of the $M(N)$ curve. The more interesting, though, is the dynamic method, which has the advantage that simultaneously with solving the stability problem one can also determine the normal modes of the stars and the lifetimes of unstable configurations. For details the reader should turn to the review article /267/. Here we shall give only a brief summary of the results obtained with the dynamic method /237a/.

To find out whether a configuration is stable or unstable, its behavior is studied under a small radial perturbation $r \rightarrow r + \xi$, where $\xi = \xi(r, t)$. Subsequently the corresponding expression for the action is formed, and the Euler equations for the radial perturbation ξ are obtained from the principle of least action. In general, every equilibrium configuration is characterized by a discrete set of radial normal modes

$$\xi(r,t) = \xi_n(r)\, e^{i\omega_n t}, \tag{55.1}$$

which differ from each other by the value of the index: $n = 0, 1, 2, 3, \ldots\ldots$, with $\omega_0^2 < \omega_1^2 < \omega_2^2 \ldots$ Substituting (55.1) in the equation of motion, we find the equations for the amplitude ξ_n and the square of the frequency ω_n^2 of the n-th normal mode. The equation for ξ_n is of the second order with respect to r. A radial normal mode of vibration will be a solution of this equation which satisfies the boundary conditions

$$\xi_n(0) = 0, \quad \delta P(R) = 0 \tag{55.2}$$

(during the motion the center of the star remains fixed, and the pressure at the surface is zero). We are thus faced with the problem of finding the eigenvalues and the eigenfunctions. This is solved through trial and error. The normal mode is stable (the oscillations are periodic) if its frequency is real, i. e. $\omega_n^2 > 0$, and unstable if $\omega_n^2 < 0$, i. e. the frequency ω_n is imaginary. Instability of some mode also means instability of the configuration as a whole, since when this vibration is excited its amplitude will grow exponentially with time.

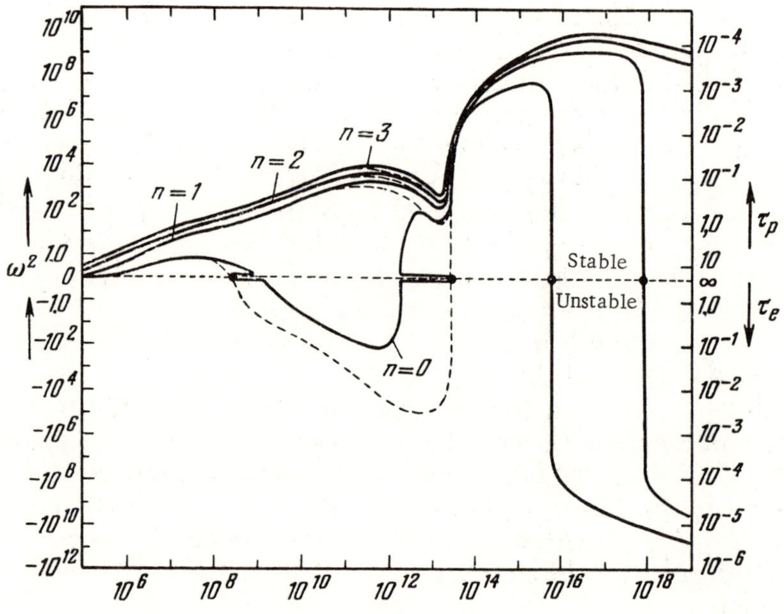

FIGURE 37. The squares of the angular frequencies ω_n^2 of the normal radial modes with $n = 0, 1, 2, 3$ as a function of the central frequency ρ_0. On the right along the ordinate are given the corresponding pulsation periods τ_p and the times τ_e during which the amplitude of unstable modes increases by a factor of e. The frequency is measured in sec^{-1}, the time in seconds, and the density in $g \cdot cm^{-3}$.

In /237a/ the dynamic method was used to calculate the squares of the frequencies of the first four normal modes of the radial pulsations of baryon configurations. A visual picture of these results can be formed from the $\omega_n^2 = \omega_n^2(\rho_0)$ curves of Figure 37.

When $\omega_n^2 > 0$ the mode is stable. When the corresponding mode becomes excited, the configuration performs harmonic oscillations. Such oscillations obviously do not disturb the configuration's stability but, on the contrary, should attenuate with time, if the possible pulsation-energy losses are taken into account. If, on the other hand, $\omega_n^2 < 0$, then the pulsation amplitude increases exponentially with time:

$$\xi(x, t) = \xi_n(x)\, e^{|\omega_n| t},$$

and therefore the configuration ultimately breaks up. On the right-hand side of Figure 37 are given the oscillation periods $\tau_p = 2\pi/\omega_n$ and the times $\tau_e = 1/|\omega_n|$ during which the pulsation amplitudes of unstable modes increase by a factor of e. At the points marked by heavy dots there is a change in the configuration stability, and they correspond to extrema in the $M(\eta)$ curve of Figure 20. A certain difference between the ρ_0 values of Figures 20 and 37 is due to the use of different equations of state in calculating the parameters of the baryon configurations.

Thus, returning to Figure 20, one can say on the basis of the above that the branch with $\eta < -1.36$ to the left of the first maximum, and the branch $-0.5 \lesssim \eta \lesssim 0.8$ of the white-dwarf configurations, represent completely stable configurations, and the configurations with $-1.36 \lesssim \eta \lesssim -0.6$ and $\eta \gtrsim 0.8$ are unstable. The numbers τ_e on the right-hand ordinate axis of Figure 37 point to the lifetimes of the unstable stars.

According to Figure 37, the period of the fundamental mode for white dwarfs is typically of the order of 10 sec, whereas for neutron configurations (in the ideal-gas version) it is $\sim 10^{-3}$ sec. When the equation of state includes the contribution of nuclear forces (the real-gas version), the fundamental period of neutron stars decreases to $\sim 0.3 \cdot 10^{-3}$ sec. When $\rho_0 \gtrsim 10^{19}$ g·cm^{-3}, more and more modes become unstable.

The eigenfunctions corresponding to the frequencies of Figure 37 were calculated in /237a/. It was shown that for stable white dwarfs and neutron stars the fundamental vibrational mode is very close to uniform, $\xi_0(r) \sim r$, and the harmonics are close to sinusoidal. For the majority of unstable configurations, the eigenfunctions of the normal modes are different from uniform or sinusoidal. In the same work the pulsation energy was also calculated for stable neutron stars (ideal-gas version) in the fundamental mode. The pulsation energy was found to be

$$E_p \approx 10^{52}\left(\frac{\xi}{r}\right)^2, \tag{55.3}$$

where ξ/r is the relative oscillation amplitude; the energy is measured in ergs. In the case of configurations with a real neutron gas (i.e. when the equation of state includes the energy of nuclear interactions between nucleons), the frequency of the fundamental mode is increased by a factor of about 3, and the energy by a factor of $\sim 3^2$; consequently,

$$E_p \approx 10^{53}\left(\frac{\xi}{r}\right)^2. \tag{55.4}$$

For comparison, let us point out that the total internal energy of the sun is $M_{\odot}c^2 = 1.8\cdot 10^{54}$ erg. As we can see, the pulsation energy of baryon stars is very high. The fact that the pulsation energy of neutron stars (which, it was assumed, are formed in outbursts of supernovae) may be very high was first established in /202a, 184a, 149a/ on the basis of Newtonian considerations.

The importance of baryon-star pulsations in astrophysics depends on their damping rate. There are several channels for energy losses, which lead to the suppression of radial oscillations. Let us enumerate the most important of these: 1) modified URCA processes

$$n + n \rightarrow n + p + e + \bar{\nu},\quad n + p + e \rightarrow n + n + \nu, \tag{55.5}$$

which take place during raising and lowering of the Fermi level /184a, 184b, 237a, 192a/; these are reactions analogous to (55.5), which proceed with the participation of μ^- and Σ^- particles /270/; 2) shock waves, generated in the atmosphere of a pulsating star /184a, 149b/; 3) hydromagnetic waves, generated by the pulsations in the magnetic field of the star /149a, 149b/; 4) electromagnetic waves, emitted by the oscillating magnetic field and the surrounding plasma /149b/; 5) coupling between normal radial and nonradial modes as a result of fast rotation and subsequent damping through the emission of gravitational waves caused by the nonradial pulsations /277a, 275a/. The relative importance of the above processes has not yet been exactly determined. Rough calculations have shown, however, that if during the first few hours after its formation a baryon star does not succeed in ridding itself of a substantial part of its angular momentum, then, owing to gravitational emission (resulting from the presence of nonradial pulsations), the radial oscillations will be damped to a negligible amplitude after a time of approximately one day. The damping time is expressed in terms of the eccentricity e of the rotating configuration by

$$\tau \sim \frac{1}{e^2}\text{sec}. \tag{55.6}$$

We would like to emphasize once again that gravitational waves cannot be emitted from purely radial vibrations of a spherical mass distribution (in analogy with the pulsations of a spherical distribution of electrical charges). These waves can be emitted only during nonradial oscillations (quadrupole radiation). According to /275a, 277a/ the gravitational emission is very fast, that is, the nonradial pulsations of baryon stars are quenched within ~1 minute, leaving only the radial oscillations.

TABLE 14. Some characteristics of superdense cold configurations rotating with the maximum angular velocity

$\rho(0)$, g·cm^{-3}	$M/M_\odot$	$10^{-50}E_r$, erg	τ_{min}, sec	$\rho(0)$, g·cm^{-3}	$M/M_\odot$	$10^{-50}E_r$, erg	τ_{min}, sec
$1.98\cdot10^{6}$	0.500	0.0624	23,0	$8.78\cdot10^{14}$	0.984	298	$7.34\cdot10^{-4}$
$5.41\cdot10^{7}$	1.06	0.457	583	$1.14\cdot10^{15}$	1.18	467	$6.45\cdot10^{-4}$
$7.04\cdot10^{8}$	1.26	0.930	2.17	$1.65\cdot10^{15}$	1.39	7940	$5.28\cdot10^{-4}$
$2.09\cdot10^{9}$	1.27	1.12	1.42	$2.44\cdot10^{15}$	1.51	1050	$4.58\cdot10^{-4}$
$7.29\cdot10^{9}$	1.25	1.33	0.873	$1.16\cdot10^{16}$	1.39	1310	$2.94\cdot10^{-4}$
$3.60\cdot10^{10}$	1.17	1.47	0.619	$3.53\cdot10^{16}$	1.17	924	$2.65\cdot10^{-4}$
$1.06\cdot10^{11}$	1.08	1.49	0.324	$6.80\cdot10^{16}$	1.08	739	$2.68\cdot10^{-4}$
$2.45\cdot10^{11}$	0.993	1.37	0.251	$1.21\cdot10^{17}$	1.04	643	$2.76\cdot10^{-4}$
$2.11\cdot10^{12}$	0.951	1.09	0.300	$2.27\cdot10^{17}$	1.03	576	$2.91\cdot10^{-4}$
$2.63\cdot10^{13}$	0.899	0.491	0.965	$5.47\cdot10^{17}$	1.05	580	$3.04\cdot10^{-4}$
$5.97\cdot10^{13}$	0.733	0.164	4.80	$1.61\cdot10^{18}$	1.08	623	$3.08\cdot10^{-4}$
$7.10\cdot10^{13}$	0.236	0.0138	10.7	$2.70\cdot10^{19}$	1.10	660	$3.00\cdot10^{-4}$
$8.69\cdot10^{13}$	0.125	0.0025	0.209	$2.70\cdot10^{20}$	1.09	643	$3.02\cdot10^{-4}$
$2.84\cdot10^{14}$	0.259	5.19	$2.53\cdot10^{-3}$	$5.68\cdot10^{20}$	1.09	642	$3.02\cdot10^{-4}$
$5.51\cdot10^{14}$	0.639	9.45	$1.01\cdot10^{-3}$	∞	1.09	660	$3.00\cdot10^{-4}$

For the following section we also need to have some idea concerning the rotation periods τ_r of baryon stars. Superdense stars are expected to have high rotational angular velocities, if one assumes that celestial bodies of equal mass also have approximately equal angular momentum. But the angular velocity Ω cannot be arbitrarily high. Its maximum value is determined by the condition that at the equator the centrifugal force $\rho R\Omega^2$ be equal to the Newtonian attraction $K\rho M/R^2$, i. e.

$$\Omega_{\max} = \left(\frac{KM}{R^3}\right)^{1/2}. \tag{55.7}$$

When $\Omega > \Omega_{\max}$, the centrifugal force prevails over the centripetal force, and outflow of matter from the equator sets in. In Sections 39 and 46 we have seen that the rotation, even at an angular velocity $\Omega = \Omega_{\max}$, results in no significant changes in the values of baryon-configuration parameters. In Table 14 are presented the minimum values of the period $\tau_r = 2\pi/\Omega_{\max}$, and the rotation energy E_r calculated for $\Omega = \Omega_{\max}$ /81/. The energy is measured in units of 10^{50} erg. As we see, for white dwarfs $\tau_{\min} > 1$ sec, and for baryon configurations $\tau_{\min} < 1$ sec. Longer rotation periods are naturally also possible.

56. PULSARS

The theory of superdense celestial bodies presented above is based on well-founded physical principles. Therefore, from the theoretical point of view, there is no doubt as to the actual existence of such stars. But the conclusions of even the most perfect theory also have to be confirmed through observation. There is no room for doubt concerning the existence of white dwarfs, since numerous stars of this type have been discovered through direct observation in recent decades.

The question of baryon stars remained unsolved until the beginning of 1968. The first hints concerning the existence of such stars came in reports on the discovery of point sources of cosmic X-ray radiation. This discovery was made in extra-atmospheric observations, carried out with the help of rockets /165/. But indirect evidence leaving no room to doubt the existence of baryon stars was obtained only with the discovery of pulsating sources of radio waves. These puzzling objects were called *pulsars*. The report on the discovery of the first pulsar, CP 1919, was received in February 1968 /201a/. Following this, many more new pulsating radio-wave sources were discovered, and by the end of 1970 there were data on approximately 55 objects of this type (see Table 15) /282/. A substantial part of the literature on this subject can be found in /36a, 281/.

Pulsars are powerful sources of radio signals, which are repeated at a fixed time interval with remarkable accuracy. The pulse repetition period for different objects is not the same. For the known pulsars the periods P vary within the interval $0.33 \lesssim P \lesssim 4$ sec. The width (duration) W of the pulses is about 5 — 50 msec, i. e.

approximately one order of magnitude lower than the period itself. Table 15 contains a list of the pulsars, as well as some of their more important characteristics. These celestial objects are labeled by Latin letters and four digits (see second column). Usually the first letter indicates the place of discovery of the source, and the second is the first letter of the word "pulsar." Sometimes the notation PSR is used, which consists of three letters out of this word. The numbers give the value of the right ascension α (the first two digits indicate hours, the last two, minutes). For some pulsars a second coordinate is also given, namely the declination in degrees. As an example, AP 1237 + 25 designates a pulsar discovered in Arecibo, with $\alpha = 12$ hr, 37 min, $\delta = +25°$, and PSR 0628-28 is a pulsar $\alpha = 0.6$ hr, 28 min, $\delta = -28°$. In the new unified notation, the letters indicating the place of discovery are omitted (see first column).

The profile of the radio pulses is roughly the same for all sources, but the detailed shape and amplitude can vary. The amplitude of the pulses is not constant, but undergoes fluctuations with widely varying time intervals (within an interval of the order of days, hours, or even minutes). The changes are noticeable even from pulse to pulse. Thus, owing to a strong weakening with time of pulses from CP 191, not all pulses could be detected (for example, on the 3.7 m wave the pulses were recorded for about a minute, and this was followed by silence during two to three minutes). The physical nature of these amplitude fluctuations is not altogether clear. The variations in the flux of pulsating radiation may be caused by fluctuations in the dispersive matter of the interstellar environment, or possibly by the emission mechanism in the source itself, or maybe by both. The pulse size also varies with the frequency, and the radiation is not always confined to a single pulse. Thus, for example, in the case of CP 0950, beside the main pulse with width $W \approx \approx 15$ msec, there is also a comparatively weak intermediate pulse which appears 100 msec earlier. The width of this forerunner of the main pulse is approximately 10 msec, and the energy flux associated with it is roughly 1.5% of the total flux.

The radiation from pulsars is strongly polarized (in the case of CP 0950 it reaches 95%), which obviously strongly indicates the presence of intensive magnetic fields in the region where the radiation is generated. The polarization is usually elliptic, but it may also be linear or circular. Within each pulse the polarization changes with frequency. The polarization is also dependent on time, varying with a period of the order of one day (there are cases where it changes from pulse to pulse).

For all pulsars enumerated in Table 15, to the exclusion of NP 0531, the radiation is concentrated in the radio-frequency region of

TABLE 15. Table of pulsars

New symbol	Old symbol	Period p, sec	Pulse width, msec	Dispersion, parsec·cm^{-3}	Distance, parsec	$\overline{n}_e$, cm^{-3}	Variation of period, nsec/day	$P:\left(\frac{dP}{dt}\right)$, yr
P 0031	MP 0031	0.943	25:	12	210	0.06		
P 0254	MP 0254	0.448	10	10	240	0.04		
P 0329	CP 0329	0.714518603	7	26.75	2000	0.01	0.177	1.1E7
P 0450	MP 0450	0.5497	20:	10	330	0.03		
P 0527	NP 0527	3.74591	148	49.3	2000	0.02		
P 0531	NP 0531	0.03309932	3	56.5	2000	0.03	36.518	2.5E3
P 0628—28	PSR 0628—28	1.2444	50:	34.4	330	0.03	8.640	
P 0736	MP 0736	0.3749	40	100	1000	0.10	1.728	
P 0808	CP 0808	1.292241305	90	5.77	190	0.03	0.002	6.0E8
P 0818	MP 0818	1.238	25:	25:	830	0.03		
P 0823+26	AP 0823+26	0.53062	13	19.4	630	0.03	0.146	9.9E6
P 0833—45	PSR 0833—45	0.0892	2	50	500	0.10	10.08	2.4E4
P 0834	CP 0834	1.2737633	38	12.9	430	0.03	0.587	5.9E6
P 0835	MP 0835	0.765	20	120.0	1200	0.10		
P 0904+77	PSR 0904+77	1.57905	80					
P 0940	MP 0940	0.662	30	145.0	2300	0.06		
P 0943	PP 0943	1.0977	24:	17,5	300	0.06		
P 0950	CP 0950	0.253065038	21	2.98	290	0.01	0.020	3.3E7
P 0959	MP 0959	1.4365	50	90.0	2300	0.04		
P 1133	CP 1133	1.188011019	43	4.87	210	0.02	0.326	9.9E6
P 1154	MP 1154	0.400		270.0	2700	0.10		
P 1237+25	AP 1237+25	1.38245	60	8.5	200	0.05		
P 1240	MP 1240	0.388	60:	220	2200	0.10		
P 1359	MP 1359	0.690	20:	20:	670	0.03		

P 1426	MP 1426	0.788	10	60	1500	0.04		
P 1449	MP 1449	0.180	5:	90	1700	0.05		
P 1451—68	PSR 1451—68	0.2633	25	12	400	0.03	0.259	
P 1508	HP 1508	0.739677616	20	19.6	250	0.08	0.442	4.6E6
P 1530	MP 1530	1.368	25:	20	670	0.03		
P 1541+09	AP 1541+09	0.74839	100:	35.0	1200	0.03		
P 1604	MP 1604	0.421	10:	10	330	0.03		
P 1642—03	PSR 1642—03	0.38765	20	40.0	1300	0.03		
P 1706	MP 1706	0.653	20	10.0	330	0.03		
P 1727	MP 1727	0.829	30	140	1500	0.09		
P 1747	MP 1747	0.742	20	40	1500	0.03		
P 1749—28	PSR 1749—28	0.5625533	4,7	50.88	1600	0.03	6.20	2.5E5
P 1818	MP 1818	0.5980	20:	70	2400	0.02		
P 1845	IP 1845	0.597731	20	141.9				
P 1857	MP 1857	0.6118	35:	35	1200	0.03		
P 1858+03	IP 1858	0,655444	120	402				
P 1911	MP 1911	0,825	20	75.0	750	0.10	0.12	3.2E7
P 1919	CP 1919	1.337301133	40	12.55	420	0.03		
P 1929+10	PSR 1929+10	0.2265	10	8	270	0.03	0.526	1.8E6
P 1933+16	IP 1933+16	0.358764		143.0	1430	0.10		
P 1944	MP 1944	0.440	35:	35	1200	0.03		
P 1946	IP 1946	0.71735		100				
P 1953+29	IP 1953	0.426676	20	20				
P 2003+31	IP 2003	2.111206	25	225				
P 2015+28	AP 2015+28	0.557954	28	14.2	470	0.03	0.011	1.4E8
P 2022	IP 2022	0.52921		10:	330	0.03		
P 2045—16	PSR 2045—16	1.9615639	100	11.40	380	0.03	1.018	5.2E6
P 2113	IP 2113	1.0147777		100:	1000	0.10		
P 2218+47	PSR 2218+47	0.538461	30	43.8	2500	0.02		
P 2303+30	AP 2303+30	1.575869	30	46	460	0.10		
P 2319+60	IP 2319	2.256483	140	96				

Note. The symbol EN in the last column means $\cdot 10^N$.

the spectrum, mainly in the meter range. To develop a theory, it is important to know the distribution of energy with frequency within a pulse. This problem has not yet been solved — there are only preliminary data, indicating that the differential energy spectrum drops exponentially with exponent $n \approx 1$. There is, however, no doubt that the energy spectrum is strongly decreasing, since no signals could be observed at high frequencies (the centimeter region and below). In some cases the energy spectrum seems to be cut off abruptly at high frequencies.

The pulse width displays a dependence on frequency, diminishing with increasing frequency as $W \sim \nu^{-n}$, where $n \approx 0.25$ and 0.21 for CP 1153 and NP 0527, respectively. The sharp rise and decay of the pulses (roughly 1.5 msec for NP 0531) point to a strictly directional bunching radiation mechanism, provided that the object is rotating (see below).

The periods of pulsars have been measured with high accuracy. It was found that in all sources the period increases slowly with time. A definite correlation is observed between the period P and its rate of lengthening $\frac{dP}{dt}$: in going from a source with short P to a source with long P the derivative diminishes (see "Variation of period" column). Thus, for example, in NP 0531 the period is $P \approx$ ≈ 0.0331 sec and $\frac{dP}{dt} \approx 36.5$ nsec/day (nsec = nanosecond = 10^{-9} sec), whereas in CP 1919 $P \approx 1.29$ and $\frac{dP}{dt} \approx 0.002$ nsec/day.

The width of a radio pulse enables one to form a fairly accurate idea regarding the dimensions of the emitting region. Their lower limit is obviously defined by the time interval between rays emitted from the center and edges of a section of the celestial structure, i. e. $W \approx R/c$, where R is the radius of a section of the emitting region. Since $W \approx 0.03$ sec, it follows from this that the linear dimensions of the emitting region are of the order of 900 km (in the case of NP 0531 $W \approx 0.003$ sec and $R \approx 90$ km). It is clear even from this simple estimate that the phenomenon of pulsating radio waves is due to supercompact celestial bodies.

It is of great importance to know the distance to the pulsars. On the average, pulsars are at distances of several hundred to several thousand parsec from the sun. Let us remember that the diameter of the Galaxy is $d \approx 30$ kparsec, and its thickness in the central region is approximately 5 kparsec and in the periphery ~ 1 kparsec. The sun is at a distance $r \approx 8$ kparsec from the center in the northern hemisphere, 14 parsec from the galactic plane. The observed pulsars are thus members of our Galaxy. From this it certainly does not follow that they do not also exist in other galaxies. There is no doubt that pulsars do exist in other galaxies, but their detection against the background noise of the whole system involves major difficulties. Knowledge of the distance and the amount of energy contained in the pulses enables one to calculate the absolute flux of radio emission, i. e. the average radio luminosity of the sources. The radio luminosity of pulsars is of the order of 10^{28}—10^{31} erg·sec^{-1}. If the short-wave part of the emission (which one has not yet been able to detect, except for NP 0531) is also taken into account, then the total luminosity may obviously amount to 10^{32} erg·sec^{-1}. For comparison, let us remember that the luminosity of the sun is $L_{\odot}=3.86\cdot10^{33}$ erg·sec^{-1}. As we see, pulsars are very powerful sources of radio emission.

Since a fairly large number of pulsars have already been discovered, one can consider the question of their spatial distribution. The statistics on 55 examples shows that they clearly tend to cluster near the galactic plane.

Two pulsars, NP 0531 and NP 0527, were found in the Crab nebula /179a/. The period of the first is the shortest, and that of the second the longest, of all the sources listed in Table 15. The pulsar NP 0531 should be discussed in somewhat more detail, since it has some outstanding characteristics which have placed it at the center of interest of the specialists. It has a period $P\approx0.0331$ sec and a pulse width $W\approx3$ msec, and, as has already been mentioned, is the fastest pulsating source of radiation. It is of great interest to identify the pulsars with known or unknown celestial bodies which are observable in the optical region. The efforts at optical identification succeeded only in the case of NP 0531. It was found that it coincides with an earlier observed weak star located at the center of the Crab nebula. This is the same star which is considered to be the remnant of a supernova outburst in the year 1054 (this event was mentioned in Chinese records). The cloud of matter, ejected during the explosion of the central massive body, is visible to this day (the radius of the cloud is of the order of 3 parsec, and the mass is $M\approx0.3\,M_{\odot}$). It moves away from the center at a rate of 1000 km·sec^{-1}.

The Crab nebula has repeatedly been at the center of the attention of astrophysicists. In it the presence of synchrotron emission in the optical frequency region was first discovered. Continuing observation of the synchrotron emission indicates that, even at present, streams of relativistic electrons are constantly being generated in the nebula, and these, moving in a magnetic field, emit magnetobremsstrahlung (i. e. synchrotron emission). This conclusion is inescapable, since the lifetime of the streams of relativistic particles generated during the supernova explosion is considerably shorter than the time elapsed since then (which is 916 years). As a result of the magneto-bremsstrahlung, the particles should have lost their energy long ago, and the relativistic streams should have disappeared. In 1963, X-ray emission from the central region of the Crab nebula was observed, once again drawing to it the attention of astrophysicists. It was found that a star located at the center of the nebula emits, beside fast-pulsating radio waves, a powerful pulsating radiation in both the optical and X-ray regions. The energy of the quanta emitted by this source, which starts in the radio region, reaches values in excess of 30 kev. Measurements have shown that the periods of the optical and X-ray pulsations coincide, and these are emitted in precise synchronization. This important fact leaves no room to doubt that radio, optical, and harder pulsating radiation are all emitted by the same source, simultaneously.

The pulsating fluxes arriving at the earth from NP 0531 are approximately equal to $6 \cdot 10^{-14}$, $6 \cdot 10^{-14}$, and 10^{-7} $\mathrm{erg \cdot cm^{-2} \cdot sec^{-1}}$ for the radio, optical, and X-ray regions of the spectrum, respectively. Since the distance to the Crab nebula is approximately 2.02 kparsec = $= 6.25 \cdot 10^{21}$ cm, one can calculate from these data the absolute energy fluxes (luminosities) of the pulsar:

$$I \approx \begin{cases} 3 \cdot 10^{31} \ \mathrm{erg \cdot sec^{-1}} & \text{radio,} \\ 3 \cdot 10^{31} \quad '' & \text{optical,} \\ 7.4 \cdot 10^{36} \quad '' & \text{X-ray.} \end{cases}$$

The pulsed optical energy flux amounts to about 10% of the total optical flux from the star identified as a pulsar, and the pulsed X-ray radiation is roughly 15% of the total X-ray flux emitted by the whole nebula.

Another pulsar which has aroused much interest is PSR 0833-45, because it is, after NP 0531, the pulsating radio source with the next shortest period, namely P = 0.0892. It stands out among the remaining pulsars also because, like NP 0531, it is evidently related to a supernova explosion. In fact, it is located at the position of Vela-X (Vela constellation), which is considered to be a supernova explosion (where a fairly faint gaseous envelope can still be seen). Comparatively weak trails indicate that here the outburst occurred considerably earlier than the supernova of the year 1054. The search for optical and X-ray emission from this pulsar yielded no positive results.

The examples of objects in the Crab nebula and Vela-X suggest that the pulsars are the remnants of supernova explosions (here there are again two conflicting points of view, the first claiming that there was a collapse of an ordinary star, the second believing in an explosion of a superdense celestial body). The absence of a gaseous envelope or any other visible traces in other pulsars lead to the interpretation that either the explosion in question occurred long ago, so that the matter ejected in the process has become scattered in the interstellar space, or that the pulsar moved away from the area of the nebula. If this is the case, then one can go one step further and state that there should be a direct correlation between the period and age of pulsars (the pulsation periods should be short for young sources and long for old ones). Accordingly, one can assume that there should exist pulsars with periods shorter than in NP 0531. Thus, if one assumes that the observed objects represent different evolutional stages of initial events of nearly the same type, then this leads to the idea that there are certain retardation and damping processes which cause the lengthening of the pulsar periods with time. This is directly supported by measurements showing that the pulsar periods do in fact increase slowly with time, while the rate of increase $\frac{dP}{dt}$ diminishes in going from short-period to long-period pulsars (i. e. the rate $\frac{dP}{dt}$ diminishes with time). There is also a correlation between the periods and powers — the luminosity of the pulsed emission of NP 0531 is considerably higher than that of other pulsars, which means that young structures should emit more powerful fluxes than old ones.

We can now discuss the nature of pulsars. The comparatively short periods and pulses clearly indicate that the phenomenon of pulsed emission is directly related to superdense stars with dimensions $R \lesssim \omega c$. The most serious and logical analysis of the related facts has been carried out by Gold /189b/. Considering the strict periodicity of the pulsars, he arrived at the conclusion that the phenomenon in question is in some way produced by the rotation of a baryon star. At present this idea is generally recognized, but at the time there were other opinions too. Thus, the literature contained the versions of rotating white dwarfs, pulsating white dwarfs, and simultaneously rotating and pulsating white-dwarf or baryon stars. It has now become clear that the versions related to white dwarfs do not apply. The short periods encountered in NP 0531 and PSR 0833-45 are excluded in the case of white dwarfs. In fact, in white dwarfs the periods for radial pulsations are $P > 0.25$ sec, and for rotation $P_r > 0.3$ sec (see Table 14). The periods of the fundamental modes of radial pulsations for typical white dwarfs are of the order of 10 sec, whereas in the case of baryon configurations they are about 0.1—0.001 sec (see Figure 37). Thus the measured periods of pulsars correspond primarily to the periods of rotation and radial pulsations of baryon stars. On the other hand, the width of the pulses, and their sudden appearance and disappearance, point to the emission beam being narrow and directional, which can be more reasonably explained if one assumes that rotation, and not pulsation, of baryon stars is involved. It is as if there were a radiating spot on the star's envelope, from which, as a result of rotation, a beam alternately arrives at the earth and disappears. Such spots might, for example, be magnetic poles. In fact, the presence of an intense field may be the reason why the emission is concentrated in certain directions in the star's atmosphere. There are also additional arguments in favor of rotation as opposed to pulsation.

In /179a/ a large number of successive pulses from AP 2015+28 and CP 1919 were analyzed, showing what appeared to be a second and shorter period P_2 in the variation of the radiation intensity, with $P_2 \ll P_1$, where P_1 is the main period, and $P_2 \approx 10$ msec. In contrast with P_1, the period P_2 was found to vary appreciably. Based on this, the authors of /36a, 189a/ proposed a model in which the baryon stars simultaneously rotate and pulsate, ascribing the period P_1 to rotation, and P_2 to radial pulsations. However, the presence of a second period was subsequently refuted.

Many theories have been proposed to explain the mechanism of the pulsed radio and optical radiation, but none of these were able to encompass all aspects of the observed phenomenon, and some even proved to be clearly incorrect. It can be stated that, to the moment of writing these lines (November 1970), a theory for the pulsar effect has not yet been developed. We do, however, have the reasonable qualitative explanations of Gold /189b/, which we shall discuss below. The idea of rotating baryon stars can be expected to yield several results, namely: 1) there should exist a direct correlation between the period P and the width W of the pulses (during slower rotation, W should obviously increase); 2) short periods P should correspond to young objects; 3) the most likely places for observing pulsars are the sites of previous supernova explosions; 4) because of slowing-down of the rotational motion (caused by energy losses), the periods should become longer with the passage of time. All these conclusions are supported by observational data.

If in the Crab nebula we are really dealing with a baryon star, then from the mass $M \approx M_\odot$, radius $R \approx 7$ km, and angular velocity $\omega \approx 200\ \mathrm{sec}^{-1}$ ($P \approx 0.0331$ sec) we obtain the value $\sim 10^{49}$ erg for the rotation energy. On the other hand, the observed slowing-down of the rotation ($\frac{dP}{dt} = 36.518\ \mathrm{nsec \cdot day^{-1}}$, the decrease in the number of cycles is one per 2400 years) enables one to estimate the value of the energy loss. It is approximately $2 \cdot 10^{38}\ \mathrm{erg \cdot sec^{-1}}$, which agrees with the number given earlier for the total emission flux from the pulsar in the Crab nebula. It is entirely inconceivable to suppose that this powerful radiation is due to the emission of electrons which move at random in a magnetic field, since this would require particles with an energy of 10^{21} ev. Therefore in the atmosphere of the baryon star there should exist some coherent mechanisms for particle emission, which may arise during correlated motion of bunched particles (the intensity will then be proportional to the square of the density of the number of particles). It is assumed that the baryon star should have a very strong magnetic field, namely 10^{12} G, or even more, at the surface itself. Gold assumes that the powerful magnetosphere of the star rotates with it. The radial dimensions of the plasma in the magnetosphere reach distances $r \approx c/\omega$, where ω is the angular velocity of rotation. The particle bunches, executing a spiral motion along the field lines, move toward the surface of the magnetosphere. When they reach a distance where their velocity attains the value $v \approx c$, a stream of particles is ejected from the magnetosphere's surface. This ejection is accompanied by optical and X-ray emission. The main part of the radio emission occurs in the direction of the tangent to the rotation circle just before the ejection of the relativistic gas.

The absence of axial symmetry in the magnetic field (skew rotator) or in the particle supply, together with the strictly directional bunched character of the emission from the relativistic moving stream during the beam's rotation, is what causes the pulsation effect, as in a beacon. Goldreich examined certain aspects of a fast-rotating magnetosphere and the mechanisms for generating high-energy particles in it. There is also a theory advocating a laser mechanism for the generation of radiation. It is assumed that the radiation generated at the surface is enhanced when it passes through the plasma, this being at the expense of the energy of the corresponding plasma excitations, as a result of induced transitions. There are also other theoretical versions, which we do not have the possibility of discussing here. In our opinion, the most plausible is Gold's assumption that there are well-defined streams of fast particles, which are necessary for the generation of electromagnetic waves in a cooperative manner, as required for explaining the observed intensities of pulsars. In all the different theories there seems to be no doubt regarding the presence of a strong magnetic field in the plasma surrounding the star. The presence of such a field is essential even for explaining the fact that the emission is polarized.

Let us now have a look at how the distances of pulsars are determined. We have already mentioned that optical identification has been achieved for NP 0531 only, therefore we know its distance, which is $s \approx 2.02$ kparsec. In the case of the other objects, which have not been optically identified, the distance is determined from the frequency dispersion of the propagation time of the radio waves through interstellar space. The index of refraction of the waves in the plasma is given by

$$n = 1 - \frac{\nu_0^2}{2\nu^2}.$$

Here $\nu_0^2 = n_e e^2/\pi m_e$ is the plasma frequency, and n_e is the electron density in the interstellar space in the direction of propagation of the beam. The waves propagate with the group velocity $v_g = \frac{d\omega}{dk}$, where $k = \frac{\omega}{v} = \frac{n\omega}{c}$ is the wave number (the number of waves over a distance of 2π) and $v = \frac{c}{n}$ is the phase velocity of the waves. Using the expression given above for the index of refraction, we find

$$\frac{1}{v_g} = \frac{dk}{d\omega} = \frac{1}{c}\left(n + \omega\frac{dn}{d\omega}\right) = \frac{1}{c}\left(n + \frac{\nu_0^2}{\nu^2}\right) = \left(1 + \frac{\nu_0^2}{2\nu^2}\right).$$

The time of arrival of a monochromatic signal is

$$t(\nu) = \int_0^s \frac{ds}{v_g} = \frac{1}{c}\int_0^s \left(1 + \frac{\nu_0^2}{2\nu^2}\right) ds,$$

where s is the distance between the source and the receiver. As we can see, the propagation time of the signal between given points depends on the frequency; it increases with decreasing frequency. From this we find that the difference between the arrival times of signals with frequencies ν_1 and ν_2 is

$$t(\nu_1) - t(\nu_2) = \frac{e^2}{2\pi m_e c}\left(\frac{1}{\nu_1^2} - \frac{1}{\nu_2^2}\right)\int_0^s n_e ds.$$

The integral $\int_0^s n_e ds$ is the number of electrons in a cylinder with a base of 1 cm^2 in the direction of observation of the pulsar, and is called the measure of dispersion (MD). Thus

$$\mathrm{MD} = \int_0^s n_e ds = \frac{2{,}4 \cdot 10^{-4}\,[t(\nu_1) - t(\nu_2)]}{\frac{1}{\nu_1^2} - \frac{1}{\nu_2^2}}. \tag{56.1}$$

Here the time is measured in seconds, the distance in parsecs (1 parsec $=3.086\cdot 10^{18}$ cm), and the frequency in megahertz.

From (56.1) it is apparent that the measure of dispersion of every pulsar can be obtained through direct measurement of the lag of the waves. Therefore, if we know the average density of electrons in the direction of observation, then we can calculate the distance from the source

$$s=\frac{\mathrm{MD}}{n_e}. \tag{56.2}$$

The average electron density is not the same for all pulsars. It is high if the line of sight passes through regions with enhanced ionization (clouds of ionized hydrogen, called H II regions), or regions containing stars of early spectral classes. The blue supergiants are able to ionize the hydrogen around them at very large distances, e. g. to about 1000 parsec. Thus, to determine the distance of pulsars more or less accurately, it is necessary to conduct in each case a thorough investigation of the objects which are near or on the line of sight. If such an analysis is not carried out, then the distance determined from the measure of dispersion may in some cases contain substantial errors. The electron density averaged over all sources presented in Table 15 is approximately 0.045 cm^{-3} /282/. Let us note that part of the measure of dispersion is due to the source itself. As an example, let us take the pulsar NP 0531 in the Crab nebula, for which the measure of dispersion is 57 parsec$\cdot$$\mathrm{cm}^{-3}$. The nebula around NP 0531 has a mass $M\approx 0.3M_{\odot}$, a radius $R\approx 0.5$ parsec, and an average electron density $\bar{n}_e\approx 40$ cm^{-3}. Consequently, the contribution of the nebula itself to the measure of dispersion is 20 parsec$\cdot$$\mathrm{cm}^{-3}$. With the present techniques, one can observe pulsars with a measure of dispersion MD $\lessapprox 270$ parsec$\cdot$$\mathrm{cm}^{-3}$.

The measurement of the polarization of pulsars can serve to determine the intensity of the magnetic fields in the interstellar space. To this end, one employs the Faraday effect of the rotation of the polarization plane. The angle through which the plane of polarization is rotated is given by

$$\theta=\frac{7.3\cdot 10^{10}}{\nu^2}\int_0^s n_e H_{\parallel}\,ds\approx\frac{7.3\cdot 10^{10}}{\nu^2}\bar{H}_{\parallel}\cdot\mathrm{MD}, \tag{56.3}$$

where $\bar{H}_{\parallel}$ is the average magnetic field intensity in the direction of observation of the pulsar, H is measured in gauss, ν in megahertz, and the distance in parsecs. As an example, the measure of dispersion of the source CP 1919 is 13 parsec$\cdot$$\mathrm{cm}^{-3}$, consequently the rotation angle of the plane of polarization for the radioastronomical frequency band $\nu = 151.5$ Mhz is $\theta \approx 4.1 \cdot 10^7 \overline{H}_{\parallel}$ radian.

If it is true that the pulsars are born after explosion of supernovae, then from the observed frequency of appearance of the latter (one supernova in 12 years for objects in the galactic disk, and one per year for the halo) one can estimate the total number of pulsars formed during the whole period of existence of the Galaxy. The result obtained for this is of the order of 10^9. The numbers in the last column of Table 15 give an idea of the age of the pulsars.

SUMMARY

In baryon configurations, the metric of the part of space occupied by the mass distribution, and to distances of the order of a radius from the surface, is strongly curved. Therefore the theory for these celestial bodies should be based on the relativistic theory of gravitation. The powerful gravitational field of baryon stars is an aid to their observation (the red shift, and the strong deflection of light from a star located in the direction of observation).

Of primary interest is the dependence of the mass on the number of baryons, $M(N)$. When the particle densities exceed the nuclear density by an order of magnitude, the function is zigzag-shaped, i. e. it consists of many branches. To a single number of baryons N correspond two, and in some regions even more, values of the mass (see Figure 30). The stable configurations are of course those belonging to the lower branch of the $M(N)$ curve, and the rest should be metastable. The configurations on the upper branches, when subjected to some kind of perturbation, should sooner or later convert to a state on the lower branch. The difference in mass between branches of the $M(N)$ curve reaches up to several percent of the star's mass, therefore the transition from metastable to stable state will be accompanied by the release of an enormous amount of energy (up to several percent of the internal energy), i. e. it will be of the nature of a powerful cosmic explosion. The multi-valuedness of the function is a relativistic effect. In this problem, the parameter $q(0) = P(0)/\rho(0)$ is of major importance. The effect in question appears only when q is close to unity, i. e. only when the density is supernuclear. This effect cannot exist at lower densities.

In the densest baryon configurations ($n > 10^{40}$ particle$\cdot$cm^{-3}) the gravitational mass defect $m_p N - M$ assumes an anomalous value, i. e. it becomes negative (see Figures 30 and 33). This is a new effect peculiar to the general theory of relativity, which does not exist in the Newtonian theory of gravitation. It is brought about by the catastrophic violation of the additivity of the internal energy in the corresponding configurations. The configurations having an anomalous mass defect lie on the upper branches of the $M(N)$ curve. They are unstable not only with respect to a transition to the lower branch of the $M(N)$ curve, but also with respect to scattering of the particles into infinity, i. e. dissociation into a diffuse state. It is at the same time important to bear in mind that the particles cannot leave the star singly, since they are in a bound state, their total energy being negative (of course without counting their rest energy). Hence a transition to a diffuse state can be accomplished only through an explosion. The energy released in such a process is of the order of the internal energy of the star.

It is of great interest to construct a model of celestial bodies with mass far above the solar mass, which at the same time have an anomalous mass defect. These could serve as models for prestellar bodies in the cosmogonic concept of V. A. Ambartsumyan. However, the treatment of such bodies within the framework of Einstein's theory of gravitation meets with serious difficulties, owing to the presence of a singularity in the Schwarzschild solution at $r=2kM/c^2$. In our opinion, this difficulty cannot be evaded through the introduction of a Lemaître-like nonstationary metric, since the metric then has a singularity of a physical nature, namely the appearance of velocities which exceed the velocity of sound: $\frac{\partial\sqrt{g_{ik}}}{\partial t}>c$.

As seen before, not all baryon configurations are stable. The instability of configurations on the upper branch of the $M(N)$ curve is obvious, and this we have already discussed (see Figure 30). However, the configurations on the lower $M(N)$ branch are also not all stable. The only stable ones are those with $\eta<-1.36$ ($\rho_0\lesssim 7\cdot 19^9$ g·cm^{-3}) and $-0.5\lesssim\eta\lesssim 0.8$ (the density corresponding to $\eta=0.8$ is $\rho_0=4\cdot 10^{15}$ g·cm^{-3}). The former are white dwarfs, the latter baryon stars (see Figure 20). The configurations with $-1.36\lesssim\eta\lesssim-0.6$ and $\eta\gtrsim 0.8$ are unstable. When subjected to a perturbation, these may disintegrate or convert into configurations on the stable sections. At the same time, only radial oscillations constitute a danger to equilibrium. Nonradial pulsations attenuate relatively fast (within about one minute), owing to the emission of gravitational waves. An idea of the orders of magntidue of the lifetimes of unstable configurations can be obtained from Figure 37. The energy of the fundamental radial oscillation is substantial; for configurations consisting of a real baryon gas it is approximately /267/

$$E_p\approx 10^{53}\left(\frac{\xi}{r}\right)^2\text{erg}.$$

Chapter IX

THE GENERALIZED EINSTEIN THEORY OF GRAVITATION

57. GENERAL CONSIDERATIONS

In Chapters VI—VII it was shown that, in the Newton-Einstein gravitation theory, large masses cannot be achieved by configurations consisting of degenerate matter. The masses of such configurations are of the order of the solar mass. We have also seen that the role of the equation of state in this problem is very limited. Through reasonable changes in the equation of state one can change the mass and radius only by a relatively small factor; in fact, there are no equations of state that can possibly lead to an increase in their order of magnitude. Even the temperature does not introduce anything basically new — first of all, high temperatures are unattainable owing to the neutrino energy losses, and then, regardless of this, in the final count the role of the temperature is reduced to changing the form of the equation of state. Furthermore, we saw that the possibilities are strongly limited even in the case of nonstationary models of superdense celestial bodies. In the relativistic theory, the main difficulty regarding an increase in the configuration's mass is due to the presence of a singularity in the Schwarzschild solution. Actually, though, even in the nonrelativistic gravitation theory one does not obtain large masses, although here there are no singularities, except the trivial one at $r=0$. In fact, in the problem at hand there is no significant difference between the relativistic and nonrelativistic variants of the theory of gravitation. The difficulty (if this is the appropriate word to use here) arising in trying to obtain large values for the mass of superdense stellar configurations is an inherent property of the current theory of gravitation. This has already been verified from the estimates given in Section 36.

The physical reason behind the inability to obtain large masses lies in the fact that the gravitational force acting on unit mass increases in proportion to the density, therefore when the latter increases the star's radius should shrink accordingly, so as to maintain the hydrodynamic equilibrium by enhancing the pressure gradient to the value required to compensate for the above force. But the shrinking of the radius opposes an increase in the star's mass, since the latter is proportional to the cube of the radius. This is why large masses cannot be obtained for superdense configurations. No noticeable effects are obtained even when rotation is taken into consideration.

One might think that there is nothing wrong in the fact that the masses of stars consisting of degenerate matter are of the order of the solar mass. As a matter of fact, this could be the nature of things. However, even curiosity alone can drive us to ponder this question and seek ways to overcome the limitation, which is strange in our eyes. But the question is not merely that of pure academic curiosity — there are also more serious reasons which compel us to doubt and search. During the last two decades, a series of important astrophysical phenomena were discovered, which have not yet found a consistent and discrepancy-free explanation. First of all we have in mind the highly vigorous processes generated in the centers of galaxies, and the recently discovered quasars, which are compact celestial bodies, comparable in mass with galaxies, but without a star population. The critical analysis of the new phenomena is based on the cosmogonic concept of V. A. Ambartsumyan /3—13/, which is in direct agreement with the idea of superdense prestellar bodies. It is the wish to find models for the superdense prestellar bodies which forces us to seek ways to avoid the above-mentioned restriction. We should, however, be clearly aware of the fact that these restrictions can be avoided only by changing the equations of the present theory of gravitation. Any infringement on this sacred theory should obviously be undertaken with due caution, so as not to enter into conflict with known astronomical facts. The motion of planets in the solar system can be forecast many centuries in advance with enormous accuracy. We shall return to this question later on, but meanwhile let us examine the possibilities available to us.

Let us return to the system of measurement units (34.1). According to (34.5) and (34.6), a mass unit in it is

$$\beta = 2\sqrt{2\pi}\,\frac{1}{m_n^2}\left(\frac{hc}{k}\right)^{3/2}. \tag{57.1}$$

In these units, the field equations yield values of the order of 0.1 for the mass of baryon configurations. As we can see, the mass measurement unit is inversely proportional to the square of a particle's mass (i. e. the mass of a neutron). The star's substance consists mainly of these particles (the situation, of course, does not change when we consider neutrons only instead of baryons). From this it follows that if the particles' mass is hypothetically diminished by a factor of n, then the mass of the configuration increases by a factor of n^2 (naturally the value of the star's mass in the units (34.1) does not change, only the measure becomes larger). Thus, for example, if in configurations of a degenerate particle gas the numerically predominant constituent were not baryons but leptons, say electrons, then we would obtain a mass of the order of $0.1\left(\frac{m_n}{m_e}\right)^2 \sim 10^6 M_\odot$. Our world is so constructed there there are both baryons and leptons, and in degenerate matter (owing to the Pauli principle and the energy minimum) the number of the former is far greater than the number of the latter. We can certainly not exclude the possibility that, at densities far above nuclear, when the internal structure of the particles is affected, there appear new, unknown, and more primary elementary particles, whose mass is smaller than m_n. However, regardless of the truth concerning this matter, this path still does not lead to a solution of our problem. The point is that if the mass of the configuration (consisting of degenerate matter) is increased while the mass of the main elementary particle is decreased, then this is accompanied by marked decrease in the average density, as a result of which the model ceases to be superdense. We thus repeat, that the reason why we are not able to construct models for celestial bodies which are supermassive and superdense at the same time is not to be found in either the particles or the equations of state, but in the gravitation theory itself.

From (57.1) it is apparent that there are also other possibilities for increasing the mass of baryon configurations. Thus, one could assume that at the fantastic densities discussed so far, our customary laws of physics should undergo some changes or refinements. As an example, the simplest thought which comes to mind is the idea of some change, maybe small, in the universal constants e, h, c, and k. By increasing h and c or decreasing the gravitational constant k, we could also obtain larger values for the mass of baryon configurations (and also of white dwarfs). Dirac repeatedly considered the question of possible variability of the constants e, h, c, and k. As far back as 1937 /168, 169/, he considered this problem in connection with cosmology, and reached the conclusion that the gravitational constant decreases slowly with time. This idea latter appears in other studies /209, 140, 106/ as well. Recently, when analyzing the future prospectives for the development of physics, Dirac stated that the fine-structure constant e^2/hc is a true constant, while the possibility of small changes in two of the quantities e, h, and c should not be excluded; but he was inclined to think that variability of the velocity was unlikely. From the viewpoint of pure logic, no harm will be done if it turns out that our concept of the existence of a certain complex of universal constants, on which the physics of our day is based, is only an approximation. In fact, it should be kept in mind that the values of the universal constants were determined for our own epoch in the development of the universe. The most logical assumption is that if the state of the universe changes with time, then there should also be a corresponding change in the universal constants, and thereby in some of the laws of nature.

In this connection, periodic changes and refinements of the values of the universal constants of physics, made as the experimental techniques progress, are among the basic problems of science in general.

Starting with Dirac's concept of the variability of the gravitational constant, Jordan and his co-workers established a new variant of the gravitation theory, called by them the generalized theory of gravitation /80—90/. This theory was subjected to sharp criticism, and has not been accorded proper attention to this day. The main reason for the attacks was the violation of a series of fundamental laws (including the law of conservation of energy) in Jordan's formulation. As we shall show below, when the theory is more correctly formulated there is no violation of the conservation laws.

In this chapter we shall present the fundamentals of the generalized variant of the theory of gravitation, and in the last chapter, on the basis of it, we shall develop a theory for baryon configurations /102—107, 2/. As will be seen, here we can have superdense configurations with masses of the order of the galactic mass or higher.

The new theory has to be formulated in such a way that it will not conflict with the numerous facts of astronomy. It seems to us that investigations in this direction are justified even if in the final count they turn out to be fruitless. In fact, the development of science usually does not proceed along a previously planned smooth path, since several research directions always exist. The Jordan theory is one of these possible directions for the development of the theory of gravitation. It should be developed to a point where its conclusions can be compared with the observational data.

The work invested therein will not be useless even if, in the worst case, it turns out to be entirely invalid, since one of the possibilities will then have been dealt with, and one will be able to focus one's efforts on other directions. We would like to emphasize once more the tentative character of the theory to be presented in this and the following chapter. Its scientific significance can be determined only after comparison of the predicted effects with observational facts.

58. THE FIELD EQUATIONS

The basic feature of the generalized theory of gravitation is that the gravitation constant varies as a function of the space-time coordinates, but otherwise it does not differ from regular theory. In accordance with this, the field is here characterized by 11 independent functions, namely the components of the metric tensor g_{ik} and the gravitation scalar $\chi = 8\pi k/c^2$. The theory is based on Jordan's variational principle

$$S = \frac{c}{2} \int \chi^{\eta} \left(R - \frac{2\chi}{c^2} \Lambda - \zeta g^{ik} \frac{\chi_i \chi_k}{\chi^2} \right) \sqrt{-g}\, d\Omega, \tag{58.1}$$

where R is the Riemannian scalar curvature of space, Λ is the Lagrangian density for the substance $\chi_i = \frac{\partial \chi}{\partial x^i}$, and η and ζ are dimensionless constants of the new theory. This postulated expression for the action S, in addition to the well-known standard properties, also satisfies the following requirement: when χ=const it coincides with the corresponding expression in Einstein's gravitation theory, and moreover, it contains no new dimensioned constants. The square of the four-dimensional interval will be written in the form

$$ds^2 = g_{ik}\, dx^i\, dx^k. \tag{58.2}$$

Upon variation of the action in the usual manner we find

$$\begin{aligned}\delta S = \int \chi^{\eta} \Bigg\{ \Bigg[& R_{ik} - \frac{1}{2} R g_{ik} - \frac{\chi}{c^2} T_{ik} - (\zeta - \eta^2 + \eta) \frac{\chi_i \chi_k}{\chi^2} + \\ & + \eta \frac{\chi_{i;k}}{\chi} - \eta g_{ik} \frac{\chi^l_{;l}}{\chi} + \frac{\zeta - 2\eta(\eta - 1)}{2} g_{ik} \frac{\chi_l \chi^l}{\chi^2} \Bigg] \delta g^{ik} + \\ & + \frac{\eta}{\chi} \Bigg[R - \frac{2(\eta + 1)}{\eta c^2} \chi \Lambda + \frac{\zeta(\eta - 2)}{\eta} \frac{\chi_l \chi^l}{\chi^2} + \\ & + \frac{2\zeta}{\eta} \frac{\chi^l_{;l}}{\chi^2} \Bigg] \delta \chi \Bigg\} \sqrt{-g}\, d\Omega = 0,\end{aligned} \tag{58.3}$$

where ; signifies covariant differentiation, $\chi^l = g^{lk}\chi_k$, and the energy-momentum tensor is related to the Lagrangian density Λ through the relationship (8.4). In the case of a continuous medium, in the absence of dissipative processes,

$$T_{ik} = (P + \rho)\, u_i u_k - P g_{ik}. \tag{58.4}$$

From (58.3) one obtains the following set of equations:

$$R_{ik}-\frac{1}{2}Rg_{ik}+[\eta(\eta-1)-\zeta]\frac{\chi_i\chi_k}{\chi^2}+\eta\frac{\chi_{i;k}}{\chi}-$$
$$-g_{ik}\left[\eta\frac{\chi^l_{;l}}{\chi}-\frac{\zeta-2\eta(\eta-1)}{2}\frac{\chi_l\chi^l}{\chi^2}\right]=\frac{\chi}{c^2}T_{ik},\qquad(58.5)$$

$$R-\frac{2(\eta+1)}{\eta c^2}\chi\Lambda+\frac{\zeta(n-2)}{\eta}\frac{\chi_l\chi^l}{\chi^2}+\frac{2\zeta}{\eta}\frac{\chi^l_{;l}}{\chi}=0.\qquad(58.6)$$

The first reduces to the Einstein equation when χ=const. As to the second equation, when χ=const it does not exist, for the simple reason that in this case $\delta\chi=0$, and therefore the coefficient of $\delta\chi$ in (58.3) drops out altogether. To perform a valid transition to the regular Einstein theory, it is necessary to set $|\zeta|=\infty$ together with χ=const.

Let us note that in the variational principle (58.3) the scalar R can be replaced by the quantity

$$G=g^{ik}\left[\Gamma^m_{il}\Gamma^l_{km}-\Gamma^l_{ik}\Gamma^m_{lm}+\frac{\eta}{\chi}\left(\Gamma^l_{ik}\chi_l-\Gamma^m_{im}\chi_k\right)\right],\qquad(58.7)$$

which no longer contains the second derivatives of g_{ik}, but differs from it by a divergent term. During variation of the action, the integral of this term in (58.3) vanishes. Therefore, although G is not a scalar, $\int\chi^\eta G\sqrt{-g}\,d\Omega$ is a scalar and coincides with $\int\chi^\eta R\sqrt{-g}\,d\Omega$.

Equations (58.5) and (58.6) can be brought to a more convenient form. To this end, we multiply the first of these by g^{ik} and sum over the recurring indices; then, with the aid of the second equation, we find

$$\frac{\chi^l_{;l}}{\chi}+(\eta-1)\frac{\chi_l\chi^l}{\chi^2}=\frac{\eta\chi}{c^2(3\eta^2-2\zeta)}\left(T-2\frac{\eta+1}{\eta}\Lambda\right).\qquad(58.8)$$

We now multiply (58.6) by $g_{ik}/2$ and combine it with (58.5); considering (58.8), we then obtain

$$R_{ik}+\eta\frac{\chi_{i;k}}{\chi}-[\zeta-\eta(\eta-1)]\frac{\chi_i\chi_k}{\chi^2}=$$
$$=\frac{\chi}{c^2}\left(T_{ik}+\frac{\zeta-\eta^2}{3\eta^2-2\zeta}g_{ik}T\right)-\frac{\eta(\eta+1)}{c^2(3\eta^2-2\zeta)}\chi g_{ik}\Lambda.$$

59. THE CONSERVATION LAWS

The equations of motion can be obtained from the following variational principle:

$$\delta S_m = \int \chi^{\eta+1} \Lambda \sqrt{-g}\, d\Omega = 0. \tag{59.1}$$

Here the variation is applied to the particle trajectories:

$$x'^k = x^k + \xi^k,$$

where ξ^k are infinitely small quantities. At the same time, the gravitational field, i. e. the quantities g_{ik} and χ, is considered fixed. However, variation of a trajectory is associated with a transition from one point in space to another, infinitely close to it; thus the functions $g^{ik}(x^i)$ and $\chi(x^i)$ should obviously also change. For the change in these quantities we have

$$\delta g^{ik} = \xi^{i;k} + \xi^{k;i}, \quad \delta\chi = -\chi_k \xi^k. \tag{59.2}$$

Considering these last relationships, we obtain from (59.1)

$$\begin{aligned}\delta S_m = &\int \left[-(\eta+1)\chi^{\eta}\sqrt{-g}\,\Lambda\chi_l\xi^l + \chi^{\eta+1}\delta\sqrt{-g}\,\Lambda)\right] d\Omega = \\ &= \int \left[(\chi^{\eta+1}T^k_l)_{;k} - (\eta+1)\chi^{\eta}\Lambda\chi_l\right]\xi^l\sqrt{-g}\,d\Omega = 0.\end{aligned}$$

Hence from the arbitrariness of the variation of the coordinates ξ^l it follows that

$$(\chi^{\eta+1}T^k_i)_{;k} = (\eta+1)\chi^{\eta}\Lambda\chi_i. \tag{59.3}$$

These are the sought equations of hydrodynamics.

According to the general idea of the relativistic theory of gravitation, equations (59.3) are not new independent relationships, but should also follow directly from the field equations. To prove this, in (58.5) we first raise the index k and multiply by χ^{η}, and then differentiate covariantly with respect to k. Afterward we multiply (58.6) by $0.5\,\eta\chi^{\eta-1}\chi_k\,\delta^k_i$ and combine the result with the previous one, ending up with the relationship (59.3).

When an electromagnetic field is present, under the integration sign in (58.1) we have to add the following two terms:

$$\frac{2\chi}{c^3} A_k j^k - \frac{\chi}{8\pi c^2} F_{ik}F^{ik},$$

where j^k is the current density, A_k is the four-dimensional potential of the electromagnetic field, and $F_{ik}=A_{k;i}-A_{i;k}$. In order to derive the equations of the electromagnetic field, we have to fix all quantities characterizing the gravitational field and the particle trajectories, subjecting the potentials A_k to a virtual change. As a result we obtain the second pair of Maxwell equations

$$\frac{1}{\sqrt{-g}}\frac{\partial}{\partial x^k}\left(\sqrt{-g}\,\chi^{\eta+1}F^{ik}\right)=-\frac{4\pi}{c}\chi^{\eta+1}j^i. \tag{59.4}$$

From this follows directly the equation of continuity

$$\frac{1}{\sqrt{-g}}\frac{\partial}{\partial x^i}\left(\sqrt{-g}\,\chi^{\eta+1}j^i\right)=0. \tag{59.5}$$

Thus the generalized theory of gravitation contains the two dimensionless parameters η and ζ. As we shall presently see, the situation regarding the choice of the numerical value of η is good; there exists only one possibility from which η is defined in a natural manner. As to the numerical value of ζ, here the situation is not yet clear. We saw earlier that when $|\zeta|\to\infty$ the generalized theory coincides with the regular theory. The quantity $|\zeta|$ then has no upper limit. The lower limit can be obtained by considering the three Einstein effects, requiring agreement with experiment within the limits of the accuracy of the measurements (see Section 60). An accurate value of ζ can be found only after comparing the results of the new theory with the experimental facts. Meanwhile, let us examine the method of obtaining the numerical value of η.

From (59.3) and (59.4) it is apparent that when $\eta\neq-1$ the laws of conservation of energy, electric charge, and baryon number are not fulfilled in the form known to us so far. Jordan, apparently for the purpose of obtaining certain results in cosmology, or perhaps for other reasons unknown to us, chose $\eta=1$, which led to violation of the mentioned fundamental conservation laws. This was the main cause for criticism of the new theory. As we can see, though, when $\eta=-1$ the conservation laws assume their usual form

$$\left.\begin{aligned} T^k_{i;k}&=0,\\ \frac{1}{\sqrt{-g}}\frac{\partial}{\partial x^i}\left(\sqrt{-g}\,j^i\right)&=0.\end{aligned}\right\} \tag{59.6}$$

The above argument had been overlooked earlier, for the reason that equation (59.3) was not formulated from this general viewpoint, but was derived for the particular value $\eta=1$, chosen beforehand. It is likewise important to note that only at $\eta = -1$ does the integral (58.1) assume the dimension of action. This is certainly an important aspect of the problem of the correct choice of η. In a later work of Jordan /231/, and also in works of Dicke and Brans /140—142, 171—179/, it was postulated that $\eta = -1$, but the question of the conservation laws was not examined. This was studied in /106/.

Thus, on the basis of the arguments presented above, we shall assume $\eta = -1$. Equations (58.8) and (58.9) then assume the form

$$\frac{\chi_{;k}^{k}}{\chi} - 2\frac{\chi_k \chi^k}{\chi^2} = \frac{\chi T}{c^2 (2\zeta - 3)}, \tag{59.7}$$

$$R_i^k - \frac{\chi_{;i}^{k}}{\chi} - (\zeta - 2)\frac{\chi_i \chi^k}{\chi} = \frac{\chi}{c^2}\left(T_i^k + \frac{\zeta - 1}{3 - 2\zeta} T \delta_i^k\right). \tag{59.8}$$

For the scalar curvature we have

$$R = \frac{2\zeta}{c^2 (3 - 2\zeta)} \chi T + \zeta \frac{\chi_i \chi^i}{\chi^2}. \tag{59.9}$$

60. THE CENTRALLY SYMMETRICAL GRAVITATIONAL FIELD IN FREE SPACE

The solution for a centrally symmetrical field in free space has been known for a long time. It is called the Heckmann solution, and its derivation for the case $\eta = 1$ is given in /209/, where it is also shown how, through simple substitution, one can from this solution obtain the solutions for arbitrary values of η. We shall only present the finished result for the case $\eta = -1$ (it is obtained from the solution for $\eta = 1$ by replacing χ by $1/\chi$). The validity of the solution given below can be confirmed through direct substitution in the corresponding field equations. Thus, in the empty space around a centrally symmetrical mass distribution we have the following Heckmann solution:

$$\left.\begin{aligned} r &= \frac{r_0}{\sqrt{\tau}\,(\tau^{-h} - \tau^{h})}, \\ e^{-\lambda/2} &= \frac{1}{2h}\left[(0.5 + h)\,\tau^{h} - (0.5 - h)\,\tau^{-h}\right], \\ e^{\nu} &= \tau^{1/B}, \\ k &= k_0 \tau^{-\beta_0/B}. \end{aligned}\right\} \tag{60.1}$$

Here $k = \frac{c^2 \chi}{8\pi}$, τ is a variable parameter, k_0, r_0, and β_0 are constants obtained during integration, and

$$h^2 = \frac{1}{4} - \frac{\beta_0}{2} \frac{1+\beta_0\zeta}{(1+2\beta_0)^2}, \quad B = 1 + 2\beta_0. \tag{60.2}$$

The four-dimensional interval is assumed to be written in the Schwarzschild form $ds^2 = e^\nu c^2dt^2 - e^\lambda dr^2 - r^2\,(d\vartheta^2 + \sin^2\vartheta\, d\varphi^2)$. When $\beta_0 = 0$ we have $h^2 = 1/4$, and (60.1) reduces to the Schwarzschild solution. When β_0 is negative $h^2 > 1/4$, and when it is positive $h^2 < 1/4$. For a certain reason (see below) we shall be interested in the case $\beta_0 < 0$, $h > 0.5$ (it should be noted that the Heckmann solution is invariant with respect to changing h into $-h$). For $r_0 > 0$ we have $0 < \tau < 1$, and $r(\tau)$ rises monotonically: $r(0) = 0$ and $r(1) = \infty$. When $\tau \longrightarrow 1$ the metric tends asymptotically toward the Euclidean metric: $r \longrightarrow \infty$, $e^\lambda \to e^\nu \to 1$, $k(r) \longrightarrow k_0$. From the requirement that at sufficiently large distances the generalized theory coincide with the regular theory, we arrive at the conclusion that k_0 is the universal constant Newtonian gravitation.

In /140/ the solution of the field equations for free space ($\eta = -1$) was found anew in isotropic coordinates. It was obtained from (60.1) by means of a coordinate transformation, and is therefore not a new solution. For further details regarding the solution (60.1), one should consult Jordan's monograph /209/.

At sufficiently large distances, where $\varepsilon = 1 - \tau \ll 1$, one can expand the functions $1/r$, e^λ, e^ν, and k in (60.1) in powers of ε. With accuracy to terms of the order of ε, we have for $1/r$ and e^ν

$$\frac{1}{r} \approx \frac{2h}{r_0}\varepsilon, \quad e^\nu \approx 1 - \frac{\varepsilon}{B} = 1 - \frac{r_0}{2hB} \cdot \frac{1}{r}.$$

Comparison of this expansion with $e^\nu = 1 - 2kM/c^2r$ from the Schwarzschild solution yields $\varepsilon = Br_g/r$, where $r_g = 2k_0M/c^2$ and

$$r_0 = \frac{4hk_0}{c^2} BM = 2hBr_g. \tag{60.3}$$

If after this we proceed to compare the functions e^λ (or $k(r)$ and k_0) and require coincidence at finite r, we obtain $\beta_0 = 0$, which signifies complete coincidence between (60.1) and the Schwarzschild solution. Thus there is complete agreement between the two solutions only at $r = \infty$. If we now consider that the nonrelativistic theory of gravitation becomes valid already at finite, but sufficiently large, distances from the source, then we arrive at the conclusion that the generalized theory of gravitation can be made to agree with it only at infinity. Thus at distances r which are arbitrarily large but still finite, the generalized theory does not, strictly speaking, coincide with Newton's gravitation theory. From this it follows that Newton's variant of the theory should also be developed with a variable gravitational constant $k(r, t)$.

Now, following Jordan /209/, we shall endeavor to determine the constant β_0. Let us consider a highly rarefied ($\rho \longrightarrow 0$) spherical cloud with a fairly large radius R. Neglecting the curvature of space, we obtain from (59.7)

$$\chi'' + \frac{2}{r}\chi' - \frac{2\chi'^2}{\chi} = -\frac{\chi^2 \rho(r)}{c^2(2\zeta - 3)}.$$

We now multiply this equation by r^2/χ^2 and integrate between zero and the distance r, obtaining

$$\frac{d}{dr}\frac{1}{\chi} = \frac{1}{4\pi(2\zeta - 3)} \frac{M(r)}{r^2},$$

where $M(r)$ is the mass contained in a sphere with radius r. Now we integrate this equation from R to ∞:

$$\frac{1}{\chi(R)} = \frac{1}{\chi(\infty)} - \frac{1}{4\pi(2\zeta - 3)} \frac{M}{R}.$$

Finally, substituting here $\chi = 8\pi k/c^2$, we find

$$k(R) = k_0\left(1 + \frac{1}{2\zeta - 3}\frac{r_g}{R}\right).$$

Let us compare this expression with the Heckmann solution for $k(r)$. Performing an expansion of $k(r)$ in (60.1), we find

$$k(R) = k_0\left(1 + \beta_0 \frac{r_g}{R}\right).$$

From comparison between the last two expressions it follows that

$$\beta_0 = \frac{1}{2\zeta - 3}. \tag{60.4}$$

Thus all constants in the Heckmann solution are expressed through the parameter ζ, whose value can only be found from comparison between the results of the new theory and experiment. Some information concerning ζ can be obtained from the three well-known effects of the general theory of relativity (deflection of rays of light, the red shift, and the precession of the perihelion of planetary orbits), which are strongly supported by observation. These effects have been studied in the general theory of relativity /66/. It is found that the red shift does not contribute anything new, and the two other effects impose a certain restriction on the absolute value of ζ. These enable one to find the lower limit of $|\zeta|$. Thus, if one requires that the angular shift in the perihelion of the orbit of a test body rotating around a second massive celestial body agree with the result of the general theory of relativity, where χ is considered constant, with an accuracy of 2% (this is the accuracy in measuring the precession of the perihelion of Mercury), then one obtains the condition $|4\beta_0/3| \leqslant 0.02$. This yields the condition $|\zeta| \geqslant 30$ for ζ. Actual numerical calculations for the configurations of celestial bodies have been carried out for

$$\zeta = -30, \tag{60.5}$$

which corresponds to $4\beta_0/3 \approx -0.02$. Investigations carried out within the framework of the nonrelativistic generalized theory of gravitation (see Section 64) showed that for positive values of ζ the gravitational scalar $k(r)$ has a divergent solution inside the configurations. Therefore it was subsequently assumed in calculating models of stellar configurations that ζ is negative, in both the relativistic and nonrelativistic variants of the theory. In the relativistic variant the question of the sign of ζ has not been properly investigated, and should be examined. It is not impossible that a nondivergent solution can be obtained even for positive ζ.

Using (60.4), we can express all integration constants in the solution (58.1) through the parameter ζ:

$$\left.\begin{aligned} &h = (4\zeta^2 - 10\zeta + 7)^{1/2}/2(1 - 2\zeta),\\ &B = (2\zeta - 1)/(2\zeta - 3), \quad \beta_0/B = 1/(2\zeta - 1). \end{aligned}\right\} \tag{60.6}$$

Assuming $\zeta = -30$, we find for the numerical values of these quantities

$$\left.\begin{aligned} &\zeta = -30, \quad h = 0.51234, \quad B = 0.96825,\\ &r_0 = 0.9925\, r_g, \quad \beta_0/B = -\frac{1}{61}. \end{aligned}\right\} \tag{60.7}$$

An exceptionally important fact should be noted, namely that in the generalized theory of gravitation a centrally symmetrical field, as is evident from (60.1), does not have the singularity characteristic of the Schwarzschild solution. This remarkable property permits the construction of models for static superdense configurations with very large mass.

As we can see from (58.1), in the variational principle the transition to Einstein's theory of gravitation is achieved through the substitution $|1/\zeta| \to 0$. In the field equations, though, besides $|\zeta| \to \infty$ we also have to postulate $k = \text{const}$. However, it is also of interest to perform a thorough study of the behavior of the Heckmann solution for a point mass when only $|\zeta| \to \infty$. The cases of positive and negative ζ require separate treatment, but here we shall examine the case $\zeta = -30$ only. Let us return to (60.1), with the numerical values of (60.7) for the constants. It is easily noticed that $e^{\lambda(r)}$ is zero when $r=0$, then it increases monotonically with r, reaching a maximum in the vicinity of $r \approx r_0 \approx r_g$, after which it drops asymptotically to the value $e^{\lambda(\infty)} = 1$. The maximum of e^{λ} increases with $|\zeta|$ and when $\zeta \to -\infty$ it becomes infinite. At the same time the value of r_0 approaches the gravitation radius $r_g = 2k_0M/c^2$, reaching it at the limit. When $r \geqslant r_0$, the asymptotic behavior of $e^{\lambda(r)}$ is given by the Schwarzschild solution $1/(1 - r_g/r)$. Inside the gravitational radius, though, $e^{\lambda(r)} \to 0$ at every point. As to the quantity $e^{\nu(r)}$, it is a monotonically rising function of r for finite values of ζ, becoming $1 - r_g/r$ outside the gravitation radius and zero inside it when $\zeta \to -\infty$.

Let us establish a direct relationship between k and r, for this purpose eliminating the parameter τ. Assuming $|\zeta| \gg 1$, we find

$$\frac{r_0}{r} \approx k^{-3/4} - k^{-2\zeta}.$$

When $-\zeta$ becomes infinite, we have

$$k(r) = \left(\frac{r}{r_g}\right)^{4/3} \quad \text{for} \quad r \leqslant r_g,$$

$$k(r) = 1 \quad \text{for} \quad r \geqslant r_g.$$

For every finite value of ζ, the quantity $k(r)$ is a monotonically rising smooth function. Later on we shall see that in the nonrelativistic generalized theory of gravitation the range of distances where $k(r)$ is substantially different from k_0 shrinks to a point when $|\zeta| \to \infty$.

Thus, whereas when $|\zeta| \to \infty$ the solution (60.1) becomes the Schwarzschild solution outside the gravitation radius, inside the gravitational sphere the character of this solution does not correspond to the existing concepts of the general theory of relativity. In our opinion, this result could serve as a basis for a different interpretation of the singular region $r < r_g$ in the regular Einstein theory of gravitation.

The exterior solution contains the mass M of the celestial body. This is the so-called active mass, measured by an observer through its gravitational effect at large distances, where Newton's law applies. Owing to the introduction of the scalar field of gravitation, described by the function $k(r)$, the concept of mass needs refining. A definition of the mass can be obtained by starting out from Gauss's theorem, which is valid for a static field /66/:

$$\int R_0^0 \sqrt{-g}\, dV = -4\pi M$$

(the system of units used is $c=k_0=1$). For a field with central symmetry we obtain from (59.8)

$$R_0^0 = 8\pi k\left(T_0^0 - \frac{\zeta-1}{3-2\zeta}T\right) - \frac{k'\nu'}{2k}.$$

Substituting this expression in Gauss's formula, we find

$$M = \frac{2}{3-2\zeta}\int k\,[(3-\zeta)\,\rho + 3\,(1-\zeta)\,P]\sqrt{-g}\, dV + \\ + \frac{1}{8\pi}\int \frac{k'\nu'}{k}\sqrt{-g}\, dV. \qquad (60.8)$$

For constant k and $|\zeta| \to \infty$, this expression for the mass becomes the well-known formula of Tolman

$$M = \int (\rho + 3P)\sqrt{-g}\, dV. \qquad (60.9)$$

The second term in (60.8) can be interpreted as the energy of the scalar field $k(r)$.

61. THE EQUATIONS FOR A SPHERICALLY SYMMETRICAL STATIC MASS DISTRIBUTION

The fact that the exterior solution contains no singularities arouses the hope of obtaining nontrivial results when investigating the interior problem. Let us consider a field with central symmetry and static mass distribution. We write the square of the four-dimensional interval in the form

$$ds^2 = e^{\nu}c^2dt^2 - e^{\lambda}dr^2 - r^2\,(d\vartheta^2 + \sin^2\vartheta\, d\varphi^2),$$

where ν and λ are unknown functions of r; outside the mass distribution these are defined by the solution (60.1). For the particular case under consideration we obtain from (59.7) and (59.8)

$$\left.\begin{aligned}
&\frac{\chi''}{\chi}+\frac{\chi'(\nu'-\lambda')}{2\chi}+\frac{2\chi'}{r\chi}-\frac{2\chi'^2}{\chi^2}=\frac{\chi e^{\lambda}T}{c^2(3-2\zeta)},\\
&\frac{\nu'\lambda'}{4}-\frac{\nu'}{r}-\frac{\nu''}{2}-\frac{\nu'^2}{4}+\frac{\chi'\nu'}{2\chi}=\\
&\qquad=-\frac{1}{c^2}\chi e^{\lambda}\left(T_0^0+\frac{\zeta-1}{3-2\zeta}T\right),\\
&\frac{\nu'\lambda'}{4}+\frac{\lambda'}{r}-\frac{\nu''}{2}-\frac{\nu'^2}{4}+\frac{\chi''}{\chi}-\frac{\chi'\lambda'}{2\chi}+(\zeta-2)\frac{\chi'^2}{\chi^2}=\\
&\qquad=-\frac{1}{c^2}\chi e^{\lambda}\left(T_1^1+\frac{\zeta-1}{3-2\zeta}T\right),\\
&\frac{\lambda'-\nu'}{2r}+\frac{e^{\lambda}-1}{r^2}+\frac{\chi'}{r\chi}=-\frac{1}{c^2}\chi e^{\lambda}\left(T_2^2+\frac{\zeta-1}{3-2\zeta}T\right).
\end{aligned}\right\}\tag{61.1}$$

According to (58.4)

$$T_1^1=T_2^2=T_3^3=-P,\quad T_0^0=\rho,\quad T=\rho-3P. \tag{61.2}$$

In (61.1) we rewrite the fourth equation without changes, and replace the remaining equations by the following combinations: (I+ II + III)/ /2+IV, (III—II—I)/2—IV; then, using (61.2), we arrive at the system

$$\left.\begin{aligned}
&\frac{\chi''}{\chi}+\frac{\zeta-4}{2}\frac{\chi'^2}{\chi^2}+\frac{2\chi-r\chi'}{2r\chi}\lambda'+\frac{2\chi'}{r\chi}+\frac{1}{r^2}(e^{\lambda}-1)=\\
&\qquad=\frac{1}{c^2}\chi e^{\lambda}\rho,\\
&\frac{\lambda'}{2r}-\frac{\nu'}{2r}+\frac{\chi'}{r\chi}+\frac{e^{\lambda}-1}{r^2}=\frac{\chi e^{\lambda}}{c^2}\cdot\frac{\zeta P+(1-\zeta)\rho}{3-2\zeta},\\
&\frac{2\chi-r\chi'}{2r\chi}\nu'+\zeta\frac{\chi'^2}{2\chi^2}-\frac{2\chi'}{r\chi}+\frac{1}{r^2}(1-e^{\lambda})=\\
&\qquad=\frac{1}{c^2}\chi e^{\lambda}P.
\end{aligned}\right\}\tag{61.3}$$

In place of a fourth independent linear combination of equations (61.1), it is practical to use the equation of hydrodynamics $T^k_{i;k}=0$, which is actually contained in them. In the present case it reduces to

$$P'+\frac{P+\rho}{2}\nu'=0. \tag{61.4}$$

The advantage of the set (61.3) and (61.4) over the previous one is that it does not contain the second derivative. In this system we can also eliminate ν'. To this end, we solve the third equation

(61.3) for ν', and substitute the result in (61.4) and in the second equation we find λ' and substitute it in the first. Subsequently, converting to the variable $k=c^2\chi/8\pi$, we perform some simple transformations and arrive at the system

$$
\left.\begin{aligned}
\frac{dP}{dr} &= -r\frac{P+\rho}{2-rk'/k}\left[8\pi kPe^{\lambda}+\frac{2k'}{rk}-\frac{\zeta}{2}\left(\frac{k'}{k}\right)^2+\frac{e^{\lambda}-1}{r^2}\right],\\
\frac{d\lambda}{dr} &= 16\,\pi rke^{\lambda}\left[\frac{\zeta P+(1-\zeta)\rho}{3-2\zeta}+\frac{P}{2-rk'/k}\right]-\\
&\qquad -2\frac{e^{\lambda}-1}{r}\cdot\frac{1-rk'/k}{2-rk'/k}+(2-\zeta)\frac{r(k'/k)^2}{2-rk'/k},\\
\frac{d}{dr}\left(\frac{1}{k}\frac{dk}{dr}\right) &= 8\pi ke^{\lambda}\Big\{(\rho-P)-\\
&\qquad -\frac{2-rk'/k}{3-2\zeta}\left[\zeta P+(1-\zeta)\rho\right]\Big\}-\frac{e^{\lambda}+1}{r}\frac{k'}{k}.
\end{aligned}\right\}\qquad (61.5)
$$

Here we changed to the system of units (34.1). These equations represent a system of differential equations of the fourth order for P, λ, and k. To complete the definition of the interior problem, it remains to add the equation of state $\rho\,(P)$ and the boundary conditions.

For the boundary conditions, we should obviously require the vanishing of the pressure P and continuity of the function λ, k, k' at the surface of the configuration. From (60.1) we thus have for the boundary conditions at the surface

$$
\left.\begin{aligned}
&P(R)=0,\quad R=\frac{4hBM}{\sqrt{\tau_0}\left(\tau_0^{-h}-\tau_0^{h}\right)},\\
&\lambda(R)=2\ln 2h-2\ln\left[(h+0.5)\tau_0^{h}+(h-0.5)\tau_0^{-h}\right],\\
&k(R)=\tau_0^{-\beta_0/B},\qquad k'(R)=-2\beta_0\frac{M}{R^2}e^{\lambda(R)/2}\tau_0^{-0.5-\beta_0/B},
\end{aligned}\right\}\qquad (61.6)
$$

where $\tau_0\equiv\tau(R)$. The values of the constants h, B, β_0 are given by (60.6) and (60.7).

The interior problem is not yet well-posed. In fact, to integrate the system of differential equations (61.5), one also needs the conditions at the center of the mass distribution. From physical considerations it is evident that, in analogy with the regular theory of gravitation, here also the basic parameter determining the configurations of degenerate gaseous masses is the pressure $P(0)$ at the center. Consequently, one of the conditions at the center is the given value of $P\,(0)$, but this alone is not sufficient to solve the problem. To undertake the integration, one also has to know the values of the functions $\lambda(r)$, $k(r)$, and $k'(r)$ at the center. In contrast to $P(0)$, the quantities $\lambda(0)$, $k(0)$, and $k'(0)$ cannot be assigned arbitrarily, since for every configuration with given mass M and radius R these should be completely defined quantities, i. e. unknown functions of the central pressure. Therefore, before we proceed to discuss the integration procedure for equations (61.5), we should learn what the conditions are at the center.

62. THE CONDITIONS AT THE CENTER OF THE CONFIGURATION

In order to study the conditions at the center, it is convenient to convert to the variables

$$s = 8\pi k r^2, \qquad f = rk'/k, \qquad E = e^{\lambda}. \tag{62.1}$$

This substitution enables one to reduce (61.5) to a system of equations whose order is lower by one:

$$\frac{dP}{ds} = \frac{P+\rho}{f+2} \cdot \frac{\psi}{s}, \tag{62.2}$$

$$\frac{dE}{ds} = \frac{2E}{s(f+2)} (sE\varphi + 1 - f - E - \psi), \tag{62.3}$$

$$\frac{df}{ds} = \frac{E}{s(f+2)} (s[\rho - P + (f-2)\varphi] - f). \tag{62.4}$$

Here the following notations have been introduced:

$$\left.\begin{aligned} \psi &= \frac{1}{f-2}\left(sEP + 2f - \frac{1}{2}\zeta f^2 - 1 + E\right), \\ \varphi &= \frac{\rho - \zeta(\rho - P)}{3 - 2\zeta}. \end{aligned}\right\} \tag{62.5}$$

In (62.2)—(62.5) passage to the limit of the corresponding equations of the theory having a constant χ is accomplished by setting $f=0$ and sending $|\zeta|$ to infinity. In fact, the equations thus obtained are

$$\left.\begin{aligned} \frac{dP}{ds} &= \frac{P+\rho}{4s}(1 - E - sEP), \\ \frac{dE}{ds} &= \frac{E}{2s}(1 - E + sE\rho), \end{aligned}\right\} \tag{62.6}$$

which, with the variables $s=8\pi r^2$ and $E=e^{\lambda}=1/(1-2u/r)$, coincide with equations (40.12). The conditions at the surface for these equations also become the corresponding conditions for the system (40.12), since at $|\zeta|\to\infty$ the Heckmann solution becomes the Schwarzschild solution.

We shall study the conditions at the center of the configuration for the case $\zeta<0$, under the assumption that the pressure $P(0)$ is finite. As is apparent from (62.2), the central pressure can have a finite value when $s\to 0$ only if at the same time $|f|\to\infty$ or $\psi(0)=0$. Otherwise, if $[(P+\rho)\psi/(f+2)]_{s=0}\neq 0$, then $\left(s\frac{dP}{ds}\right)_{s=0}\neq 0$, and the function $P(s)$ will diverge at the star's center at least at a logarithmic rate. Of the two alternatives, the first has to be excluded for the following reason: if one assumes that $|f|\to\infty$ when $s\to 0$, then $P(0)$ can remain finite only if $\psi(0)$ is finite. From the first expression (62.5) it can be seen that, when $\lim\limits_{s\to 0}|f(s)|\to\infty$ the function $\psi(s)$ does not approach infinity only if $\lim\limits_{s\to 0}E(s)\to 0.5\,\zeta f^2(0)<0$. But the function $E(s)=e^{\lambda(s)}$ cannot have a negative value, therefore the assumption $|f(s)|_{s\to 0}\to\infty$ is incorrect, and $f(0)$ has to be finite. Then, if $f(0)$ is finite, $E(0)$ also has to be finite, as follows from the expression for ψ and the requirement that the central pressure be finite. Thus the second alternative has to take place, namely $\psi(0)=0$, which means that

$$2f(0)-\frac{\zeta}{2}f^2(0)-1+E(0)=0. \qquad (62.7)$$

Let us now examine equation (62.3) for small s:

$$s\,\frac{dE}{ds}=\frac{2E(0)}{f(0)+2}\,[1-f(0)-E(0)].$$

If the right-hand side of the equation is not zero, then the function $E(s)$ diverges at the point $s=0$, and this is inadmissible (since then $\psi(0)$, and consequently also $P(0)$, would become infinite). Therefore we should have

$$E(0)\,[1-f(0)-E(0)]=0. \qquad (62.8)$$

From (62.7) and (62.8) one obtains the following solutions for $f(0)$ and $E(0)$ for the case $\zeta=-30$:

$$\left.\begin{array}{l} \text{I. } E(0)=0\text{: } f(0)=\frac{2}{\zeta}\left(1-\sqrt{1-\zeta/2}\right)=\frac{1}{5}, \\ \text{II. } E(0)=0\text{: } f(0)=\frac{2}{\zeta}\left(1+\sqrt{1-\zeta/2}\right)=-\frac{1}{3}, \\ \text{III. } E(0)=1\text{: } f(0)=0, \\ \text{IV. } E(0)=1-\frac{2}{\zeta}=\frac{16}{15}\text{: } f(0)=\frac{2}{\zeta}=-\frac{1}{15}\,. \end{array}\right\} \qquad (62.9)$$

We can now consider the problem of a gravitational static distribution of degenerate mass to be correctly formulated — we have the differential equations (61.5), and the boundary conditions (61.6) at the surface and (62.9) at the center. We see that, in contrast to the boundary conditions at the surface, at the center there are four different possibilities. However, the existence of many alternatives at the center is not an indication of the existence of the same number of solutions. During numerical calculations one may find that some of these possibilities are incompatible with the boundary conditions at the surface. So far, only alternatives I and III have been thoroughly investigated. Especially interesting results were obtained by integrating equations (61.5) with the first set of conditions at the center (see Chapter X), which will henceforth be the only one discussed here. It should be noted that, in addition to the solutions with finite central pressure (i.e. with conditions (62.9)), there also exists a whole class of solutions with infinite pressure at the center, but with a finite value of the total mass of the configuration.

Let us now examine the behavior of the solutions of the system (62.2)—(62.5) in the vicinity of the point $s=0$ for alternatives I and II. Since at $s=0$ the quantities ψ and E become zero, for small s equations (62.3) and (62.4) assume the form

$$\frac{dE}{ds}=\frac{2(1-f_0)}{f_0+2}\cdot\frac{E}{s}, \quad \frac{df}{ds}=-\frac{f_0}{f_0+2}\frac{E}{s}, \quad f_0\equiv f(0).$$

Integration of these equations yields $E(s)$ and $f(s)$ in the vicinity of the origin in the first approximation in s^l:

$$\left.\begin{array}{r} f(s)=f_0-\frac{f_0}{2(1-f_0)}E, \\ E=Ds^l, \quad l=\frac{2(1-f_0)}{2+f_0}, \end{array}\right\} \qquad (62.10)$$

where D is an integration constant. The expansion is in powers of s^l. The pressure $P(s)$ satisfies the condition $P'(0)=0$. In the numerical integration of the system (62.2)—(62.4), we start with the behavior (62.10) of the functions E and f for small s and a choice of the "free parameter" D (trial method), striving to satisfy all of conditions (61.1) at the surface, and alternatives I or II of conditions (62.9) at the center. The expressions for the functions $P(s), E(s)$, and $f(s)$ at small s for the last two alternatives of the central conditions (62.9) will not be needed below, therefore we shall not present them here.

Finally, we shall give the exterior solution in terms of the variables (62.1). This is more easily obtained, not from (60.6) by means of a transformation, but by solving the system (62.2)—(62.5) for the exterior region. Substituting in these $\rho = P = 0$, we find

$$\frac{dE}{df} = 2E\frac{f-1}{f(f-2)} + \frac{2(1-f)+(2-\zeta)f^2}{f(f-2)}.$$

The solution of this equation satisfying the boundary conditions at infinity $s=\infty$, $f(\infty)=1$, and $E(\infty)=1$ is

$$E(f) = 1 + (2-\zeta)f - \frac{3}{2}(1-\zeta)f^2. \qquad (62.11)$$

In accordance with this solution, in integrating the differential equations (62.2)—(62.4) from the center, it is convenient to use the following expression for the conditions at the surface:

$$E(F) = 1 + (2-\zeta)F - \frac{3}{2}(1-\zeta)F^2, \qquad (62.12)$$

where $F=f(R)$ is the value of the function $f=rk'/k$ at the surface. We shall not discuss here the procedure for integrating the differential equations for the interior region. This will be done in the last chapter.

63. THE NEWTONIAN VARIANT OF THE GENERALIZED THEORY OF GRAVITATION

For the sake of completeness we would like to dwell briefly also on the Newtonian variant of the generalized theory of gravitation. The field equations could be obtained from (59.7) and (59.8) through a suitable limiting procedure. However, to obtain an independent and logically consistent picture, we consider it advisable to start out from the corresponding variational principle. Here the gravitational field is characterized by two scalar functions, namely the Newtonian potential φ and the gravitational scalar $\chi=8\pi k/c^2$. The variational principle can obviously be written in the form

$$\delta S=\delta \int\left[\frac{\rho v^2}{2}-P-\rho\varphi-\frac{1}{c^2}\frac{(\nabla\varphi)^2}{\chi}+c^2\zeta\frac{(\nabla\chi)^2}{\chi^3}-\zeta\frac{\dot{\chi}^2}{\chi^3}\right]\times dVdt=0. \quad (63.1)$$

All notations are as before, and ζ is the same dimensionless parameter as in the relativistic theory. The expression for the Lagrangian density appearing under the integration sign in (63.1) is obtained from (58.1) through the suitable limiting procedure. The quantities ρ, P, v, φ and χ are functions of the coordinates and the time. A dot over χ signifies a time derivative. Inside the integral we omitted the term which is proportional to $\dot{\varphi}^2$, since it is of a higher order of smallness (we should remember that φ is of the order of $1/c^2$). Generally speaking, in weak fields the dominant term in (63.1) is the one before last, the first four terms being small in comparison. However, these are the first nonvanishing terms in the expansion of the relativistic Lagrangian of the matter and the field in powers of v/c and φ/c^2. Keeping these terms is essential, since it is from these that one obtains the equations of motion and field through virtual variations of the trajectory and the potential. Moreover, if we neglect them we obtain $\chi=\text{const}$.

The variation is performed with respect to $\mathbf{r}$, φ, and χ independently, and leads to the following equations

$$\left.\begin{aligned}\frac{d}{dt}(\rho v)+\nabla P+\rho\nabla\varphi&=0,\\ \operatorname{div}\left(\frac{\Delta\varphi}{\chi}\right)&=\frac{1}{2}c^2\rho,\\ \operatorname{div}\left(\frac{\nabla\chi}{\chi^{3/2}}\right)-\frac{1}{c^2}\frac{d}{dt}\left(\frac{\chi}{\chi^{3/2}}\right)&=\frac{1}{2\zeta c^4}\frac{(\nabla\varphi)^2}{\chi^{1/2}}.\end{aligned}\right\} \quad (63.2)$$

The first of these is the equation of hydrodynamics. The second becomes the Poisson equation when $\chi=\text{const}$. As to the last equation, in the case $\chi=\text{const}$ it is invalid, for the simple reason that then $\delta\chi=0$. As before, in order to effect here a transition to the regular Newtonian theory of gravitation, we should, together with $\chi=\text{const}$, also postulate $|\zeta|=\infty$.

In the static case, the system (63.2) has to be supplemented by the equation of state $\rho(P)$ and the boundary conditions. In the nonstatic problem, though, besides the equation of state one also has to assign the distribution of energy sources, the equations of energy transport, and, finally, the initial conditions.

Let us examine a static centrally symmetrical mass distribution. From (63.2) we have

$$P' = -\rho\varphi', \tag{63.3}$$

$$\frac{d}{dr}\left(\frac{r^2\varphi'}{\chi}\right) = \frac{1}{2}\rho r^2, \tag{63.4}$$

$$\chi^{1/2}\frac{d}{dr}\left(\frac{r^2\chi'}{\chi^{3/2}}\right) = \frac{1}{2\zeta}r^2\varphi'^2. \tag{63.5}$$

Here we have returned to the system of units (34.1).

Outside the mass distribution

$$\frac{d}{dr}\left(\frac{r^2\varphi'}{\chi}\right) = 0, \tag{63.6}$$

$$\chi^{1/2}\frac{d}{dr}\left(\frac{r^2\chi'}{\chi^{3/2}}\right) = \frac{1}{2\zeta}r^2\varphi'^2. \tag{63.7}$$

For the boundary conditions to the last equations we take

$$\chi(\infty) \equiv \chi_0 = 8\pi, \quad \varphi' \to \frac{M}{r^2} \text{ as } r \to \infty, \tag{63.8}$$

where M is the active gravitational mass of the celestial body.

From equation (63.6) we have

$$\varphi' = \frac{\chi(r)}{8\pi}\cdot\frac{M}{r^2}. \tag{63.9}$$

Substituting this expression for φ' in (63.7) and integrating, we obtain

$$\chi(r) = \frac{\chi_0 r^2}{r^2 + 2C_1 r + C_2}. \tag{63.10}$$

Here C_2 is an integration constant which is dependent on the parameter ζ, and

$$C_1 = \frac{1}{2}(4C_2 + M^2/\zeta)^{1/2}. \tag{63.11}$$

The dependence of C_2 on ζ should be such that $C_2 \to 0$ when $|\zeta| \to \infty$; in this case also $C_1 \to 0$, and from (63.10) it then follows that $\chi = \text{const}$.

Substituting (63.10) in (63.9) and integrating the latter, we obtain the following expressions for the potential outside the mass distribution:

$$\varphi(r) = -2\sqrt{-\zeta}\arctan\frac{M}{2\sqrt{-\zeta}(r+C_1)} \quad \text{for} \quad \zeta<0, \tag{63.12}$$

$$\varphi(r) = -\sqrt{\zeta}\ln\frac{r+C_1+M/\sqrt{4\zeta}}{r+C_1-M/\sqrt{4\zeta}} \quad \text{for} \quad \zeta>0. \tag{63.13}$$

The potential is normalized so that it vanishes at infinity. At large distances, both formulas give the expansion

$$\varphi(r) = -\frac{M}{r} + C_1\frac{M}{r^2}\,. \tag{63.14}$$

Expression (63.10) has to be matched to the Heckmann solution (60.1) for χ. Thus, requiring that the two solutions coincide at large distances, we find

$$C_1 = \frac{M}{3-2\zeta}\,. \tag{63.15}$$

We now introduce the new notations

$$C_1 = \alpha M, \qquad C_2 = \beta M^2, \tag{63.16}$$

where

$$\alpha = \frac{1}{3-2\zeta} = \begin{cases} -0.0175 & \text{for} \quad \zeta = +30, \\ 0.0159 & \text{for} \quad \zeta = -30, \end{cases}$$

$$\beta = \alpha^2 - \frac{1}{4\zeta} = \begin{cases} -0.00802 & \text{for} \quad \zeta = +30, \\ -0.00858 & \text{for} \quad \zeta = -30. \end{cases}$$

From equations (63.3)—(63.5) we eliminate the potential φ and introduce the new function

$$m(r) = -8\pi r^2 P'/(\chi\rho),$$

which has the meaning of the mass contained in a sphere of radius r. As a result, we arrive at the system of equations

$$\left.\begin{aligned} & m' = 4\pi\rho r^2, \quad P' = -\frac{1}{8\pi}\frac{\chi\rho m}{r^2}, \\ & \chi'' + \frac{2}{r}\chi' - \frac{3}{2}\frac{\chi'^2}{\chi} = \frac{1}{128\pi^2\zeta}\cdot\frac{\chi^3 m^2}{r^4}, \quad \rho = \rho(P). \end{aligned}\right\} \tag{63.17}$$

The last of these is the equation of state.

Solution (63.17) of the interior problem has to be matched to the exterior solution. For this purpose, let us find the behavior of the functions $\chi(r)$ and $\chi'(r)$ at the surface of the configuration. The derivative φ' should obviously be continuous for all r. From (63.9) it then also follows that $\chi(r)$ is continuous at the surface. In the Newtonian theory, φ'' has a discontinuity of $\chi_0\rho/2$ at the surface. From the requirement of a smooth transition between the new and old theories it follows that, even in the case of a variable χ, the second derivative of the potential should have a discontinuity of $\chi\rho/2$. This condition leads to the continuity of χ', as obtained from equation (63.4). Thus from expressions (63.10) one obtains the following boundary conditions at the surface of the configuration:

$$\left.\begin{aligned}\chi(R) &= \frac{8\pi R^2}{R^2 + 2\alpha M + \beta M^2},\\ \chi'(R) &= \frac{16\pi MR(\alpha R + \beta M)}{(R^2 + 2\alpha M + \beta M^2)^2},\end{aligned}\right\} \tag{63.18}$$

where R is the radius of the configuration, obtained from the equation $P(R)=0$.

Let us also formulate the conditions at the center. In addition to the trivial condition $m(0)=0$, we should obviously also have the numerical values of the quantities

$$P(0), \quad \chi(0), \quad \chi'(0). \tag{63.19}$$

However, because of the existence of the boundary conditions (63.18), there are two relationships between the parameters $P(0)$, $\chi(0)$, and $\chi'(0)$. Hence the quantities $\chi(0)$ and $\chi'(0)$ are functions of the central pressure, and cannot be assigned in an arbitrary manner. Under these circumstances, integration from the center of the mass distribution involves some difficulties, and can be carried out, when the central pressure $P(0)$ is given, through many repeated trials. For a given $P(0)$, the correct values of $\chi(0)$ and $\chi'(0)$ are those which satisfy conditions (63.18) at the surface. We thus arrive at the conclusion that the mass, radius, function $\chi(r)$, and the structure of the configuration are determined by one parameter only, namely the pressure at the center.

The integration can also be carried out from the surface (we are referring to numerical integration, since the system (63.17) has no analytical solution). In this case, the integration should be started with the following values of the functions:

$$r=R, \quad P(R)=0, \quad m(R)=M, \quad \chi(R), \quad \chi'(R), \tag{63.20}$$

where, according to (63.18), the values of $\chi(R)$ and $\chi'(R)$ are determined through M and R. Since the configurations are defined by a single parameter, M and R are dependent on each other and cannot be assigned arbitrarily. The question which arises is, how to determine the mass M for a given radius R. To answer this, let us see what happens if we carry out the integration from the surface to the center $r=0$ for randomly chosen R and M. The only problem which can arise during this is, obviously, the nonvanishing of the function $m(r)$ at the center. By fixing R and integrating for a series of values M, we can ultimately find the true value of M, which corresponds to the given R and satisfies $m(0)=0$. With the aid of computers this can be carried out automatically, obtaining converging results if the initial value of M is successfully chosen.

SUMMARY

In the theory of gravitation with a variable interaction constant k, the field equations are defined by equations (59.7) and (59.8). These contain only the one constant dimensionless unknown parameter ζ, whose numerical value can be determined only after the nontrivial predictions of the new theory are compared with astronomical data. Passage from the new theory to the limit of the old is accomplished by setting $k=\text{const}$ and $|\zeta|\to\infty$.

Since the existing theory of gravitation describes the motion of the planets of the solar system with high accuracy, the new variant of the theory obviously has to be matched to it. From the requirement that the motion of test bodies (Mercury) coincide with the proper motion within observational accuracy, one can establish the lower limit for the absolute value of the parameter $|\zeta|$; it was found that $|\zeta|>30$.

The laws for the conservation of energy, and electric, baryon, and lepton charge, are not violated in the new variant of the theory. Their formulation is exactly the same as before (see (59.6)).

The exterior solution (60.1), found by Heckmann for a centrally symmetrical mass distribution, is analogous to the Schwarzschild solution. It has no singularities at $r\neq 0$, and this is the most significant aspect of the new variant of the theory.

When $|\zeta|\to\infty$, the Heckmann solution for a point mass coincides with the Schwarzschild solution outside the gravitation radius r_g, and differs from it at $r<r_g$. Inside the Schwarzschild sphere, $\lim\limits_{|\zeta|\to\infty} e^{\nu}=\lim e^{\lambda}=0$. The result obtained for the

gravitational scalar is: $k(r) = (r/r_g)^{4/3}$ for $r < r_g$, and $k(r) = 1$ for $r > r_g$.

The Newtonian variant of the generalized theory of gravitation is also of definite interest. The exterior solution for a centrally symmetrical field is here given by equations (63.10), (63.12)—(63.14). The interior problem in the case of a centrally symmetrical mass distribution is defined by the system of differential equations (63.17).

Chapter X

CONFIGURATIONS OF DEGENERATE GASEOUS MASSES IN THE GENERALIZED THEORY OF GRAVITATION

64. THE GRAVITATIONAL SCALAR IN THE NONRELATIVISTIC CASE

In integrating the differential equations (63.17) for a spherically symmetrical mass distribution, it is convenient to convert to the variables

$$y=\frac{r}{M},\quad \mu=\frac{m}{M}. \tag{64.1}$$

The system (63.17) then assumes the form

$$\left.\begin{aligned}&\mu'=\lambda\rho y^2,\quad P'=-\frac{k\rho\mu}{y^2},\\&k''+\frac{2}{y}k'-\frac{3}{2}\frac{k'^2}{k}=\frac{1}{2\zeta}\frac{k^3\mu^2}{y^4},\ P=P(\rho),\end{aligned}\right\} \tag{64.2}$$

where $\lambda=4\pi M^2$, and the primes signify differentiation with respect to y. Let us denote $\frac{M}{R}=W$, and call this the compactness of the configuration. In the new variables, the radius of the configuration is $Y=\frac{R}{M}=\frac{1}{W}$. In accordance with (63.18) and (63.20), the boundary conditions at the surface are

$$\left.\begin{aligned}&P(Y)=0,\quad \mu(Y)=1,\\&k(Y)=\frac{1}{1+2\alpha W+\beta W^2},\\&k'(Y)=\frac{2W(\alpha+\beta W)}{(1+2\alpha W+\beta W^2)^2},\end{aligned}\right\} \tag{64.3}$$

and one has to require that at the center of the mass distribution

$$\mu(0)=0. \tag{64.4}$$

To each value of the compactness W should correspond a particular configuration. Accordingly, in integrating from the surface, one fixes the value of W, and the true value of the parameter $\lambda = 4\pi M^2$ which corresponds to it is found from the condition $\mu(0)=0$ at the center by the method of repeated trial integrations.

First let us discuss the integration of the last equation of the system (64.2). This equation admits a fairly accurate approximate integration, which is important for forming a clear picture of the behavior of the gravitational scalar inside the configuration.

Near the center $\mu(y) \sim y^3$, therefore the right-hand side of the equation

$$k'' + \frac{2}{y}k' - \frac{3}{2}\frac{k'^2}{k} = \frac{1}{2\zeta}\frac{k^3\mu^2}{y^4} \tag{64.5}$$

is a small quantity of the order of y^2. Discarding this term, we obtain the equation

$$\left(\frac{k'y^2}{k^{3/2}}\right)' = 0,$$

which is easily integrated:

$$k(y) = \frac{ay^2}{(y+b)^2}\,. \tag{64.6}$$

The constants a and b will be determined from the boundary conditions (64.3) for k and k', if the integration of (64.5) is continued numerically to the surface of the configuration. A check made during the numerical integration indicated that, although the relative magnitude of the right-hand side of equation (64.5) increases as the surface is approached, nevertheless, it remains to the end quite small compared to the terms on the left. Therefore solution (64.6) is a good approximation, all the way to the surface, for all values of the compactness, i. e. for all cases. Analysis shows that the relative departure of the approximate solution from the precise numerical calculation does not exceed

$$1/[160|\zeta|(\alpha+\sqrt{\beta})^2] \approx 20\% \quad \text{for} \quad |\zeta| = 30.$$

Thus, since the approximate solution is adequate through the entire interior region, we can determine the constants a and b from the boundary conditions (64.3). We have

$$a = \frac{1 + 2\alpha W + \beta W^2}{(1 + \alpha W)^2}, \quad b = M\frac{\alpha + \beta W}{1 + \alpha W}. \tag{64.7}$$

When $\zeta < 0$, the constants α and β (see (63.16)), and consequently also a and b, are positive. As a result, the gravitational scalar $k(r)$ is a smooth function, increasing from the value of $k(0) = 0$ at the center to $k(\infty) = 1$ at infinity. The exterior solution (63.10) can also be approximated by equation (64.6), thereby extending it to the entire range of distances $0 < r < \infty$.

When $\zeta > 0$, $b < 0$, since α and β are negative. In this case, introducing the notation $b = -y_0$, solution (64.6) can be written in the form

$$k(y) = \frac{ay^2}{(y - y_0)^2}, \tag{64.8}$$

where $y_0 > 0$. Thus, when ζ is positive, the function $k(r)$ diverges somewhere between the center and the surface of the configuration. According to the second equation (64.2), the pressure and its derivative also diverge at this point. For this reason we shall henceforth disregard the case $\zeta > 0$, and investigate only the case of negative ζ. The calculations of static configurations were carried out for $\zeta = -30$.

According to (64.6), for all configurations the gravitational scalar behaves in the vicinity of the center as

$$k(y) = ay^2. \tag{64.9}$$

We shall show that when the compactness is large, (64.9) can serve as a good solution all the way to the surface. In the first place, this is directly apparent from the exterior solution at the surface. In fact, according to (64.3), for $W > \sqrt{|\zeta|}$

$$Y = \frac{1}{W}, \quad k(Y) \approx \frac{1}{\beta W^2} = \frac{Y^2}{\beta}, \quad k'(Y) \approx \frac{2}{\beta W^2},$$

i.e. up to the surface itself we have a solution of the form $k \approx ay^2$. Then, from matching of the solutions, one obtains $a = \frac{1}{\beta}$ and $k = \frac{y^2}{\beta}$. The validity of this result can also be proven through a term-by-term analysis of equation (64.5). On its left-hand side, each term is of the order of $1/\beta \sim |\zeta|$, whereas its right-hand side is very small when $W \gg \sqrt{|\zeta|}$. In fact, the ratio between this term and each of the terms on the left-hand side of the equation is of the order of

$$\left|\frac{\beta}{2\zeta}\frac{k^3\mu^2}{y^4}\right| \approx |\zeta|\frac{r^2m^2}{M^4} \leqslant |\zeta|\frac{R^2}{M^2} = \frac{|\zeta|}{W^2} \ll 1.$$

Thus, for compact configurations with $W \gg 1$, the following approximate solution for the gravitational scalar is valid throughout the interior region:

$$k(r) \approx \frac{r^2}{\beta M^2}\,. \tag{64.10}$$

In the case of a model for configurations consisting of an incompressible liquid, one can also integrate the two remaining equations of the system (63.17), if $W \gg 1$. We have

$$P(r) = P_0\left(1 - \frac{r^4}{R^4}\right),$$

where $P_0 = 3W^2/(16\pi\beta M^2)$ is the pressure at the center, and $M = \frac{4}{3}\pi\rho R^3$. For comparison, let us give the result obtained from the theory with $k = \text{const}$:

$$P = P_0\left(1 - \frac{r^2}{R^2}\right),$$

where $P_0 = \frac{W}{2}$.

65. STATIC CONFIGURATIONS OF DEGENERATE GASEOUS MASSES

1. Configurations of a degenerate electron-nuclear gas. Let us see which static celestial bodies consisting of a degenerate particle gas are possible according to the Newtonian variant of the generalized theory of gravitation. First, let us examine configurations consisting of a degenerate electron-nuclear plasma (the Ae-phase). The problem involves integration of the system (64.2) with the equation of state

$$\left.\begin{aligned} \rho &= \frac{8}{3\pi}\left(\frac{m_e}{m_n}\right)^3 x_e^3\,(2 + a_1x_e + a_2x_e^2 + a_3x_e^3),\\ P &= \frac{1}{3\pi}\left(\frac{m_e}{m_n}\right)^4 [x_e\,(2x_e^2 - 3)\sqrt{1 + x_e^2} + 3\ln(x_e + \sqrt{1 + x_e^2})],\end{aligned}\right\} \tag{65.1}$$

given by (29.8). It should be remembered that x_e is the Fermi momentum of the electrons, measured in units of $m_e c$; it varies within the range $0 < x_e \lesssim 45$. The values of the constants were given following equation (29.8). The boundary conditions at the surface are defined by the system (64.3), and the condition at the center by the relationship (64.4). The pressure vanishes when $x_e = 0$, therefore at the surface of the configuration $x_e(R) = 0$.

We are here dealing with a typical boundary-value problem, the boundary conditions being given at the center and at the surface. In /104/, the integration was carried out from the surface. A value was assigned to the parameter $W = \frac{M}{R}$, and the method of repeated trial integrations was then employed to seek the value of $\lambda = 4\pi M^2$ which satisfies $\mu(0) = 0$.

TABLE 16. Parameters of static configurations consisting of a degenerate electron gas and atomic nuclei, in the Newtonian variant of the generalized theory of gravitation

$x(0)$	M	R	$W = \frac{M}{R}$
0.615	$9.81 \cdot 10^{13}$	$9.81 \cdot 10^{7}$	10^{6}
1.111	$1.06 \cdot 10^{13}$	$2.66 \cdot 10^{7}$	$4 \cdot 10^{5}$
1.631	$2.16 \cdot 10^{12}$	$1.08 \cdot 10^{7}$	$2 \cdot 10^{5}$
2.707	$3.75 \cdot 10^{11}$	$3.75 \cdot 10^{6}$	10^{5}
3.637	$1.45 \cdot 10^{11}$	$2.07 \cdot 10^{6}$	$7 \cdot 10^{4}$
4.885	$5.76 \cdot 10^{10}$	$1.15 \cdot 10^{6}$	$5 \cdot 10^{4}$
9.370	$8.01 \cdot 10^{9}$	$3.20 \cdot 10^{5}$	$2.5 \cdot 10^{4}$
11.68	$4.17 \cdot 10^{9}$	$2.08 \cdot 10^{5}$	$2 \cdot 10^{4}$
18.94	$1.01 \cdot 10^{9}$	$8.08 \cdot 10^{4}$	$1.25 \cdot 10^{4}$
24.08	$5.06 \cdot 10^{8}$	$5.06 \cdot 10^{4}$	10^{4}
30.88	$2.50 \cdot 10^{8}$	$3.12 \cdot 10^{4}$	$8 \cdot 10^{3}$
43.51	$9.67 \cdot 10^{7}$	$1.61 \cdot 10^{4}$	$6 \cdot 10^{3}$
55.41	$5.11 \cdot 10^{7}$	$1.02 \cdot 10^{4}$	$5 \cdot 10^{3}$

The results are presented in Table 16 and Figure 38 (the section $\rho \lesssim 10^{13}\ \mathrm{g \cdot cm^{-3}}$). In the table were included only those data which pertain to very large values of the compactness W. The most important feature of these solutions is that, to a single value of $x_0 \equiv x_e(0)$ (i.e. the pressure P_0 or the density ρ_0 at the center), there correspond two configurations, which differ widely in their parameters. Owing to this double-valuedness of the solutions, the curves $M(\rho_0)$ for the mass and $R(\rho_0)$ for the radius consist of two branches, termed the normal and anomalous branches. The data pertaining to the normal branch are not given in Table 16. The configurations on this branch are similar in their characteristics to those obtained from the theory with a constant gravitational scalar k; thus it describes white dwarfs. The anomalous branch contains configurations with abnormally large values of the compactness W. The parameters of these configurations differ appreciably from those of white dwarfs, although the composition of matter in them is identical. For the same range of central densities, the masses of these formations are many orders of magnitude above the sun's, covering the interval $10^8-10^{14} M_{\odot}$ for $45 > x_0 > 0.6$. If x_0 is formally extended below the value of 0.6, then $M \to \infty$ for $x_0 \to 0$. The dependence of M on x_0 for $x_0 \ll 1$ is described by the following asymptotic expression:

$$M \sim x_0^{-4.5}. \tag{65.2}$$

The radii of anomalous configurations are of the order of $10^5 \lesssim R \lesssim 10^9$ km, which is considerably lower than their gravitational radius.

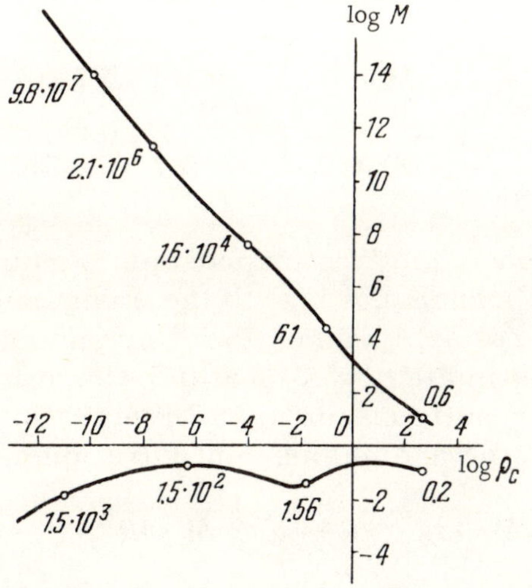

FIGURE 38. The mass M as a function of the central density ρ_0 for static configurations consisting of degenerate gaseous masses, in the generalized Newtonian theory of gravitation. The region $\rho_0 < 0.001$ corresponds to configurations of a degenerate electron-nuclear plasma. The sections of the curves with $\rho_0 > 0.001$ describe neutron configurations. The mass is measured in units of $M_{\odot}$, and the density in $6.4 \cdot 10^{15}$ g·cm^{-3}. The lower branch is close to the corresponding curve in the regular Newtonian theory. The numbers on the curves indicate the radii of the corresponding configurations in km.

The interior solution for the gravitational constant $k(r)$ is given by the approximate analytical solution (64.6). For configurations on the normal branch, nearly everywhere inside the mass distribution one can consider $k(r)$ to be a constant equal to $k_0=1$, since they are characterized by small values of the compactness, $W \ll 1$. In the case of configurations pertaining to the upper branch of $M(\rho_0)$, though, the compactness $W \gg 1$, and the approximate solution (64.10) is valid everywhere inside the configuration.

The values of the parameters of configurations consisting of a degenerate electron gas can be estimated in the following manner. In equations (64.2) and (65.1) we replace all quantities by their averages, using expression (64.3) for the gravitational scalar at the surface. After some simple algebraic operations we obtain

$$\left.\begin{aligned} R &\approx \frac{8.5 \cdot 10^2}{x_0}, \quad M \approx 0.06 \quad \text{for } W \lesssim 10, \\ R &\approx \frac{1.3 \cdot 10^8}{x_0^2}, \quad M \approx \frac{2 \cdot 10^{14}}{x_0^3} \quad \text{for } W \gtrsim 10, \end{aligned}\right\} \tag{65.3}$$

for $x_0 \gtrsim 1$, and

$$\left.\begin{aligned} R &\approx \frac{9 \cdot 10^7}{x_0^{2.5}}, \quad M \approx \frac{7 \cdot 10^{13}}{x_0^{4.5}} \quad \text{for } W \gtrsim 10, \\ R &\approx \frac{1.2 \cdot 10^3}{x_0^{0.5}}, \quad M \approx 0.17 \cdot x_0^{1.5} \quad \text{for } W \lesssim 10, \end{aligned}\right\} \tag{65.4}$$

for $x_0 \approx 1$. Of the above, the first expression pertains to configurations on the normal branch, the last to the anomalous branch, and the second and third do not apply, since there are no such points on the $x_0(W)$ curve. The estimates (65.3) and (65.4) are in fairly good agreement with the results of numerical integration. According to the above estimates, for anomalous configurations

$$M \sim x_0^{-9/2}, \quad R \sim x_0^{-5/2}, \quad M \sim R^{9/5}. \tag{65.5}$$

2. Neutron configurations. The equation of state for a degenerate neutron gas has the form

$$\left.\begin{aligned} \rho &= \frac{1}{4\pi}(\operatorname{sh} t_n - t_n), \\ P &= \frac{1}{12\pi}\left(\operatorname{sh} t_n - 8 \operatorname{sh}\frac{t_n}{2} + 3t_n\right). \end{aligned}\right\} \tag{65.6}$$

The definition of t_n is given by (32.1). Here also, integration of the system (64.2) was conducted from the surface, with the boundary conditions (64.3), for a series of fixed W, using the method described above. The radius of the surface is here determined by the condition $t_n(R)=0$. The results /103/ are presented in Table 17 and Figure 38.

All parameters of the neutron configurations, too, are double-valued functions of the central density ρ_0 (or, it can also be said, of the parameter $t_n(0)$). Let us remember that, in the regular theory, the mass, radius, and other characteristics of neutron stars are single-valued functions of ρ_0. In our case, the $M(\rho_0)$ and $R(\rho_0)$ curves consist of two branches — the lower (regular), and upper (anomalous). The lower branch pertains to configurations with masses of the order of the sun's, and radii of the order of 10 km. In their properties, these configurations differ but little from those obtained from the regular theory of gravitation; in this section the $M(\rho_0)$ for the two cases lie close together (see Table 8). The upper branch, on the other hand, describes configurations with abnormally high values of the compactness W, and it has no analogue in the regular theory. On the anomalous branch, the mass of neutron configurations varies from the lowest value to $\sim 10^8 M_\odot$, and the radius also varies from very small values to values of the order of 10^6 km.

TABLE 17. The most important parameters of neutron configurations according to the Newtonian variant of the generalized theory of gravitation

Normal branch				Anomalous branch			
$t_n(0)$	M	R	W	$t_n(0)$	M	R	W
1.0	0.0354	1.560	0.0227	0.342	$3.2\cdot10^7$	3280	10^4
2.2	0.101	1.005	0.1	1.743	$3.08\cdot10^4$	61.6	500
3.27	0.152	0.761	0.2	2.83	$3.47\cdot10^3$	17.36	200
3.74	0.169	0.676	0.25	4.18	$5.94\cdot10^2$	5.936	100
4.26	0.181	0.603	0.3	5.18	$2.19\cdot10^2$	3.123	70
5.55	0.185	0.463	0.4	6.44	77.24	1.545	50
5.93	0.181	0.431	0.42	8.5	20.75	0.593	35
8.41	0.126	0.301	0.42	11.11	4.46	0.178	25
8.81	0.115	0.287	0.4	13.55	1.266	0.0633	20
10.35	0.0872	0.291	0.3	22.7	$1.26\cdot10^{-2}$	$1.26\cdot10^{-3}$	10
11.96	0.0730	0.365	0.2	31.2	$1.72\cdot10^{-4}$	$3.44\cdot10^{-5}$	5
12.81	0.0739	0.434	0.17				
14.01	0.0802	0.535	0.15	33.5	$5.75\cdot10^{-5}$	$1.43\cdot10^{-6}$	4
				35.8	$1.75\cdot10^{-5}$	$5.83\cdot10^{-6}$	3
				38.4	$4.83\cdot10^{-6}$	$2.41\cdot10^{-6}$	2
				41.3	$1.197\cdot10^{-6}$	$1.197\cdot10^{-6}$	1
				41.85	$9.1\cdot10^{-7}$	$1.137\cdot10^{-6}$	0.8

When $W \gg 1$, the parameters M, R, and $t_n(0)$ can be approximated by asymptotic analytical expressions, similarly to those found for configurations of a degenerate gas of electrons and atomic nuclei. The possibility of finding such asymptotic expressions is due to the fact that, when $W \gg 1$, the gravitational scalar $k(r)$ inside the configuration diminishes strongly with increasing W. Simultaneously with this the pressure gradient also decreases, so that the $\rho(r/R)$ function becomes flattened. In the given case $t_n(0)$ decreases, and the downslope of the function $t_n(r/R)$ is strongly moderated with distance from the center. At sufficiently large W, when $t_n(0) > 1$ (this takes place when $W \gtrsim 10^4$), one can expand $\operatorname{sh} t_n$ in powers of t_n, and the system of differential equations defining the interior problem can then be written in the form

$$\mu' = \frac{1}{6} M^2 y^2 t_n^2, \quad t_n' = -\frac{16}{\beta} \frac{\mu}{t_n}. \tag{65.7}$$

In the last equation, the function $k(y)$ was replaced by the asymptotic y^2/β. It should be remembered that $y = \frac{r}{M}$. Equations (65.7) satisfy the following similitude transformation:

$$y \to \alpha y, \quad M \to \alpha^{-9/4} M, \quad t_n \to \alpha^{1/2} t_n, \quad \mu \to \mu,$$

from which it follows that

$$M \sim W^{9/4}, \quad t_n(0) \sim W^{-1/2}, \quad R \sim W^{5/4}. \tag{65.8}$$

In order to determine the proportionality coefficients, we proceed in the following manner. In (65.7) we convert to the average values

$$\bar{y} \approx \frac{1}{2W}, \quad \bar{y^2} \approx \frac{1}{4W^2}, \quad \overline{\mu(y)} \approx \frac{1}{2}, \quad \overline{t_n(y)} \approx \frac{t_n(0)}{2},$$

$$\overline{\mu'(y)} \approx W, \quad \overline{t_n'(y)} \approx -t_n(0) W, \quad \overline{t_n^3(y)} \approx \frac{t_n^3(0)}{8}, \quad \beta \approx -\frac{1}{4\zeta}.$$

If we now replace the true quantities in (65.8) by their averages, as given above, we obtain

$$t_n(0) \approx \frac{C_1}{\sqrt{W}}, \quad M \approx C_2 W^{9/4}, \quad R \approx C_2 W^{5/4} \tag{65.9}$$

with the values $C_1 \approx 60$, $C_2 \approx 0.03$, which are in fairly good agreement with the results of numerical integration, $C_1 = 34.2$ and $C_2 = 0.0325$. We thus see that, when $W \gg 1$, the parameter $t_n(0)$ is inversely proportional to $\sqrt{W}$, and the mass is proportional to $W^{9/4}$. From (65.9) one can obtain the dependence of the mass and radius on $t_n(0)$, as well as $M(R)$:

$$M \approx 3.3 \cdot 10^6 t_n^{-9/2}(0), \quad R \approx 860 t_n^{-5/2}(0), \quad M \approx 17.2 R^{9/5}. \tag{65.10}$$

The following reservation has to be made regarding the results presented in this section. The radii of configurations pertaining to the anomalous branch of $M(\rho_0)$ are smaller than their gravitational radii $r_g = 2M$. In this case relativistic effects may alter the above results appreciably. Therefore, in calculating configurations with compactness $W \gg 1$, it is necessary to start out from the corresponding equations of the general theory of relativity.

66. SUPERCOMPACT CONFIGURATIONS

From the second expression in the system of boundary conditions (61.6) we have

$$W = \frac{1}{4hB}\left(\tau_0^{0,5-h} - \tau_0^{0,5+h}\right).$$

From (60.7) we know that, when $\zeta = -30$, $h = 0.51234$ and $4hB = 1.9843$. Hence it is apparent that, when $W \ll 1$, the parameter τ_0 is very small compared to unity. Neglecting the second small term, we find

$$\tau_0 = (4hBW)^{-j}, \quad W \gg 1, \tag{66.1}$$

where $j = \frac{1}{h - 0.5} = 81.0373$. In the same approximation,

$$E(R) \equiv e^{\lambda(R)} \approx (2hj\tau_0^h)^2 \ll 1.$$

Thus, at the surface of supercompact configurations, the absolute value of the radial component of the metric tensor is very small, with $E(r) \to 0$ when $W \to \infty$. Comparison between this and condition (62.9) at the center, according to which $E(0) = 0$ (we are referring to the first two versions of the central conditions), leads one to believe that for large values of the compactness the function $E(R)$ is small throughout the internal region, diminishing monotonically with increasing W, and approaching zero in the limit $W \to \infty$. Numerical integration of equations (62.2)—(62.4) confirms this conclusion. The above enables one to find asymptotic expressions for the parameters of superdense configurations.

Let us rewrite the system of equations (62.2)—(62.5), neglecting the function $E(r)$ in them as being small compared to unity:

$$\left.\begin{aligned} \frac{dP}{ds} &= \frac{P+\rho}{f+2}\cdot\frac{\psi}{s}, \\ \frac{dE}{ds} &= \frac{2E}{s(f+2)}(sE\varphi+1-f-\psi), \\ \frac{df}{ds} &= \frac{E}{s(f+2)}\{s[\rho-P+(f-2)\varphi]-f\}, \end{aligned}\right\} \qquad (66.2)$$

where

$$\psi = \frac{1}{f-2}\left(sEP+2f-\frac{\zeta}{2}f^2-1\right), \quad \varphi = \frac{\rho-\zeta(\rho-P)}{3-2\zeta}.$$

Equations (66.2) are invariant with respect to the following similitude transformation

$$s\to\alpha s, \quad E\to\alpha^{-1}E, \quad f\to f, \quad P\to P, \quad \rho\to\rho. \qquad (66.3)$$

Configurations with $W\gg 1$ are thus similar to each other. They are characterized by identical values of the functions $f(s)$, $P(s)$, and the product $sE(s)$. In converting to the variable r this similarity breaks down, i. e. with a given r these functions will have different values for different W. We shall term configurations with $W\gg 1$ *gravitars*. They are characterized by almost identical values of the pressure $P(0)$ at the center.

In (60.1), neglecting small terms and then using (66.1), we obtain the following expressions for the integral parameters of gravitars:

$$\left.\begin{aligned} e^{\lambda(R)} &\approx C_1W^{-2hj}\approx C_1W^{-83}, \\ e^{\nu(R)} &\approx C_2W^{-j/B}\approx C_2W^{-83.7}, \\ k(R) &\approx C_3W^{j\beta_0/B}\approx C_3W^{-1.328}, \\ f(R) &= Rk'(R)/k(R)\approx -\frac{\beta_0 j}{B}\approx 1.328, \end{aligned}\right\} \qquad (66.4)$$

where $C_1=\frac{4h^2j^2}{(4hB)^{2hj}}=1.35\cdot10^{-21}$, $C_2=(4hB)^{-j/B}=1.25\cdot10^{-25}$, and $C_3=(4hB)^{j\beta_0/B}=0.4025$. The numbers given here refer to the case $\zeta=-30$. Similar asymptotic expressions can also be found for the mass and radius, if one takes into consideration that at the surface of all gravitars the product $sE(s)$ has the same value

$$s_0E(s_0)=8\pi R^2k(R)E(R)=\text{const},$$

where $s_0=s(R)$. Taking into account the expressions for $k(R)$ and

$E(R)$ given above, we find

$$M = WR, \quad R \approx C_4 W^{j(2hB-\beta_0)/2B} \approx C_4 W^{42.16}. \qquad (66.5)$$

The numerical value of C_4 can be found by comparing (66.5) with the result obtained by integrating the differential equations for some large value of W.

67. BARYON CONFIGURATIONS IN THE GENERALIZED THEORY OF GRAVITATION

In solving the internal problem, it is convenient to start with the formulation given by (62.2)—(62.5). Integration of this system is most easily carried out from the center. In accordance with (62.10), one has to start with the following conditions at the center of the configuration:

$$P(0), \quad f(0) = f_0, \quad E(s) = D \cdot s^{2(1-f_0)/(2+f_0)}, \qquad (67.1)$$

where D is an unknown constant, and, according to (62.9.I), $f_0 = -\frac{2}{\zeta}\left(1 - \sqrt{1 - \frac{\zeta}{2}}\right) = 0.2$ for $\zeta = -30$. Through trial integrations one then selects the value of D so that at the surface, where $P(s_0) = 0$, the functions $f(s)$ and $E(s)$ satisfy, with predetermined accuracy, the external solution (62.11), i.e. the equation

$$E(F) = 1 + (2-\zeta)F - 1.5\,(1-\zeta)F^2, \qquad (67.2)$$

where $F \equiv f(s_0)$. With the value obtained for $E(s_0)$ we then find the quantities $\tau_0 \equiv \tau(R)$, W, and $k(R)$ from (60.1). With the help of these quantities one can then determine the radius and the mass:

$$R = \left(\frac{s_0}{8\pi k(R)}\right)^{1/2}, \quad M = WR$$

(see the transformation equations (62.1)). With this, one can consider the problem of determining the integral parameters of the configurations as solved.

Generally speaking, the entire interior problem can also be considered solved, since during the numerical calculation of the integration procedure described above one can obviously also obtain the values of the functions $P(s)$, $f(s)$, and $E(s)$. However, to form a clear picture of the internal structure of the configuration, we have to change over to the variable r. In principle, this transition can be accomplished through the inverse of transformation (62.1). However, since the conditions at the center (after prior integration) are already known, it is more convenient to perform one more integration of the differential equations determining the internal parameters as a function of r, than it is to carry out the equally complicated transition from the independent variable s to r. Let us write out the equations for the most important functions characterizing the internal structure of the configurations:

$$\left.\begin{aligned}
\frac{dP}{dr} &= \frac{P+\rho}{r}\psi,\\
\frac{dE}{dr} &= \frac{2E}{r}(sE\varphi - E + 1 - f - \psi),\\
\frac{df}{dr} &= \frac{E}{r}\{s[\rho - P + (f-2)\varphi] - f\},\\
\frac{dk}{dr} &= f\frac{k}{r},\\
\frac{d\nu}{dr} &= -2\frac{\psi}{r},\\
\frac{d\mu}{dr} &= 4\pi\rho\sqrt{E}\,r^2,\\
\frac{dm}{dr} &= 4\pi\left\{\frac{s}{3-2\zeta}[(2-\zeta)\rho + 3(1-\zeta)P] + \right.\\
&\qquad\left. + \frac{k'\nu' r^2}{8\pi k}\right\}\sqrt{E}\,e^{\nu/2},\\
\frac{dN}{dr} &= 4\pi\sqrt{E}\,r^2 n(r),\\
\rho &= \rho(P).
\end{aligned}\right\} \qquad (67.3)$$

Let us recall that $s=8\pi k(r)r^2$, $f(r)=rk'(r)/k(r)$, $E=e^{\lambda(r)}$, $\varphi=\frac{\rho-\zeta(\rho-P)}{3-2\zeta}$, $\psi=\frac{1}{f-2}\left(sEP+2f-\frac{1}{2}\zeta f^2-1+E\right)$.

The first four equations are easily obtained from (61.5), the fifth is equation (61.4), the sixth gives the proper mass (the mass which does not include the energy of the gravitational field) $\mu(r)$ in a sphere of radius r, the eighth defines the total number of baryons $N(r)$ up to the distance r ($n(r)$ is the number density of baryons), and the seventh equation in a certain sense characterizes the total mass $m(r)$ in a sphere with radius r. Strictly speaking, a clear physical meaning is attached only to $m(R)$, which, according to (60.8), coincides with the active mass M of the exterior solution.

The conditions at the center are

$$\left.\begin{array}{lll} P(0)\equiv P_0, & E(r)\approx D(8\pi\gamma)^l\, r^{2-2f_0}, & \\ f(0)\equiv f_0, & k(r)\approx \gamma r^{f_0}, & \nu(0), \\ \mu(0)\equiv 0, & m(0)=0, & N(0)=0. \end{array}\right\} \quad (67.4)$$

Here $l=2(1-f_0)/(2+f_0)=8/11$ for $\zeta=-30$, the value of the central pressure P_0 as specified above, the value of the constant D which corresponds to the chosen P_0 is assumed known after integration of the system (62.2)—(62.4), and γ and $\nu(0)$ are unknown constants, which are determined during the second stage of the integration. The behavior of the functions $E(r)$ and $k(r)$ at $r\approx 0$ given above follows directly from (62.10) and (62.1).

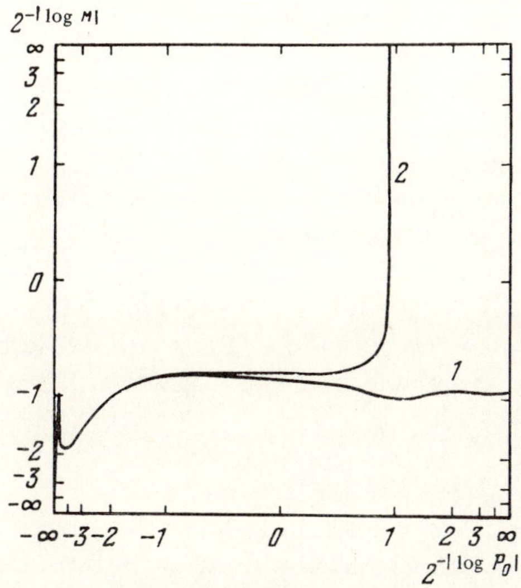

FIGURE 39. The mass as a function of central density for static configurations consisting of a real gas of baryons (equation of state (33.6)). The variable along the abscissa is $2^{-|\log P_0|}$, measured from the point $+\infty$, when $P\geqslant 1$, and from $-\infty$ when $P_0\leqslant 1$. In an analogous manner, the variable along the ordinate is $2^{-|\log M|}$. In the regular Einstein theory (curve 1) the mass of the configuration is of the order of $M_\odot$, and the central pressure can assume any values. In the generalized relativistic theory (curve 2), the pressure has an upper limit, but the mass can assume any values, since $M\to\infty$ when $P_0\to 7.52$. The figures at the points indicate the values of the parameter $q_0=\dfrac{P_0}{\rho_0}$ for the corresponding configurations [sic].

Integration of the system (67.3) with conditions (67.4) is more easily carried out from the center. For the unknown constants γ

and $\nu(0)$ one at first chooses arbitrary values γ_1 and $\nu_1(0)$ and performs a trial integration to the point where the pressure becomes zero, i.e. to the surface. The results thus obtained for the functions $k(r)$ and $\nu(r)$ are certain values $k_2(R)$ and $\nu_2(R)$, which naturally do not coincide with the true values $k(R)$ and $\nu(R)$, found during the first stage of the integration together with τ_0 and W. This is followed by a repeated integration of the system (67.3) with the same conditions at the center, but with the following values for γ and $\nu(0)$:

$$\gamma=\gamma_1 k(R)/k_2(R), \qquad \nu(0)=\nu_1(0)+\nu(R)-\nu_2(R).$$

Analysis of equations (67.3) easily shows that these new values of the constants γ and $\nu(0)$ are the correct ones, since all conditions at the surface are now satisfied. As a result of this second integration we obtain the functions $P(r), \lambda(r), f(r), k(r), \nu(r), \mu(r), m(r), N(r)$, and also the integral characteristics M, R, and $N(R)$ of the configurations, for a given value of the pressure P_0 at the center. As we can see, in the generalized theory of gravitation, too, static spherical configurations of degenerate mass are determined by a single parameter, namely the central pressure.

The method described above was used to calculate the parameters of configurations consisting of a real gas of baryons /107/, with the equation of state (33.2)—(33.6), for $\zeta=-30$. The results of these calculations are presented in Table 18. Figure 39 shows the mass M as a function of the pressure P_0 at the center. At some points on the curves are indicated the values of the so-called relativistic parameter $q_0=\frac{P_0}{\rho_0}$. The lower curve, marked curve 1, describes the function $M(P_0)$ according to the regular relativistic theory of gravitation. As we already know from the material presented in Chapter VII, static baryon configurations can here exist with any central pressure P_0, but with limited values of the mass and radius ($M\sim M_\odot$, $R\sim 10$ km); when $P_0\to\infty$, the mass $M(P_0)$ oscillates with an attenuating amplitude while it approaches a limit close to the solar mass. The upper curve corresponds to the generalized relativistic theory of gravitation.

From Table 18 and Figure 39 it is seen that the results of the new theory are not much different from those of the regular theory, significant deviations appearing only for $q_0 \gtrsim 0.5$. With increasing q_0 (and consequently also P_0), the mass and radius of static baryon configurations go up, but the central pressure cannot assume values above $P_0 \approx 7.52$ ($4.86\cdot 10^{37}$ dyne$\cdot$cm^{-2}). As to inverse function $P_0(M)$, P_0 increases with M, and asymptotically approaches the limit $P_0 \approx 7.52$ when $M\to\infty$. At the same time, the parameter q_0 tends

TABLE 18. Some of the most important characteristics of configurations consisting of a real baryon gas, in the generalized relativistic theory of gravitation

P_0	q_0	R	M	W	M_0	$10^{-58}N$	$-\nu(0)$	$-\nu(R)$	$-\ln D$	$\lambda(R)$	γ	$k(R)$
0.00223	0.04	0.930	0.050	0.054					−8			0.998
0.0159	0.11	0.827	0.122	0.148	0.136	0.143	0.741	0.346	−5	0.331	2.39	0.995
0.239	0.37	0.651	0.185	0.284	0.254	0.236	2.19	0.834	−2	0.793	1.60	0.987
1.64	0.67	0.500	0.180	0.366	0.342	0.251	3.94	1.27	0	1.194	1.32	0.980
2.34	0.72	0.489	0.188	0.385	—	—	—	—	0.5	—	—	0.978
3.05	0.75	0.490	0.200	0.407	0.485	0.310	—	1.58	1	1.475	1.21	0.975
4.55	0.81	0.552	0.232	0.445	—	—	—	—	2	—	—	0.969
5.73	0.84	0.610	0.290	0.474	1.22	0.658	6.24	2.47	3	2.211	1.08	0.961
6.93	0.86	0.972	0.491	0.510	3.73	1.89	7.50	3.56	5	2.855	0.946	0.945
7.39	0.87	2.20	1.17	0.533	21.5	10.7	9.27	5.27	8	2.881	0.772	0.920
7.48	0.87	3.91	2.12	0.543	—	—	—	—	10	—	—	0.902
7.49	0.87	16.9	9.51	0.563	$1.36\cdot10^3$	$6.72\cdot10^2$	13.4	9.35	15	−0.444	0.479	0.862
7.51	0.88	74.9	43.7	0.583	$2.63\cdot10^4$	$1.30\cdot10^4$	16.2	12.3	20	−3.33	0.341	0.823
7.52	0.88	327	197	0.604	$5.11\cdot10^5$	$2.52\cdot10^5$	19.2	15.2	25	−6.22	—	0.785
7.52	0.88	1420	890	0.626	$9.95\cdot10^6$	$4.91\cdot10^6$	22.1	18.1	30	−9.12	0.172	0.747
7.52	0.88	6180	4010	0.648	$1.92\cdot10^8$	$9.46\cdot10^7$	25.0	21.0	35	−12.01	0.123	0.716
7.52	0.88	$2.68\cdot10^4$	$1.81\cdot10^4$	0.671	$3.71\cdot10^9$	$1.83\cdot10^9$	27.9	23.9	40	−14.91	0.081	0.684

Note. P_0 and ρ_0 are the pressure and energy density at the center, $q_0=\frac{P_0}{\rho_0}$, R is the radius, M is the mass of the configuration, $W=M/R$. M_0 is the proper mass (the mass without counting gravitation), N is the number of baryons in the configuration, $\nu(0)$ and $\nu(R)$ are the values of the function $\nu(r)=\ln\, g_{00}(r)$ at the center and at the surface, $\lambda(R)=\ln|g_{rr}(R)|$ (g_{00} and g_{rr} are the components of the metric tensor), $k(R)$ is the value of the gravitational scalar at the surface, and D and γ are the constants appearing in expressions (67.4) for $e^{\lambda}=-g_{rr}$ and $k(r)$.

toward a limit which is approximately equal to 0.88. Actually, the existence of the vertical branch of $M(P_0)$ was already proven in the preceding section. In fact, there it was shown that configurations with $W \ll 1$ have almost identical central pressures, though their mass and radius may be different (gravitars). For gravitars we have the asymptotic functions (66.4) and (66.5).

The integral parameters of configurations with small compactness $W \gg 1$ correspond to the regular Einstein theory (correspondence principle). As follows from the table, such configurations also have a small value of $q_0 = \frac{P_0}{\rho_0}$. Here the relativistic effects are also insignificant. The quantity q_0 is thus an important parameter, simultaneously characterizing the effects of both the relativistic theory of gravitation and the generalized theory of gravitation. When the parameter q_0 increases, the compactness, mass, and radius also increase. Simultaneously there is also a weakening of the gravitational interaction, i. e. a decrease of the gravitational scalar $k(r)$, which promotes the maintenance of hydrostatic equilibrium for larger masses and lower pressures P_0 than in the regular Einstein theory. When the mass is much above the sun's, the decrease in the mutual attraction of masses becomes so effective that the pressure at the center of the configuration reaches saturation and almost ceases to rise with further increase of the configuration's mass. The $R(P_0)$ curve is similar to $M(P_0)$ for $W \gg 1$, as is obvious from both (66.5) and the tabulated data.

No behavior similar to that seen in Figure 39 was found in the case of configurations consisting of a degenerate ideal baryon gas. Calculations performed for $\zeta = -30$ showed that the $M(P_0)$ curves obtained in the two variants of the gravitation theory were not much different from each other, and the same was true for all other curves of the configuration parameters. The reason for this is that the separation between the parameters obtained in the two theories occurs at $q_0 \gtrsim 0.5$, whereas in the case of an ideal gas $q_0 \leqslant 1/3$. This indicates once more that the important factor here is not the central pressure, but the relativistic parameter q_0. The value of q_0 at which noticeable differences between the results of the two theories set in depends on the parameter ζ — it increases with $|\zeta|$.

In /251/ models of configurations of an ideal neutron gas were calculated for $\zeta = -2$. No noticeable deviations from the results of the regular theory of gravitation were found. However, in this paper, as was correctly pointed out in /2/, the most interesting central conditions, namely (62.9.I, II), were overlooked. In our opinion, significant effects should be obtained for such small values of $|\zeta|$ if the boundary conditions at the center and at the surface are correctly formulated.

In the case of gravitars, one can find approximate analytical expressions for the constants γ and D appearing in (67.4). Thus, from dimensional considerations, it follows that $\gamma \sim c^2 M^{-1} R^{0.8}$. Comparing this with the data in the last few rows of Table 18, we find that $\gamma \approx 0.451 \ W^{-1} R^{-0.8}$. In an analogous manner, one obtains for the constant D

$$D \approx 10^{-81} W^{-273} R^{-16/11} .$$

68. VARIOUS PROBLEMS

Comparison between the nonrelativistic (Figure 38) and relativistic (Figure 39) generalized theories reveals that, for large values of the compactness, $W = \frac{M}{R} > 1$, the difference between the results is not only quantitative, but qualitative as well. This is not surprising at all, since when $W > 1$ we are dealing with bodies whose radius is smaller than their gravitational radius $r_g = 2M$. In such cases, owing to the strong curvature of space inside the mass distribution, the Newtonian variant of the theory may even yield incorrect results, since the relativistic effects can become quite substantial. In the case of the regular theory of gravitation, we cannot say how matters stand with respect to the above, since then there are no bodies with $R < r_g$ and therefore a similar situation does not arise. A qualitative agreement can be expected only for configurations with $R \gtrsim r_g$, and this does in fact exist. However, regardless of the strong difference between the results, they still have similar traits. In both Figure 38 and Figure 39, the mass curve $M(P_0)$ consists of two branches — the normal, which does not differ much from the regular theory, and the anomalous branch, which has no analogue in the regular theory of gravitation. In both cases, the compactness continues to increase as one moves from left to right along the $M(P_0)$ curve (here we ignore the oscillations on the normal branch of the $M(P_0)$, $R(P_0)$ and $W(P_0)$ curves, which play no role in this). At the same time, in the nonrelativistic theory P_0 (and consequently also $q_0 = \frac{P_0}{\rho_0}$) at first increases up to some maximum value $P_m(0)$, and then, with further progress from left to right, it drops abruptly; i.e. starting at $P_0 = P_m$, the $M(P_0)$ curve and others bend backward in the direction of the ordinate axis, and continue to rise (see Figure 38). In the relativistic theory the quantity P_0 also has a maximum, but here it is the limiting value. For $P_0 < 1$ (in the units of (34.1)) the function $M(P_0)$ is depicted by a single line (see Figure 39), but when $P_0 \gtrsim 1$ it splits into two; one of these proceeds nearly parallel to the abscissa axis (normal branch), and the second bends gradually to the left. rising vertically at $P_0 \to P_m \approx 7.52$ (which corresponds to $q_0 \approx 0.88$). As to the inverse function $P_0(M)$, in progressing from left to right, in the nonrelativistic case P_0 at first rises, has a maximum at some value $M = M_m$ of the abscissa, then decreases monotonically with increasing M, so that $P_0 \to 0$ when $M \to \infty$. The $P_0(M)$ curve has a maximum for all equations of state (polytropic /70a/, incompressible liquid /102/, and others). The $M(P_0)$ function is described in Figure 38 by two separate lines. As we can see, these lines approach each other, and it is apparent that they should join at some point to form a single uninterrupted curve. This missing section was not plotted due to the lack of corresponding points. Calculations for this part of the $M(P_0)$ curve were not carried out owing to serious difficulties in the numerical computations. In the relativistic case, the $P_0(M)$ curve rises monotonically, approaching the line $P_0 = P_m(0) = 7.52$ when $M \to \infty$. Thus, owing to relativistic effects, the anomalous branch of the $P_0(M)$ curve of the relativistic theory turns upward, straightens out, and becomes parallel to the abscissa axis.

The similarity in the behavior of the gravitational scalar $k(r)$ is much stronger, although the dependence on r is not the same. In both cases the function $k(r)$ rises monotonically from zero at the center to the limiting value $k_0 = k(\infty)$ at infinity; when the configuration's mass M increases, the value of $k(r)$ diminishes everywhere, but its boundary values $k(0) = 0$ and $k(\infty) = k_0$ remain unchanged. The value of the gravitational scalar $k(r)$ is in both cases substantially different from the Newtonian value at distances $r \lesssim r_g$. As we can see, there is still some qualitative agreement between the results of the two variants of the generalized theory.

For polytropes with indices $n = 0$, 1.5, 3, the maximum value of q_0 is 1.51, 1.18, 1.02, respectively, in the nonrelativistic variant, and 2.07, 1.135, 0.925 in the relativistic variant (see /70a, 70b, 71/).

An important role in the appearance of the anomalous branch of the mass, radius, and other configuration characteristics is played by the parameter $q_0 = \frac{P_0}{\rho_0}$, while the value of the central pressure is less important. This is apparent even from the fact that, in the relativistic variant, the gravitar branch (anomalous branch) does not appear for all equations of state, although in all cases there is no limitation on the value of the pressure. The reason for this is precisely that not for all equations of state can q_0 reach the critical value q_m, which is necessary for the appearance of the gravitar branch. Thus, for example, in the case of an ideal baryon gas $q_0 \lesssim 1/3$; the limiting value $q_0 = 1/3$ is reached in the ultrarelativistic case, but this value is insufficient for the emergence of the anomalous branch. On the other hand, for a real baryon gas with the equation of state (33.6) we have $0 \leqslant q \leqslant 1$; thus the highest value $q_0 = 1$ is here larger than the value $q_m(0) = 0.88$ at which the anomalous branch of the mass sets in.

The results given in this chapter apply to $\zeta = -30$. It is important to find out what happens to these results when ζ is changed. A qualitative picture of this can be obtained, without performing new calculations, through careful study of the nature of the $M(q_0)$ and $R(q_0)$ curves. From Figure 39 it is seen that the two branches of the $M(q_0)$ curve (in the future we shall often refer to q_0 instead of P_0, since in the case of degeneracy there is a one-to-one relationship between them) coincide in their initial sections, and only at large q_0 the anomalous branch bends upward toward the left. This behavior is common to all equations of state which involve a gravitar branch. Further, since the results of the generalized and regular theories should coincide when $|\zeta| \to \infty$, from the above-mentioned behavior of the curves it follows that, when $|\zeta|$ is decreased, the maximum value of q_0 in the nonrelativistic theory, and its limiting value in the relativistic theory, should decrease, whereas when $|\zeta|$ is increased these should also increase, thus shortening and lengthening the coinciding sections respectively. In this manner, when $|\zeta|$ is increased, the anomalous branch of the $M(q_0)$ curve should shift to the right, in the direction of large q_0, thereby lengthening the common section of the two branches. In the nonrelativistic case, the validity of this rule can be proven directly. In fact, as was shown in /71, 70b/, in the case of the incompressible-liquid model with other conditions being equal, the pressure at the center, and consequently also its maximum value $P_m(0)$, are inversely proportional to $|\zeta|^{1/2}$, thus directly confirming the above conclusion. From the arguments presented here it follows that the critical value q_m at which large effects appear should be dependent on $|\zeta|$. When $|\zeta| < 30$, the anomalous branch of the mass curve $M(q_0)$ should be present even for small values of q_0, in particular, in the case of an ideal baryon gas. However, small $|\zeta|$ are not admissible, since at $|\zeta| < 30$ we run into the risk of conflicting with Einstein's theory of gravitation.

Gravitars have an extremely large gravitational mass defect $\Delta M_1 = M_0 - M$, where M is the active gravitational mass appearing in the external solution (60.1) and M_0 is the proper mass

$$M_0 = 4\pi \int_0^R \rho r^2 e^{\lambda/2}\, dr. \tag{68.1}$$

The reason for this is that, throughout the internal region of configurations with $W>1$, the functions $k(r)$, $e^{\lambda(r)}$, and $e^{\nu(r)}$ are very small compared to unity. Comparison between (68.1) and (60.7) shows that $M/M_0 \approx \overline{ke^{\nu/2}} \ll 1$. The absolute gravitational mass defect is also abnormally high:

$$\Delta M = Nm_p - M.$$

Here N is the number of baryons in the star. From (60.7) we have

$$M = 4\pi \int (\rho + 3P)\, k(r)\, r^2 e^{\frac{\nu+\lambda}{2}} dr + \frac{1}{2} \int \frac{k'\nu'}{k} e^{\frac{\nu+\lambda}{2}} r^2 dr + $$
$$+ \frac{12\pi}{3-2\zeta} \int (\rho - 3P)\, k(r)\, e^{\frac{\nu+\lambda}{2}} r^2 dr. \qquad (68.2)$$

All terms in this expression are positive, therefore one can state that

$$M \gtrsim \overline{V\rho \exp\left(\frac{\nu+\lambda}{2}\right)},$$

where V is the proper volume of the configuration. Consequently,

$$\frac{M}{Nm_p} \lesssim \overline{ke^{\nu/2}} \ll 1.$$

To clarify, let us give one example. We shall examine a configuration with the parameters $P_0=7.52$, $W=0.671$, $M=1.81\cdot10^4$, $M_0=$ $=3.71\cdot10^9$, and $N=1.83\cdot10^{67}$; then $M/M_0 \approx 5.10^{-5}$ and $M/Nm_p \approx 10^{-5}$. For configurations with $W \gtrsim 0.5$ the absolute mass defect is always positive, and there is no amomaly.

In the spirit of the generalized theory of gravitation, the value of the gravitational scalar far away from a celestial body should depend on the conditions existing in the universe. In particular, in the case of a uniform isotropic space the quantity k_0 should depend on the average density of matter ρ. At the present time, it is generally recognized that the state of the universe is nonstationary. If we disregard the weighty theoretical arguments (we are referring to the Friedman solution, which should undoubtedly have its analogue in the generalized theory) /69/, direct confirmation of the above is furnished by the radial scattering of metagalactic objects with velocities proportional to their distance from the observer. Consequently, ρ, and thereby k_0, are functions of the age of the universe. What can be said regarding the function $k_0(t)$? From the example of individual stellar configurations we know that, with increasing mass density ρ, the gravitational scalar $k(r)$ diminishes when its values $k(0)=0$ at the mass center and $k(\infty)=k_0$ at infinity remain fixed. On the basis of this it is possible to postulate that, at the time when the universe was compressed to a point, $k_0(0)$ was equal to zero. Since the ultrarelativistic liquid cannot be confined for long to a small volume, it must instantly expand, and while $\rho(t)$ decreases, the function $k_0(t)$ should increase continuously. With these simple considerations, it is difficult to say where the expansion process will end. Will it continue indefinitely, or will the expansion be replaced at some moment t by contraction?

Let us consider the possible results of the hypothesis that k_0 changes with time for individual celestial bodies. The system of equations (67.3) is invariant with respect to the following similitude transformation

$$\left.\begin{aligned} k(r) &\to \alpha k(r), \\ k_0(t) &\to \alpha k_0(t) \end{aligned}\right\} \tag{68.3}$$

(it should be remembered that basically the field equations and boundary conditions contain not $k(r)$, but the ratio $k(r)/k_0(t)$, although k_0 does not appear in the equations due to the fact that in the system of units used $k_0=1$). In the transformation (68.3), for a given P_0, the numerical value of the mass, radius, and other parameters remains unchanged. The change of $k_0(t)$ entails a proportional change in the function $k(r)$.

If one considers the preceding remark, it would seem to follow from this that the change of k_0 with time will have no effect on the individual bodies. However, this is not so. Needless to say, in the system of units (34.1) used, where $k_0=1$, although the numerical values of the quantities do not change there is still a change in their scale of measurement. Thus, according to (34.5) and (34.6), the unit of distance is $\alpha \sim k_0^{-1/2}$ and unit mass $\beta \sim k_0^{-3/2}$; consequently, the scales for distance and mass decrease and increase when k_0 is increased or decreased respectively. For fixed numerical values of the configuration parameters (M, R, etc.), in the first case this leads to a decrease of M and R, and in the second, to their increase, which is entirely reasonable (because strengthening of the gravitational attraction should be accompanied by a decrease in the mass of the celestial body). Certain physical quantities, whose units of measurement do not contain $k_0(t)$, remain altogether unchanged, e. g., the pressure, energy density, etc. As to the stability of the configurations, it is not disturbed by the variation of k_0.

At the surface of gravitars $g_{00}(R) = e^{\nu(R)} \ll 1$, therefore the electromagnetic radiation emitted by them undergoes an improbably large red shift:

$$\omega(\infty) = \omega(R)\, e^{\frac{\nu(R)}{2}} \ll \omega(R). \qquad (68.4)$$

In the presence of an extended atmosphere the situation may, of course, change appreciably, since $g_{00}(r)$ increases fast with distance. The deflection of light rays passing close to a gravitar is also very large. In our opinion, both effects can be exploited to verify the theory of variable k. The role of the atmosphere, though, is of major importance. It is necessary to develop models for hot stars with sources, and only after this can one consider verification of the new theory of gravitation.

69. ENERGY SOURCES

In the cosmogonic concept which considers the initial stage to be that of the diffuse state of matter, various aspects of stellar energy sources have been treated extensively. Two basic forms of stellar energy enter into this, namely gravitational and nuclear. Under certain conditions, as a result of the gravitational attraction, diffuse cosmic clouds of gas and dust gradually become compressed, thus forming stars. At the same time, the problem of how the gas was formed is evaded. Compression of the cloud is accompanied by conversion of gravitational energy into kinetic energy of the gas particles. As a result the celestial body heats up, and at a certain stage of the compression, when the temperature reaches tens of millions of degrees, an effective thermonuclear synthesis of light nuclei sets in. The nuclear energy sources suffice to cover the star's energy losses and to maintain it in a stationary thermal state for several billion years. After exhaustion of the nuclear energy, the Kelvinian period of evolution begins. The star cools off during a period of the order of several tens of millions of years. During the cooling-off period the star's dimensions continue to shrink, while part of the gravitational energy is converted to thermal energy, which partially covers the energy losses from the surface. At the end of the Kelvinian period of evolution there begins a catastrophic compression of the celestial body, called collapse. The collapse of the body can continue to the end, i. e. until it becomes locked inside the singular Schwarzschild sphere. For a suitable value of the mass (approximately equal to the sun's mass), or, if it is not appropriate, but during its rapid compression the star succeeds in discarding its excess mass, the process of the dying of the star may end with the formation of a celestial body pertaining to the stable segments of the $M(\eta)$ curve (see Figure 20), i. e. the formation of a white dwarf or a baryon star. In this approach, formation of the unstable baryon configurations with $-1.36 < \eta < -0.6$ and with $\eta > 0.8$ is excluded. Most astrophysicists are inclined to believe that the phenomenon of supernova outbursts is in fact the collapse of a star.

Let us now examine the problem of energy sources according to the cosmogonic concept in which the initial stage is assumed to be the superdense state of matter. It should be added that, generally speaking, in this approach it is not required that the prestellar matter be in a superdense state. It is simply postulated that the development of matter in the universe is from dense to rarefied states, for all we see are outbursts accompanied by ejection of mass, expansion of the system of bodies, dispersion of matter, etc. In this concept, the gravitational energy is obviously deprived of the important role which it has in the diffuse-state concept. The nuclear energy source, on the other hand, is here also highly significant.

Let us list the processes, related to sources of nuclear energy, which could in principle be of importance in the superdense-state cosmogony.

1. The ordinary thermonuclear reactions, which occur at $T \gtrsim 10^7$ degrees.

2. Cold nuclear reactions, occurring in a cold electron-nuclear plasma (the Ae-phase) at sufficiently high densities (see Section 19). When the average distance between particles becomes smaller than the radius of the atomic K shell, light nuclei can, through the tunnel effect, combine with each other or with heavier nuclei, at the same time releasing energy related to the mass defect. Such reactions are possible not only in the Ae-phase, but in the Aen-phase and in a baryon gas as well, if the equilibrium is disturbed as a result of expansion. The nuclear reactions continue until such time as the expansion stops and a new stable equilibrium compound is formed, corresponding to the new value of the substance's density. These are called pycnonuclear reactions.

3. Fission reactions of heavy atomic nuclei. From the material presented in Chapter III, we know that at sufficiently high densities there is fusion of atomic nuclei through the tunnel effect, and their enrichment with neutrons owing to the neutronization effect. At densities of 10^9—10^{13} $g \cdot cm^{-3}$, the number of neutrons in the nuclei is several times higher than the number of protons. Let us now assume that for some reason the celestial body begins to expand. As

a result of the reduction in density the Fermi-energy level of the electron gas is lowered, disturbing the equilibrium between components and leading to the conversion of neutrons into protons $n \rightarrow p+e+\bar{\nu}$. This destroys the stability of the nuclei, and they disintegrate into lighter nuclei through fission. This process will be accompanied by the release of thermal energy (the kinetic energy of fission fragments). The situation in this case is actually the same as in uranium atomic bombs, where an enormous amount of thermal energy is released as a consequence of a chain reaction involving fission of plutonium nuclei.

4. In our opinion, an especially important role in the cosmogony of superdense celestial bodies belongs to the energy $\Delta\varepsilon=\rho(A, Z_{min})-\rho(A_{min}, Z_{min})$, which is plotted in Figure 8. Let us recall briefly what kind of energy this is. In Section 20 we examined two states of thermodynamic equilibrium of a degenerate electron-nuclear plasma (Ae-phase). The first is the state formed through β-decay processes at a fixed mass number A of the atomic nucleus (the minimum of the energy density ρ is with respect to Z only). Compared to the cosmogonic time intervals, thermodynamic equilibrium is in this case established very fast, even instantaneously. The second is the state during which the substance consists of the most stable atomic nuclei (with maximum nuclear binding energy at a given ρ). In this case the internal-energy density $\varepsilon=\rho-n_n m_n-n_p m_p$ (n_n and n_p are the number of neutrons and protons per unit volume, respectively) and the total-energy density ρ have an absolute minimum, i. e. a minimum not only with respect to Z, but with respect to A as well. At medium densities, the most stable nuclei are those of the iron group with $A=56$. At high densities, when neutronization of nuclei becomes significant, the mass number A for which ρ has an absolute minimum is dependent on density $A_{min}=A(\rho)$.

In contrast to the first case, the state with an absolute energy minimum $\rho(A_{min}, Z_{min})$ is reached very slowly. Here, it seems that the relaxation times can even be of the order of a billion years, and this statement is in fact justified. Under the given conditions, the most stable nucleus can be formed through pycnonuclear reactions, the synthesis of light nuclei, or through fission of heavy nuclei. Obviously, formation of the most stable nucleus at once, in a single act of synthesis or fission, is a highly improbable event. Therefore the establishment of the most stable state of thermodynamic equilibrium will no doubt be of the nature of an extremely slow stochastic process.

During the slow approach of the Ae- and Aen-plasmas to the state of highest stability there is also a gradual conversion of the excess internal energy $\Delta\varepsilon$ to thermal energy. The above mechanism for gradual release of nuclear energy operates only in a degenerate electron-nuclear plasma (Ae-phase). It may be of major importance not only for white dwarfs, but also for baryon stars, since matter in the envelope of the latter is also in the Ae- and Aen-phases.

In the problem of energy sources, a certain role may evidently also be played by the protons, neutrons, and electrons emitted by radioactive nuclei.

From what we said above it follows that the picture we have of dwarfs and baryon configurations, as well as that of dead celestial bodies, could turn out to be incorrect. Apparently, these celestial bodies continue to live a vigorous life owing to the slow depletion of the internal-energy reserves available in the Ae- and Aen-phases. In particular, we cannot exclude the possibility that the pulsar phenomenon is due precisely to the release of the above-mentioned energy $\Delta\varepsilon$.

SUMMARY

According to the regular theory of gravitation, at least for static configurations, the compactness obeys

$$W=\frac{M}{R}<0.5.$$

Because of the Schwarzschild singularity at $r=r_g=2M$, there are no stars with $W>0.5$. In the generalized theory of gravitation, configurations with arbitrarily high compactness $W>0.5$ are admissible, this being due to the absence of singularities in the solution for a point mass.

When $W<0.5$, the star parameters calculated from the regular and generalized variants of the theory of gravitation differ little from each other. For large W, the curve $M(q_0)$, which represents the dependence of the mass on the relativistic parameter $q_0=P_0/\rho_0$, consists of two branches (see Figure 39) — the normal branch, which is not much different from the corresponding curve in the regular Einstein theory, and the anomalous curve, which has no analogue in this theory. The anomalous branch contains configurations with mass $M_\odot \lesssim M<<\infty$.

An important factor here is the parameter q_0, whose value indicates the relativistic nature of the configuration. Celestial bodies can have large masses only for q_0 which are close to a certain value q_m. If for the relevant equation of state q_0 does not attain the value

q_m (simply because in this case q_0 cannot assume such values), the anomalous branch of the mass does not appear. Thus, for example, in the case of an ideal baryon gas $0 < q_0 < 1/3$, whereas $q_m \approx 0.88$, therefore (for the chosen value $\zeta = -30$, of course) there is no anomalous branch. In the case of a real gas, however, it does not exist, since the required value of q_0 is reached: $0 < q < 1$.

To avoid misunderstandings, we judge it necessary to warn that the generalized theory of gravitation presented in the last two chapters is only one of the conceptually plausible directions for developing the theory. It is still too early to consider its confirmation through astronomical facts. It is quite probable that it has nothing in common with reality. However, this version should not be rejected offhand, without waiting for its comparison with facts. Since this theoretical version exists, it should be investigated. With time this theory may be shown to be entirely incorrect, but one should not reject the attractive and reasonable idea on which the generalized theory of gravitation is based. We are referring to Dirac's idea concerning the variability of the gravitational constant k. From the viewpoint of pure logic there is nothing wrong with having a variable k, and what is more, who can state that everywhere, including the region $r > r_g$, k has the same value as outside the mass distribution? Nowhere has this been shown to be the case, and such a statement would certainly be a postulate in a certain sense. Needless to say, a theory with variable k can always be constructed so as not to conflict with the existing theory of gravitation.

Thus, in contrast to the first eight chapters, the material presented in the last two chapters is of a controversial nature. These were written with the sole purpose of raising doubts as to the reality of the singularity appearing in the Schwarzschild solution, and to indicate one of the theoretical possibilities for its removal. We hope that the concrete results thus obtained will in the near future lead to a decision on the fate of this theory.

Note added in proof. In this book, not much attention was paid to the theory of hot superdense celestial bodies, since, at the time of writing and publication of the book, these had not been sufficiently studied. In the last two years, substantial work has been accomplished in this area:

1. It has been shown that the residual nuclear energy available in white dwarfs can be released at the necessary rate only through exchange of neutrons between nuclei. Models for hot white dwarfs with such sources have been worked out. For lifetimes of the order of cosmogonic times, the values obtained for the luminosity are $L \sim (0.001 - 0.1) L_\odot$ (Saakyan, G. S. and R. M. Avakyan. — Astrofizika, 1972; Astron. Zh., 1971).

2. It has been established that, of all known sources of internal energy in superdense stars, the most powerful ($10^{48} - 10^{51}$ erg) is the added gravitational energy $kM^2\Delta R/R^2$, which is due to distention

of the star during rotation, if sufficiently fast (Saakyan, G.S., D.M. Sedrakyan, E.V. Chubaryan, R.M. Avakyan, and G.G. Arutyunyan. — Astrofizika, 1972).

3. It has been found that there is a possibility for the existence of an extended atmosphere in some baryon stars and white dwarfs: $T \sim 1/r$, $\rho \sim 1/r^3$ (Saakyan, G.S., D.M. Sedrakyan, and E.V. Chubaryan. — Astrofizika, 1972).

BIBLIOGRAPHY

1. Abrikosov, A. A. — Vop. Kosmog. 3 (1954), 11.
2. Avakyan, R. M. and M. A. Mnatsakanyan. — Astrofizika 5 (1969), 169.
3. Ambartsumyan, V. A. — Doklady AN ArmSSR 13 (1951), 129.
4. Ambartsumyan, V. A. — Soobshch. Byurakan. Observ. 15 (1954).
5. Ambartsumyan, V. A. Nonstationary Stars. — Erevan, 1957. (Russian)
6. Ambartsumyan, V. A. — Izv. AN ArmSSR (Seriya Fiz. -Mat. Nauk) 11 (1958), 9.
7. Ambartsumyan, V. A. — Rev. Mod. Phys. 30 (1958), 944.
8. Ambartsumyan, V. A. — Proc. Solvay Conference, p. 241. — Brussels, 1958. [Original not traced.]
9. Ambartsumyan, V. A. Questions in the Evolution of the Universe. — AN ArmSSR, Erevan, 1958. (Russian)
10. Ambartsumyan, V. A. — Scientific Works, Vol. II, Erevan, 1960. (Russian)
11. Ambartsumyan, V. A. — Astrophys. J. 66 (1961), 536.
12. Ambartsumyan, V. A. — Questions in Extragalactic Astronomy. — Transactions IAU, XI. 1961. [Russian ?].
13. Ambartsumyan, V. A. Nonstationary Phenomena in the Galaxy. — AN ArmSSR, Erevan, 1968. (Russian)
14. Ambartsumyan, V. A. and G. S. Saakyan. — Astron. Zh. 37 (1960), 193.
15. Ambartsumyan, V. A. and G. S. Saakyan. — Astron. Zh. 38 (1961), 785.
16. Ambartsumyan, V. A. and G. S. Saakyan. — Astron. Zh. 38 (1961), 1016.
17. Ambartsumyan, V. A. and G. S. Saakyan. — Vop. Kosmog. 9 (1963), 91.
18. Ambartsumyan, V. A. and G. S. Saakyan. — Astrofizika 1 (1965), 7.
19. Ambartsumyan, V. A., L. V. Mirzoyan, G. S. Saakyan, S. K. Vsekhsvyatskii, and V. V. Kazyutinskii. Questions in Contemporary Cosmogony.— "Nauka," 1972. (Russian)
20. Arutyunyan, G. G. and D. M. Sedrakyan. — Uchen. Zap. Erevan. Gos. Univ. 3 (1969), 23.
21. Arutyunyan, G. G. and D. M. Sedrakyan. — Astrofizika 7 (1971).
22. Bethe, H. A. and P. Morrison. Elementary Nuclear Theory, 2nd ed. — New York, Wiley, 1956.
23. Bisnovatyi-Kogan, G. S. — Astron. Zh. 43 (1966), 89.
24. Bisnovatyi-Kogan, G. S. and Ya. M. Kazhdan. — Astron. Zh. 43 (1966), 761.
25. Vartanyan, Yu. L. and G. S. Saakyan. — Izv. AN ArmSSR 15 (1962), 83.
26. Vartanyan, Yu. L. — Izv. AN ArmSSR 16 (1963), 137.
27. Vartanyan, Yu. L. — Astrofizika 1 (1965), 183.
28. Vartanyan, Yu. L. — Astrofizika 2 (1966), 45.
29. Vartanyan, Yu. L. — Astrofizika 3 (1967), 4.
30. Vartanyan, Yu. L. — Astrofizika 4 (1968), 373.
31. Vartanyan, Yu. L. and A. V. Ovsepyan. — Astrofizika 4 (1968), 643.
32. Vartanyan, Yu. L.— Astrofizika 6 (1970), 167.
33. Vartanyan, Yu. L. and A. V. Ovsepyan. — Astrofizika 6 (1970), 601.

34. Vartanyan, Yu. L. and A. V. Ovsepyan. — Astrofizika 7 (1971), 107.
35. Vartanyan, Yu. L. — Candidate Dissertation. Erevan, 1964. (Russian)
36. Ginzburg, V. L. and D. A. Kirzhnits. — ZhETF 47 (1964), 2006.
36a. Ginzburg, V. L., V. V. Zheleznyakov, and V. V. Zaitsev. — UFN 98 (1969), 201.
37. Guseinov, O. Kh. — Trudy Shemakh. Astr. Observ. 7 (1966).
38. Guseinov, O. Kh. — Candidate Dissertation. Baku, 1967. (Russian)
39. Dzhelepov, B. S. and L. K. Peker. Decay Schemes of Radioactive Nuclei. — Moscow, "Nauka," 1966. (Russian)
40. Dzhelepov, B. S., L. K. Peker, and V. O. Sergeev. Decay Schemes of Radioactive Nuclei. — Moscow, AN SSSR, 1963. (Russian)
41. Dmitriev, N. A. and S. A. Kholin. — Vop. Kosmog. 9 (1963).
42. D'yachenko, V. F., Ya. B. Zel'dovich, V. S. Imshennik, and V. V. Paleichik. — Astrofizika 4 (1968), 159.
42a. Doroshkevich, A. G., Ya. B. Zel'dovich, and I. D. Novikov. — ZhETF 49 (1965), 170.
43. Zel'dovich, Ya. B. — ZhETF 33 (1957), 991.
44. Zel'dovich, Ya. B. — ZhETF 37 (1959), 569.
45. Zel'dovich, Ya. B. — ZhETF 41 (1961), 1609.
46. Zel'dovich, Ya. B. — ZhETF 42 (1962), 641.
47. Zel'dovich, Ya. B. — ZhETF 42 (1962), 1667.
48. Zel'dovich, Ya. B. — ZhETF 43 (1962), 1561.
49. Zel'dovich, Ya. B. — ZhETF 43 (1962), 1982.
50. Zel'dovich, Ya. B. — Vop. Kosmog. 9 (1963), 57.
51. Zel'dovich, Ya. B. and I. D. Novikov. — UFN 84 (1964), 377.
52. Zel'dovich, Ya. B. and I. D. Novikov. — UFN 86 (1965), 477.
53. Zel'dovich, Ya. B. and I. D. Novikov. Relativistic Astrophysics. — Moscow, "Nauka," 1967. (Russian)
54. Zel'dovich, Ya. B. and O. Kh. Guseinov. — Doklady AN SSSR 162 (1965), 791.
55. Zel'dovich, Ya. B. — Astron. Zh. 45 (1968), 913.
56. Zel'dovich, Ya. B. and I. D. Novikov. — In: Einshteinovskii sbornik 1 (1966).
57. Zel'dovich, Ya. B. — UFN 86 (1965), 303.
58. Zel'dovich, Ya. B. — Vop. Kosmog. 9 (1963), 240.
59. Zel'dovich, Ya. B. — Doklady AN SSSR 67 (1964), 155.
60. Zel'dovich, Ya. B. — Astron. Tsirk. 250 (1963).
61. Ivanenko, D. D. and D. F. Kurdgelaidze. — Astrofizika 1 (1965), 479.
62. Kaplan, S. A. — Uchen. Zap. Leningrad. Gos. Univ. Ser. Fiz. -Mat. 15 (1949), 109.
63. Kaplan, S. A. and I. A. Klimishin. — Tsirk. Astron. Observ. Leningrad Gos. Univ. 17 (1953), No. 27.
64. Kaplan, S. A. and G. A. Lukanov. — Astron. Zh. 42 (1965), 299.
65. Kirzhnits, D. A. — ZhETF 38 (1960), 503.
66. Landau, L. D. and E. M. Lifshits. Field Theory. — Moscow, "Nauka," 1967. (Russian)
67. Landau, L. D. and E. M. Lifshits. Statistical Physics. — Moscow, "Nauka," 1964. (Russian)
68. Landau, L. D. and E. M. Lifshits. Mechanics of Continuous Media. — Moscow, Gostekhizdat, 1953. (Russian)
69. Mandzhos, A. V. — Vestn. Kiev. Univ. 10 (1968), 104.
69a. Mandzhos, A. V. — Astrofizika 5 (1969), 645.
70. Mnatsakanyan, M. A. — Doklady AN ArmSSR 49 (1969), 33.
70a. Mnatsakanyan, M. A. — Astrofizika 5 (1969), 645.
70b. Mnatsakanyan, M. A. — Candidate Dissertation. Erevan, 1969. (Russian)
71. Mnatsakanyan, M. A. and R. M. Avakyan. — Astrofizika 4 (1968), 646.
72. Moszkowski, S. A. Models of the nucleus. — In: Stroenie atomnogo yadra (Structure of the Atomic Nucleus). Moscow, Inostrannaya Literatura, 1959. [Russian; original not traced.]

73. Novikov, I. D. and L. M. Ozernoi. — Doklady AN SSSR **150** (1963), 1019.

74. Novikov, I. D. — Soobshch. Gos. Astron. Inst. Shternberg **132** (1964).

75. Papoyan, V. V. — Doklady AN ArmSSR **47** (1968), 21.

76. Papoyan, V. V., D. M. Sedrakyan, and E. V. Chubaryan. — Astrofizika **3** (1967), 41.

77. Papoyan, V. V., D. M. Sedrakyan, and E. V. Chubaryan. — Soobshch. Byurakan. Observ. **39** (1968), 101.

78. Papoyan, V. V., D. M. Sedrakyan, and E. V. Chubaryan. — Soobshch. Byurakan. Observ. **40** (1969), 86.

79. Papoyan, V. V., D. M. Sedrakyan, and E. V. Chubaryan. — Astrofizika **5** (1969), 95.

80. Papoyan, V. V., D. M. Sedrakyan, and E. V. Chubaryan. — In: Sbornik trudov V Mezhdunarodnoi konferentsii po gravitatsii i teorii otnositel'nosti. Tbilisi, 1970.

81. Papoyan, V. V., D. M. Sedrakyan, and E. V. Chubaryan. — Astrofizika **5** (1969). 415.

82. Papoyan, V. V., D. M. Sedrakyan, and E. V. Chubaryan. — Doklady ArmSSR **49** (1969), 237.

82a. Papoyan, V. V. — Candidate Dissertation. Erevan, 1969. (Russian)

82b. Petrov, A. Z. New Methods in the General Theory of Relativity. — Moscow, "Nauka," 1966. (Russian)

83. Pinaeva, G. V. — Astron. Zh. **41** (1964), 25.

84. Podurets, M. A. — Astron. Zh. **41** (1964), 28.

85. Rylov, Yu. L. — ZhETF **40** (1961), 1755.

86. Saakyan, G. S. — ZhETF **38** (1960), 843.

87. Saakyan, G. S. — ZhETF **38** (1960), 1593.

88. Saakyan, G. S. — Izv. AN ArmSSR **14** (1961), 117.

89. Saakyan, G. S. and D. M. Sedrakyan. — Izv. AN ArmSSR **14** (1961), 109.

90. Saakyan, G. S. — Astron. Zh. **39** (1962), 1014.

91. Saakyan, G. S. — Izv. AN ArmSSR (Ser. Fiz. -Mat. Nauk) **15** (1962), 123.

92. Saakyan, G. S. — Doctoral Dissertation. Moscow, Fiz. Inst. Lebedev, 1962. (Russian)

93. Saakyan, G. S. — Astron. Zh. **40** (1963), 82.

94. Saakyan, G. S. — Astron. Zh. **41** (1964), 193.

95. Sahakian, G. S. and Ju. L. Vartanian. — Nuovo Cim., Serie X, **30** (1963), 82.

96. Saakyan, G. S. and Yu. L. Vartanyan. — Soobshch. Byurakan. Observ. **33** (1963), 55.

97. Saakyan, G. S. and Yu. L. Vartanyan. — Nuovo Cim. **27** (1963), 1497.

98. Saakyan, G. S. and Yu. L. Vartanyan. — Astron. Zh. **41** (1964), 193.

99. Saakyan, G. S., Yu. L. Vartanyan, and E. V. Chubaryan. — In: Problemy gravitatsii, p. 205. Tbilisi, 1965.

100. Saakyan, G. S. and Yu. L. Vartanyan. — Astrofizika **3** (1967), 503.

101. Saakyan, G. S. and E. V. Chubaryan. — Soobshch. Byurakan. Observ. **34** (1963), 99.

102. Saakyan, G. S. and M. A. Mnatsakanyan. — Astrofizika **3** (1967), 311.

103. Saakyan, G. S. and M. A. Mnatsakanyan. — Astrofizika **4** (1968), 181.

104. Saakyan, G. S. and M. A. Mnatsakanyan. — Soobshch. Byurakan. Observ. **40** (1968), 98.

105. Saakyan, G. S. and M. A. Mnatsakanyan. — Tezisy dokladov Mezhdunarodnoi konferentsii po gravitatsii i teorii otnositel'nosti, p. 196. Tbilisi, 1968.

106. Saakyan, G. S. and M. A. Mnatsakanyan. — Astrofizika **4** (1968), 567.

107. Saakyan, G. S. and M. A. Mnatsakanyan. — Astrofizika **5** (1969), 555.

108. Sedrakyan, D. M. and E. V. Chubaryan. — Astrofizika **4** (1968), 239.

109. Sedrakyan, D. M. and E. V. Chubaryan. — Astrofizika **4** (1968), 481.

110. Seidov, Z. F. — Astrofizika **3** (1967), 189.

111. Seidov, Z. F. — Soobshch. Shemakh. Astr. Observ. **3** (1964), 25.

112. Seidov, Z. F. — Izv. AN AzerbSSR (Ser. Fiz. -Mat. Tekhn. Nauk) **5** (1968), 93.

113. Seidov, Z. F. — Soobshch. Shemakh. Astr. Observ. **5** (1965).

114. Seidov, Z. F. and O. Kh. Guseinov. — Soobshch. Shemakh. Astr. Observ. **2** (1966), 13.
115. Harrison, B. K., K. S. Thorne, M. Wakano, and J. A. Wheeler. Gravitation Theory and Gravitational Collapse. — Chicago, Univ. of Chicago Press, 1965.
116. Fowler, W. A. and F. Hoyle. Neutrino Processes and Pair Formation in Massive Stars and Supernovae. — Chicago, Univ. of Chicago Press, 1964.
117. Fok, V. A. Theory of Space, Time and Gravitation. — Fizmatgiz, 1961. (Russian) [English translation: Pergamon Press, New York, 1959.]
118. Chandrasekhar, S. An Introduction to the Study of Stellar Structure. — New York, Dover, 1957.
119. Chubaryan, E. V. — Izv. AN ArmSSR **16** (1963), 95.
120. Chubaryan, E. V. — Doklady AN ArmSSR **38**, No. 1. (1964), 17.
121. Chubaryan, E. V. — Candidate Dissertation. Erevan, 1964. (Russian)
122. Shatsman, E. — Astron. Zh. **33** (1956), 800.
123. Schwarzschild, M. Structure and Evolution of the Stars, 2nd ed. — New York, Dover, 1965.
124. Albada, G. B. van. — Bull. Astr. Inst. Netherlands **10** (1964), 161.
125. Albada, G. B. van. — Astrophys. J. **105** (1947), 393.
126. Baade, W. and F. Zwicky. — Phys. Rev. **45** (1934), 138.
127. Baade, W. and F. Zwicky. — Astrophys. J. **88** (1938), 411.
128. Baade, W. and F. Zwicky. — Proc. Nat. Acad. Sci. **20** (1934), 259.
129. Baglin, A. — Compt. Rend. **260** (1965), 2424.
130. Baglin, A. — Compt. Rend. **258** (1964), 5801.
131. Baglin, A. — Ann. d'Astrophys. **29** (1968), 103.
132. Bahcall, J. N. and R. A. Wolf. — Phys. Rev. Lett. **14** (1965), 343.
132a. Bahcall, J. N. and R. A. Wolf. — Phys. Rev. **140B** (1965), 1452.
133. Barach-Schmidt, N., A. Barbaro-Galtieri, L. R. Price, M. Roos, A. H. Rosenfeld, P. Söding, and C. G. Wohl. Particle Properties. — In: Review of Particle Properties, USRI-8030. Berkeley, 1968.
134. Bardeen, J. M., K. S. Thorne, and D. M. Meltzer. — Astrophys. J. **145** (1966), 505.
135. Bethe, H. and J. Goldstone. — Proc. Roy. Soc. **A238** (1957), 551.
136. Beandet, G., V. Petrosian, and E. Salpeter. — Astrophys. J. **150** (1967), 979.
137. Brueckner, K. A. — Rev. Mod. Phys. **30** (1958), 561.
138. Brueckner, K. A. and I. I. Gammel. — Phys. Rev. **109** (1958), 1023.
139. Brueckner, K. A., I. I. Gammel, and I. I. Kubis. — Phys. Rev. **118** (1960), 1095.
140. Brans, C. and R. H. Dicke. — Phys. Rev. **124** (1960), 925.
141. Brans, C. — Ph. D. thesis, Princeton University. Princeton, 1961.
142. Brans, C. — Phys. Rev. **125** (1962), 388.
143. Brans, C. — Phys. Rev. **125** (1962), 2194.
144. Brill, D. R. and J. Cohen. — Phys. Rev. **143** (1966), 1011.
145. Burbidge, G. — Astrophys. J. **137** (1963), 995.
146. Cameron, A. G. W. — Astrophys. J. **130** (1959), 916.
147. Cameron, A. G. W. — Astrophys. J. **130** (1959), 884.
148. Cameron, A. G. W. — Mém. Soc. Rog. Sci. Liège (5) 3 (1960), 461.
149. Cameron, A. G. W. — Astr. J. **64** (1959), 325.
149a. Cameron, A. G. W. — Nature **205** (1965), 787.
149b. Cameron, A. G. W. — Nature **206** (1965), 1342.
150. Carson, T. R., D. F. Mayers, and W. H. Stibls. — Mon. Not. R. Astr. Soc. **140** (1968), 483.
151. Cernuschi, F. — Phys. Rev. **56** (1939), 450.
152. Chandrasekhar, S. — Mon. Not. R. Astr. Soc. **95** (1935), 222.

153. Chandrasekhar, S. — Mon. Not. R. Astr. Soc. **95** (1935), 207.
154. Chandrasekhar, S. — Astrophys. J. **140** (1964), 417.
155. Chandrasekhar, S. — Mon. Not. R. Astr. Soc. **93** (1933), 390.
156. Chandrasekhar, S. and N. R. Lebovitz. — Astrophys. J. **136** (1962), 1082.
157. Chandrasekhar, S. and P. Roberts. — Astrophys. J. **138** (1963), 809.
158. Chandrasekhar, S. — Phys. Rev. Lett. **12** (1964), 437.
159. Chandrasekhar, S. — Phys. Rev. Lett. **14** (1965), 241.
160. Chandrasekhar, S. and R. Tooper. — Astrophys. J. **139** (1964), 1396.
161. Chiu, H. I. — Ann. Phys. **26** (1964), 364.
162. Chiu, H. I. — Astrophys. J. **137** (1963), 343.
163. Chiu, H. I. — Ann. Phys. **15** (1961), 1.
164. Chiu, H. I. — Ann. Phys. **16** (1961), 321.
165. Chiu, H. I. and E. E. Salpeter. — Phys. Rev. Lett. **12** (1964), 413.
165a. Cocke, W. J., M. J. Disney, and D. J. Taylor. — Nature **221** (1969), 525.
166. Critchfield, C. L. — Astrophys. J. **96** (1942).
167. Dehnen, H. — Z. Astrophys. **68** (1968). 190.
168. Dirac, P. A. M. — Nature **139** (1937), 323.
169. Dirac, P. A. M. — Proc. Roy. Soc. (A) **165** (1938), 199.
170. Dirac, P. A. M. — Scient. Am. **208**, No. 5 (1963), 45.
171. Dicke, R. H. — Rev. Mod. Phys. **29** (1957), 355.
172. Dicke, R. H. — Rev. Mod. Phys. **29** (1957), 363.
173. Dicke, R. H. — J. Wash. Acad. Sci. **48** (1958), 213.
174. Dicke, R. H. — Science **129** (1959), 621.
175. Dicke, R. H. — Am. Scient. **47** (1959), 25.
176. Dicke, R. H. — Am. J. Phys. **28** (1960), 344.
177. Dicke, R. H. — Phys. Rev. **125** (1962), 2163.
178. Dicke, R. H. — Rev. Mod. Phys. **34** (1962), 110.
179. Dicke, R. H. — Astrophys. J. **152** (1968), 1.
179a. Drake, F. D. and H. D. Craft Jun. — Nature **220** (1968), 231.
180. Eddington, A. S. — Mon. Not. R. Astr. Soc. **100** (1940), 582.
181. Euwema, R. N. — Phys. Rev. **133** (1964), B1046.
182. Fierz, M. — Helv. Phys. Acta **29** (1956), 128.
183. Finkelstein, P. — Phys. Rev. **110** (1958), 965.
184. Finzi, A. F. — Ann. Phys. **26** (1964), 411.
184a. Finzi, A. F. — Phys. Rev. Lett. **15** (1965), 599.
184b. Finzi, A. F. — Mon. Not. R. Astr. Soc. **127** (1963), 21.
184c. Finzi, A. F. — In: High-Energy Astrophysics (ed. by L. Grutton), New York, Academic Press, 1967.
185. Fowler, W. A. — Rev. Mod. Phys. **36** (1964), 545.
186. Fronsdal, C. — Phys. Rev. **116** (1959), 778.
187. Gamow, G. and E. Teller. — Phys. Rev. **52** (1937), 929.
188. Gamow, G. and M. Schienberg. — Phys. Rev. **59** (1941), 539.
189. Gamow, G. Atomic Nuclei and Nuclear Transformations (2nd ed.), p. 234. — Oxford, 1936.
189a. Ginzburg, V. L. and V. V. Zaitsev. — Nature **222** (1969), 230.
189b. Gold, T. — Nature **218** (1968), 731; **221** (1969), 25.
190. Gombás, P. and D. Z. Kisdi. — Naturforschung **21** (1966), 2009.
191. Gomes, L., J. Valecka, and V. Weiskopf. — Ann. Phys. 3 (1958), 241.
192. Hamada, T. and E. E. Salpeter. — Astrophys. J. **134** (1961), 683.
192a. Hausen, C. J. — Nature **211** (1966), 1069.
193. Hartle, J. B. — Astrophys. J. **150** (1967), 1005.

194. Hartle, J. B. and K. S. Thorne. — Astrophys. J. 153 (1968), 807.
195. Hartle, J. B. and K. S. Thorne. — Slowly rotating relativistic stars, 111. — Preprint, 1968.
196. Hartle, J. B. and D. H. Sharp. — Astrophys. J. **147** (1967), 317.
197. Hayashi, Ch. and R. C. Cameron. — J. Phys. Soc. Japan **17**, Suppl. A-111 (1962), 53.
198. Hayashi, Ch. and R. C. Cameron. — Astr. J. **67** (1962), 577A.
199. Heckmann, O., P. Jordan, and W. Z. Fricke. — Z. Astrophys. **28** (1951), 113.
200. Heckmann, O. — Naturwissenschaften **38** (1951).
201. Heckmann, O. — Z. Astrophys. **40** (1956), 278.
201a. Hewish, A., S. J. Bell, D. H. Pilkington, P. F. Scott and R. A. Collins. — Nature **217** (1968), 709.
202. Heckmann, O. and E. Schücking. Handbuch der Physik, Vol. 53, Astrophysik IV, Sternsysteme. — 1959.
202a. Hoyle, F., J. V. Narlikar, and J. A. Wheeler. — Nature **203** (1964), 914.
203. Hoyle, F. and W. A. Fowler. — Astrophys. J. **132** (1960), 565.
204. Hoyle, F. and W. A. Fowler. — Astrophys. J. Suppl. **91** (1964), 201.
205. Hohl, H. and H. Dehnen. — Z. Astrophys. **68** (1968), 181.
206. Iton, N. — Preprint. Department of Physics, Kyoto University, 1970.
207. James, R. A. — Astrophys. J. **140** (1964), 552.
208. Johnson, M. C. — Observatory **64** (1938), 74.
209. Jordan, P. Schwerkraft und Weltall. — Braunschweig, 1955.
210. Jordan, P. and C. Müller. — Naturforschung **2a** (1947), 1.
211. Jordan, P. — Astr. Nachr. **276** (1948), 193.
212. Jordan, P. — Akad. Wiss. Lit. (Mainz) (1950), 319.
213. Jordan, P. — Rev. Mod. Phys. **34** (1962), 596.
214. Jordan, P. — Naturwissenschaften **25** (1937), 513.
215. Jordan, P. — Naturwissenschaften **26** (1938), 417.
216. Jordan, P. — Z. Phys. **132** (1952), 655.
217. Jordan, P. — Z. Phys. **133** (1952), 558.
218. Jordan, P. — Z. Phys. **157** (1959), 112.
219. Jordan, P. — Z. Phys. **201** (1967), 394.
220. Jordan, P. — Z. Astrophys. **68** (1968), 201.
221. Kerr, R. P. — Phys. Rev. Lett. **11** (1963), 237.
222. Kothari, D. S. — Proc. Roy. Soc. London (A) **162** (1937), 521.
223. Kuchowicz, B. — Rep. Inst. Nucl. Research, Warsaw, No. 384/VII, 11, 1963.
224. Kuchowicz, B. — Bull. Acad. Polon. Sci. Ser. Sci. Math. Astron. Phys. **11** (1963), 203.
225. Kuchowicz, B. — Postępy Ast. **11** (1963), 203.
226. Kuchowicz, B. Nuclear Astrophysics. Bibliographical Survey. Pts. I, II, III. — Warsaw, Nuclear Energy Information Center, 1965.
226a. Kuchowicz, B. — Nature **225** (1970), 440.
226b. Kuchowicz, B. — Acta Phys. Polonica **36** (1969), 1105.
226c. Kuchowicz, B. — Fortschr. Phys. **17** (1969), 517.
226d. Kuchowicz, B. — Phys. Lett. 30B (1969), 79.
227. Landau. L. — Phys. Z. SowjUn. 1 (1932), 285.
228. Landau, L. — Nature **141** (1938), 333.
229. Lebovitz, N. R. — In: Annual Review of Astronomy and Astrophysics (ed. by L. Goldberg), Vol. 5, 1967.
230. Levinger, J. N. and L. M. Simmons. — Phys. Rev. **124** (1961), 916.
230a. Lindquist, R. W. — Ann. Phys. 37 (1966), 487.
231. Liebes, S. Jr. — Phys. Rev. **133** (1964), B835.

232. Ludwig, G. — Arch. Math. **1** (1948), 2.
233. Ludwig, G. and C. Müller. — Ann. Phys. **2** (1948), 76.
234. Ludwig, G. — Naturforschung **2a** (1947), 482.
235. Ludwig, G. — Z. Phys. **124** (1948), 450; **125** (1949), 545.
236. Ludwig, G. Fortschritte der projektiven Relativitätstheorie. — Braunschweig, 1951.
236a. Markov, M. A. — Phys. Lett. **10** (1964), 122.
237. Marshak, R. E. — Astrophys. J. **92** (1940), 321.
237a. Meltzer, D. W. and K. S. Thorne. — Astrophys. J. **145** (1966), 514.
238. Misner, G. and H. S. Zapolsky. — Phys. Rev. Lett. **12** (1964), 635.
239. Morton, D. C. — Nature **201** (1964), 1308.
240. Nakagawa, K., T. Ohmura, H. Takebe, and S. Obi. — Progr. Theor. Phys., Osaka **16** (1956), 389.
240a. O'Mengain, E. P., N. A. Porter, J. White, D. J. Fegan, D. M. Jennings, and B. G. Lawless. — Nature **219** (1968), 1348.
241. Oppenheimer, J. R. and R. Serber. — Phys. Rev. **54** (1938), 540.
242. Oppenheimer, J. R. and H. Snyder. — Phys. Rev. **56** (1939), 455.
243. Oppenheimer, J. R. and G. H. Volkoff. — Phys. Rev. **55** (1939), 374.
244. Pacini, F. — Nuovo Cim. **37** (1965), 767.
245. Pacini, F. — Nature **209** (1966), 389.
246. Papapetrou, A. — Proc. Roy. Ir. Acad. **52** (1948), 11.
247. Reeves, H. and E. E. Salpeter. — Phys. Rev. **116** (1959), 1505.
248. Roeder, R. C. — Can. J. Phys. **44**, No. 8 (1966).
249. Rosenfeld, A. H., A. Bazbaro Galtieri, W. J. Podolsky, L. R. Price, P. Söding, Ch. G. Wohl, M. Roos, and W. J. Wollis. — Rev. Mod. Phys. **39** (1967), 1.
250. Sahakian, G. S. and Ju. L. Vartanian. — Nuovo Cim. **109** (1958), 1023.
251. Salmona, A. — Phys. Rev. **154** (1967), 1218.
252. Salmona, A. — Phys. Rev. **160** (1967), 1008.
253. Salpeter, E. E. — Astrophys. J. **134** (1961), 669.
254. Salpeter, E. E. — In: Proceedings of Symposium on Low Energy Nuclear Physics, 1—18. Bombay, 1961.
255. Salpeter, E. E. — Astrophys. J. **134** (1961), 669.
256. Salpeter, E. E. — Ann. Phys. **11** (1960), 393.
257. Sampson, D. H. — Astrophys. J. **129** (1959), 734.
258. Schatzmann, E. White Dwarfs. — Amsterdam, North-Holland, 1958.
259. Schatzmann, E. — In: Stars and Stellar Systems (ed. by H. Y. Chiu and W. F. Hoffman), Vol. VIII, 327—365. Chicago, Univ. of Chicago Press, 1965.
260. Serber, R. — Rev. Mod. Phys. **36** (1964), 649.
261. Schücking, E. — Z. Phys. **148** (1957), 72.
261a. Schutz, F. — IR, Preprint, California Inst. Technology, Pasadena, California.
262. Tolman, R. C. Relativity, Thermodynamics, and Cosmology. — Oxford, Clarendon Press, 1934.
263. Thorne, K. S. — Astrophys. J. **152** (1968), L71.
264. Thorne, K. S. — Phys. Rev. Lett. **21** (1968), 320 (gravitational radiation damping).
265. Thorne, K. S. and J. R. Ipser. White-dwarf and neutron star interpretation of pulsating radio sources. — Preprint, 1968.
266. Thorne, K. S. and D. Meltzer. — Astrophys. J. **145** (1966), 514.
267. Thorne, K. S. Relativistic stellar structure and dynamics. — Preprint, 1967.
268. Thorne, K. S. — Science **150** (1965), 1671.
269. Thorne, K. S. The rotation and pulsation of general relativistic stellar models. — Preprint, 1967.

269a. Tsuruta, S. — Nature **207** (1965), 364.
270. Tsuruta, S. and A. G. W. Cameron. — Can. J. Phys. **44** (1966), 1863.
270a. Tsuruta, S., J. P. Wright, and A. G. W. Cameron. — Nature **206** (1965), 1137.
271. Tsuruta, S. and A. G. W. Cameron. — Can. J. Phys. **43** (1965), 2056; **44** (1966), 1895.
272. Volkoff, G. M. — Phys. Rev. **55** (1939), 421.
273. Volkoff, G. M. — Phys. Rev. **55** (1939), 413.
274. Weiskopf, V. — Nucl. Phys. **3** (1957), 423.
275. Wheeler, J. A. The superdense star and the critical nucleon number. — In: Gravitation and Relativity (ed. by H. Y. Chiu and W. F. Hoffman), 195—299. New York, Benjamin, 1964.
275a. Wheeler, J. A. — In: Annual Review of Astronomy and Astrophysics (ed. by L. Goldberg), Vol. 4. Palo Alto, Calif. 1966.
276. Wildhack, W. A. — Phys. Rev. **57** (1940), 81.
277. Wolf, R. A. — Preprint (November, 1964), California Inst. Technology, Pasadena, California.
277a. Zee, A. and J. A. Wheeler. — Paper in preparation, 1966.
278. Zwicky, F. — Astrophys. J. **88** (1938), 522.
279. Zwicky, F. — Phys. Rev. **55** (1939), 726.
280. Zwicky, F. — In: Max-Planck Festschrift 1958, pp. 243—251, Berlin, Deutscher Verlag, 1959.
281. Pulsating Stars 1., "Nature" reprint (1968), with intr. by F. G. Smith and A. Hewish; Pulsating Stars 2., "Nature" reprint (1969), with intr. by T. Gold.
282. Terzian, J. — Lecture preprint, 1970.

NOTE (Abbreviations used in the Bibliography for Russian Journals):

AN: Akademiya Nauk
UFN: Uspekhi Fizicheskikh Nauk
ZhETF: Zhurnal Eksperimental'noi i Teoreticheskoi Fiziki

SUBJECT INDEX